Off-Diagonal Bethe Ansatz for Exactly Solvable Models

Yupeng Wang · Wen-Li Yang
Junpeng Cao · Kangjie Shi

Off-Diagonal Bethe Ansatz for Exactly Solvable Models

Springer

Yupeng Wang
Institute of Physics
Chinese Academy of Sciences
Beijing
China

Junpeng Cao
Institute of Physics
Chinese Academy of Sciences
Beijing
China

Wen-Li Yang
Institute of Modern Physics
Northwest University
Xi'an
China

Kangjie Shi
Institute of Modern Physics
Northwest University
Xi'an
China

ISBN 978-3-662-46755-8 ISBN 978-3-662-46756-5 (eBook)
DOI 10.1007/978-3-662-46756-5

Library of Congress Control Number: 2015935415

Springer Heidelberg New York Dordrecht London

Printed on acid-free paper

Springer-Verlag GmbH Berlin Heidelberg is part of Springer Science+Business Media
(www.springer.com)

身似别无长物
心如不系之舟
或问几何功业
一诺十年此书

Preface

Quantum integrable models play important roles in a variety of fields such as quantum field theory, condensed matter physics, and statistical physics. For decades, a number of theoretical methods have been developed for solving the eigenvalue problem of integrable models. Among them, the three typical and most popular methods are the coordinate Bethe Ansatz method proposed by H. Bethe in 1931, the $T-Q$ method proposed by R.J. Baxter in the early 1970s, and the algebraic Bethe Ansatz method proposed by the Leningrad Group in the late 1970s. These methods have been demonstrated as powerful in solving most of the known quantum integrable models. After Baxter's work on the eight-vertex model, people realized that a special class of quantum integrable models exists in which the $U(1)$ symmetry is broken and, in some cases, obvious reference states are absent. Some well-known examples are the *XYZ* spin chain (or equivalently the eight-vertex model), the quantum Toda chain, the anisotropic spin torus, and the quantum spin chains with nondiagonal boundary fields. Several methods have since been developed to approach this remarkable problem. Among them, two promising ones are Baxter's $T-Q$ method and Sklyanin's separation of variables (SoV) method, which provide efficient tools to treat quantum integrable models with functional analysis.

This book serves as an introduction to the off-diagonal Bethe Ansatz (ODBA) method, a newly developed analytic theory to approach exact solutions of quantum integrable models, especially those with nontrivial boundaries. In any sense, ODBA is not an isolated theory but one based on extensive existing knowledge. Therefore, this book also covers some main ingredients of $T-Q$ relation, algebraic Bethe Ansatz, thermodynamic Bethe Ansatz, fusion techniques and Sklyanin's SoV basis, etc. It is organized in a parallel structure to explain how ODBA works for different types of integrable models. Chapter 1 is devoted to the basic knowledge of quantum integrable models, and Chap. 2 to a comprehensive introduction of the algebraic Bethe Ansatz, the fusion techniques, and the SoV scheme. In addition, the thermodynamic Bethe Ansatz method is introduced as a tool for deriving the physical quantities. Chapter 3 focuses on the application of ODBA in the periodic *XXZ* model and the *XYZ* model, and Chap. 4 on the topological boundary problem using the anisotropic spin torus as example. Chapter 5 studies the exact solution of the spin-$\frac{1}{2}$ chain Hamiltonian with generic open boundaries, which had been a long-standing problem for over two decades. Chapter 6 is devoted to the one-dimensional Hubbard model and the super-symmetric $t-J$ model with generic integrable boundaries. Chapters 7 and 8 focus on the generalizations of ODBA to high-spin integrable models. Chapter 9 is devoted to the Izergin-Korepin model with generic boundaries, a typical integrable model beyond the A-type models. Calculations of some important physical quantities based on the Bethe Ansatz equations, especially the nontrivial boundary contributions, are given in Chaps. 2–5 and the method for retrieving the eigenstates based on the inhomogeneous $T-Q$ relations and the SoV basis is introduced with concrete examples in Chaps. 4 and 5.

In general, the authors aim to introduce topics that are under ongoing research and are developing at a stimulating pace in this fascinating field. These contents are selected for the book according to the authors' own understanding of the topics under discussion. Thus, they devote much attention to methods that work well for the nontrivial boundaries (Research on nontrivial surface effects, including edge states of the quantum Hall effect, surface states of topological insulators, open strings, and stochastic processes in nonequilibrium statistical physics, has become a trend in modern physics. The authors study this problem from the mathematical physics side.). The two-dimensional lattice models and most of the well-established knowledge on the models with periodic and diagonal boundary conditions are not included, since several excellent books have already covered these topics. This book was originally planned for around 100 pages but then was expanded to the present size, thanks to suggestions of numerous colleagues that detailed calculations should be included as much as possible to make it easy to follow the method. Although most of the results contained in this book have been rigorously proven, we still use the word "exactly" in the title as Baxter did for his book, for the reason that some results in this book are not that rigorously proven. For example, the thermodynamic limit is constructed based on reasonable physical arguments. For most models considered in this book, numerical results are provided to support the analytical ones, which is a conventional way for physicists and scientists in other fields to support their proposals, though it may not meet mathematical rigorousness. For physicists, to propose something correct is always more ambitious than to prove it.

Also, the methodology still needs to be developed and leaves some open questions, among which are: how to apply it in graded integrable models and in cyclic integrable models with nontrivial boundaries; how to retrieve the Bethe states and to derive the scalar products of high-rank quantum integrable models, etc. We expect that those issues may undergo significant progress in the near future.

The authors would like to share with you their happiness in undertaking the collaboration, which started in Fall 2012. At that time little was known about ODBA. Without the ensuing teamwork, it would have been impossible to achieve the main original results contained in this book!

We apologize if any important references are omitted. Any such error is definitely due to the limits of the authors' knowledge of the literature.

Acknowledgments Y. Wang would like to acknowledge a number of people who directly or indirectly helped to make the book possible. The first is F.C. Pu, who introduced the author into this interesting field. Under Pu's supervision, the author came to notice the problem of solving those integrable models without $U(1)$ symmetry 30 years ago when introduced to Baxter's papers and Takhtajan and Faddeev's paper about the XYZ model. As a junior graduate student, the author was unable to understand why the model could only be solved for an even number of sites. In 1997, when he became aware of the paper about the open *XYZ* model by H. Fan et al., he gradually realized that Baxter's local gauge transformation (vertex-face transformation or corner matrix) method could also be applied to the *XXZ* spin chain with nondiagonal boundary fields. The idea became clearer and clearer during his collaboration with J. Cao, H.-Q. Lin, and K. Shi. The author's thanks also go to D. Jin and L. Yu who have continually helped, supported, and encouraged him in his career. Fruitful discussions with R.J. Baxter and R.I. Nepomechie about the manuscript are especially acknowledged. Most importantly, the author would like to extend his deep gratitude to his wife Yan, who has been dedicating herself to taking care of the family so that her husband could devote more of his time to work even before the baby was coming!

W.-L. Yang, J. Cao, and K. Shi would like to express deep thanks to their supervisor B.-Y. Hou, whose strict style of scholarship has influenced them a lot in their career. They also benefited much from the academic atmosphere of Institute of Modern Physics of Northwest University founded by B.-Y. Hou.

All the authors would like to acknowledge some of the referees for their constructive comments on several of the authors' original papers. Remarkably, a question of "how to get the root distribution of the Bethe Ansatz equations" stimulated the authors to write the paper on the thermodynamic limit, while the comments of "how to prove the completeness of the solutions" and "what is the corresponding eigenstate in the homogeneous lattice case" stimulated the authors to propose the two theorems in Chap. 1 and related corollaries and to retrieve Bethe states based on the inhomogeneous $T-Q$ relations. These papers become important parts of the ODBA scheme.

All authors thank Q. Fu, G.-L. Li, K. Hao, and the students Y.-Y. Li, X. Zhang for proofreading the manuscript. The help in numerical simulations from two students, Y. Jiang and S. Cui is especially acknowledged.

This book was financially supported by the Natural Science Foundation of China, the Ministry of Science and Technology of China and the Chinese Academy of Sciences.

Beijing, February 2015 Yupeng Wang
Xi'an Wen-Li Yang
Beijing Junpeng Cao
Xi'an Kangjie Shi

Contents

Chapter 1
Overview

Quantum integrable models are exactly solvable models defined by the Yang-Baxter equation (YBE) [1, 2] or the Lax representation [3]. These models play important roles in a variety of fields such as quantum field theory, condensed matter physics and statistical physics, because they can provide solid benchmarks for understanding the many-body effects in corresponding universal classes and sometimes even yield conclusions to debates about important physical concepts. For instance, the exact solution of the two-dimensional Ising model [4] gives concrete evidence for the existence of thermodynamic phase transitions; the exact solution of the one-dimensional Hubbard model [5] clarifies the concept of the Mott insulator; while the spinon excitations obtained from the exact solution of the Heisenberg spin chain [6] elucidate how fractional charges could be generated from low-dimensional correlated quantum systems. In recent years, new applications have been found in cold atom systems, quantum information, AdS/CFT correspondence and many other aspects. For example, the Lieb-Liniger model [7, 8], the δ-potential Fermi gas model [1, 9] and the one-dimensional Hubbard model [5] have provided important benchmarks for one-dimensional cold atom systems and even fit experimental data with incredibly high accuracy [10]. On the other hand, the anomalous dimensions of operators of $\mathcal{N} = 4$ super-symmetric Yang-Mills field theory can be given by the eigenvalues of the Hamiltonians for certain integrable spin chains [11, 12].

For several decades, a number of theoretical methods have been proposed for solving the eigenvalue problem of quantum integrable models. Among them, the three typical and most popular methods are the coordinate Bethe Ansatz method proposed by Bethe [13], the $T - Q$ method proposed by Baxter [14, 15] and the algebraic Bethe Ansatz method [16–22] proposed by the Leningrad Group. Those methods have been demonstrated to be powerful in solving the eigenvalue problem of the known quantum integrable models and a great number of papers have been devoted to this topic in the literature. Among the family of quantum integrable models, there exists a large class of models that do not possess $U(1)$ symmetry and an obvious reference state is usually absent. Some well-known examples are the XYZ spin chain with an odd number of sites [17], the anisotropic spin torus [23] and the quantum spin chains with non-diagonal boundary fields [24–27]. These models have been

Y. Wang et al., *Off-Diagonal Bethe Ansatz for Exactly Solvable Models*,
DOI 10.1007/978-3-662-46756-5_1

found to possess important applications in non-equilibrium statistical physics (e.g., stochastic processes [28–33]), in condensed matter physics (e.g., a Josephson junction embedded in a Luttinger liquid [34], spin-orbit coupling systems, one-dimensional cold atoms coupled with a BEC reservoir, etc.) and in high energy physics (e.g., open strings and coupled D-Branes). Many efforts have been made [24–27, 29–32, 35–53] to approach this nontrivial problem.

Actually, Baxter's theory [2] already provided a powerful method for approaching exactly solvable models with functional analysis, allowing us to solve those models without $U(1)$ symmetry. A remarkable example is the exact solution of the eight-vertex model [2]. Another important functional analysis method to approach these models is the quantum separation of variables (SoV) method [54–57] proposed by Sklyanin, which has also been successfully applied to several nontrivial quantum integrable models. A famous example is the solution of the quantum Toda chain [54]. Nevertheless, for a long time, the Bethe Ansatz equations could only be obtained for constrained boundary conditions [24, 25, 37] or for special crossing parameters [26, 27, 35, 36] associated with spin-$\frac{1}{2}$ chains or with spin-s chains [58–61]. An analytic method for solving the integrable models with or without obvious reference state, i.e., the off-diagonal Bethe Ansatz (ODBA) method was proposed in 2013 [62]. With this method, several models without obvious reference states were solved exactly [62–71] by the construction of the inhomogeneous $T-Q$ relations, and a method to obtain the physical quantities in the thermodynamic limit was established [72] subsequently, based on the ODBA equations. Soon after that, Sklyanin's SoV method was applied to the spin-$\frac{1}{2}$ chains with generic integrable boundaries [73], and a set of Bethe states was conjectured via the algebraic Bethe Ansatz [74]. A systematic method to retrieve the Bethe-type eigenstates based on the ODBA solutions and the SoV basis is developed in [75, 76].

This chapter is a brief introduction of the integrability associated with YBE; the boundary conditions associated with the integrability; the factorizability induced by YBE and the ideas of the coordinate Bethe Ansatz; the $T-Q$ relation; and the basic ingredients of ODBA.

1.1 Integrability and Yang-Baxter Equation

The concept of integrability originated from classical mechanics, wherein a physical system is usually described by a set of differential equations (the equations of motion). The solutions of these differential equations are their integrals. In such a sense, integrable means solvable. The integrals are accompanied by some integral constants that do not depend on time and are usually called integrals of motion or conserved quantities. For a mechanical system with N degrees of freedom, if N independent integrals of motion which are in involution can be obtained, then the system is completely integrable.

A precise definition of classical integrability is given by Liouville's theorem [77]: Given a Hamiltonian system described by the coordinates $\{x_j | j = 1, \ldots, N\}$ and the momenta $\{k_j | j = 1, \ldots, N\}$, if there exists a canonical transformation $x_j, k_j \to q_j, p_j$ to make the Hamiltonian to be only a function of the canonical momenta $\{p_j\}$, the system is integrable. This is true because the following Poisson brackets hold:

$$\{k_j, x_l\} = \delta_{j,l}, \quad \{p_j, q_l\} = \delta_{j,l},$$
$$\frac{dp_j}{dt} = \{H, p_j\} = 0, \quad \frac{dq_j}{dt} = \frac{\partial H}{\partial p_j}, \tag{1.1.1}$$

which imply that the N canonical momenta are conserved quantities and the evolution of the N canonical coordinates is linear in time t. The Liouville's theorem indicates that the variables of an integrable system are in fact completely separable. However, such a separation process is usually rather nontrivial.

To show the integrability clearly, let us first consider a simple classical integrable system which might give a bridge to the quantum integrable systems: N classical indistinguishable objects moving in a straight line. Suppose each object carries a momentum k_j initially and the collisions among the objects are elastic. Consider the collision process between two neighboring objects. If the objects carry momenta k_i and k_j before the collision, and k_i' and k_j' after the collision, respectively, the conservation laws of momentum and energy require that

$$k_i + k_j = k_i' + k_j', \tag{1.1.2}$$
$$k_i^2 + k_j^2 = k_i'^2 + k_j'^2. \tag{1.1.3}$$

The above equations have two sets of solutions: (1) $k_i' = k_i, k_j' = k_j$; (2) $k_i' = k_j$, $k_j' = k_i$. Since these objects are not penetrable, only the second set of solutions is allowed, i.e., the objects exchange their momenta after the collision. If the objects are moving in a ring (periodic boundary condition), the system can be described by a parameter set $\{k_1, \ldots, k_N\}$ which does not change with the collision processes. Such phenomenon is usually called non-diffraction behavior and is a common feature of the integrable systems. Obviously, the following conserved quantities hold:

$$\mathrm{C}_n = \sum_{j=1}^{N} k_j^n, \quad n = 1, \ldots, N, \tag{1.1.4}$$

indicating that this system is completely integrable. If the objects move in an interval with boundaries, the momenta they carry are no longer a conserved set of parameters. The object carrying a momentum k_j must be reflected at the boundaries and its momentum is changed to $-k_j$ after reflection. However, the system still preserves its integrability because of the existence of the following conserved quantities:

$$\mathrm{C}_n^o = \sum_{j=1}^{N} |k_j|^n, \quad n = 1, \ldots, N. \tag{1.1.5}$$

The central point of quantum integrability lies in the conservation laws governed by the YBE. There are several ways to derive YBE. Here we adopt Yang's procedure [1]. Consider that N indistinguishable quantum particles are moving in one spatial dimension. Suppose its wave function initially takes the following asymptotic form:

$$\Psi_{in} \sim e^{i\sum_{j=1}^{N} k_j x_j}, \quad x_1 \ll x_2 \ll \cdots \ll x_N. \tag{1.1.6}$$

The first particle reaches the right end of the system from the left after scattering with all the other particles. The asymptotic wave function after this process becomes

$$\Psi_{out} \sim S_{1,23\ldots N} e^{i\sum_{j=1}^{N} k_j x_j}, \quad x_2 \ll x_3 \ll \cdots \ll x_N \ll x_1, \tag{1.1.7}$$

where $S_{1,23\ldots N}$ is the scattering matrix of particle 1 to all the other particles. If the many-body S-matrix can be factorized as the product of two-body S-matrices

$$S_{1,23\ldots N} = S_{1,N}(k_1, k_N) \cdots S_{1,3}(k_1, k_3) S_{1,2}(k_1, k_2), \tag{1.1.8}$$

we call the system a factorizable system. We note that the following inversion identity for the two-body S-matrix holds:

$$S_{1,2}(k_1, k_2) S_{2,1}(k_2, k_1) = 1. \tag{1.1.9}$$

The factorizability ensures the integrability of a quantum system. To show this point clearly, let us consider the three-particle case. There are two routes from the initial state $|1, 2, 3\rangle$ to the final state $|3, 2, 1\rangle$ as shown in Fig. 1.1. These two routes must be equivalent because of the uniqueness of the final wave function. If the system is factorizable, we have the following equation:

$$S_{1,2}(k_1, k_2) S_{1,3}(k_1, k_3) S_{2,3}(k_2, k_3) = S_{2,3}(k_2, k_3) S_{1,3}(k_1, k_3) S_{1,2}(k_1, k_2). \tag{1.1.10}$$

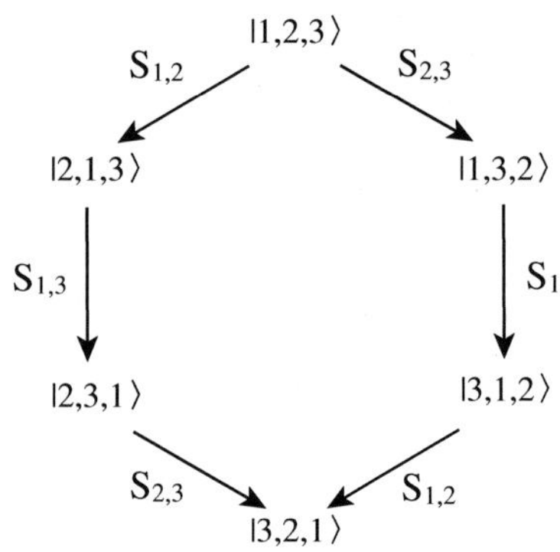

Fig. 1.1 Schematic diagram of the Yang-Baxter equation: the two routes from the initial state $|1, 2, 3\rangle$ to the final state $|3, 2, 1\rangle$ must be equivalent

This is the YBE, which was realized in [78] and first emphasized by Yang [1] in solving the one-dimensional δ-potential Fermi gas model and by Baxter [2] in constructing the $T-Q$ method for solving the two-dimensional vertex models. It was demonstrated by Yang [79] that YBE is the sufficient condition of Yang-Baxter quantum integrability with proper boundary conditions. This equation also ensures factorizability, thus constituting the cornerstone for constructing and solving the quantum integrable models. In fact, the factorizability indicates that the basic scattering process is the two-body one, and that some conserved quantities that possess the eigenvalues of Eqs. (1.1.4) or (1.1.5) always exist, because of the conservation laws of momentum and energy.

Usually, if the spectral parameters in the S-matrix are additive, i.e., $S_{i,j}(k_i, k_j) \sim R_{i,j}(k_i - k_j)$, the YBE is written as

$$\begin{aligned} R_{i,j}(u_i - u_j)R_{i,k}(u_i - u_k)R_{j,k}(u_j - u_k) \\ = R_{j,k}(u_j - u_k)R_{i,k}(u_i - u_k)R_{i,j}(u_i - u_j). \end{aligned} \tag{1.1.11}$$

Throughout this book we adopt the standard notations: for any matrix $O \subset \mathrm{End}(\mathbf{V})$, O_j is an embedding operator in the tensor space $\mathbf{V} \otimes \mathbf{V} \otimes \cdots \otimes \mathbf{V}$, which acts as O on the jth factor space and as identity on the other factor spaces; $R_{i,j}(u)$ is an embedding operator of the R-matrix in the tensor space, which acts as identity on the factor spaces except for the ith and jth ones. Moreover, we denote id as the identity operator in the corresponding space.

1.2 Integrable Boundary Conditions

There are several possible boundary conditions associated with the quantum integrability. To show them clearly, let us first introduce the procedure for constructing quantum integrable models based on YBE. In principle, given an R-matrix, we can seek solutions of the equation

$$R_{0,\bar{0}}(u-v)L_{0,n}(u)L_{\bar{0},n}(v) = L_{\bar{0},n}(v)L_{0,n}(u)R_{0,\bar{0}}(u-v). \tag{1.2.1}$$

Obviously, $L_{0,n}(u) = R_{0,n}(u - \theta_n)$ is a solution of this equation. $L_{0,n}(u)$ is usually called the Lax operator and θ_n is a site-dependent parameter (inhomogeneous parameter). Given an R-matrix satisfying YBE, we define the monodromy matrix

$$T_0(u) = L_{0,N}(u)L_{0,N-1}(u)\cdots L_{0,1}(u), \tag{1.2.2}$$

where N is the number of sites of the system. The transfer matrix of the system is defined as the trace of the corresponding monodromy matrix in the auxiliary space

$$t(u) = tr_0 T_0(u). \tag{1.2.3}$$

The concept of the transfer matrix originated from the classical statistical models [2] and was adopted later in the study of quantum integrable models.

An important step to construct and to solve quantum integrable models is the RTT relation proposed by Baxter. Since

$$[L_{0,m}(u), L_{\bar{0},n}(v)] = 0, \quad m \neq n, \tag{1.2.4}$$

from YBE (1.2.1) we have

$$\begin{aligned}
&R_{0,\bar{0}}(u-v)T_0(u)T_{\bar{0}}(v)\\
&\quad = R_{0,\bar{0}}(u-v)L_{0,N}(u)L_{\bar{0},N}(v)\cdots L_{0,1}(u)L_{\bar{0},1}(v)\\
&\quad = L_{\bar{0},N}(v)L_{0,N}(u)R_{0,\bar{0}}(u-v)\cdots L_{0,1}(u)L_{\bar{0},1}(v)\\
&\quad = L_{\bar{0},N}(v)L_{0,N}(u)\cdots L_{\bar{0},1}(v)L_{0,1}(u)R_{0,\bar{0}}(u-v)\\
&\quad = T_{\bar{0}}(v)T_0(u)R_{0,\bar{0}}(u-v).
\end{aligned} \tag{1.2.5}$$

Multiplying $R^{-1}_{0,\bar{0}}(u-v)$ from the left side of the Eq. (1.2.5) and taking the trace in the auxiliary spaces 0 and $\bar{0}$, we obtain

$$[t(u), t(v)] = 0. \tag{1.2.6}$$

Expanding $t(u)$ in terms of u

$$t(u) = \sum_{n=0}^{\infty} t^{(n)} u^n, \tag{1.2.7}$$

we readily have that the coefficients are mutually commuting

$$[t^{(m)}, t^{(n)}] = 0. \tag{1.2.8}$$

Choosing one of them or a certain combination of them as a Hamiltonian H, then $[H, t^{(n)}] = 0$ and the model is integrable. If we obtain the eigenvalues of the transfer matrix, we can obtain all the eigenvalues of the coefficients. The boundary condition for the transfer matrix defined by (1.2.2) and (1.2.3) is periodic.

In most of the cases, Eq. (1.2.1) allows c-number solution $L_{0,n}(u) = G_0$ which is independent of the spectral parameter u. This allows us to construct the following transfer matrix

$$t(u) = tr_0\{G_0 L_{0,N}(u)L_{0,N-1}(u)\cdots L_{0,1}(u)\}, \tag{1.2.9}$$

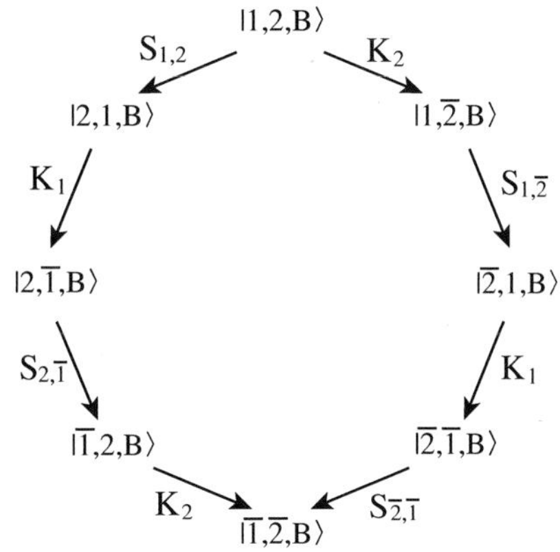

Fig. 1.2 Schematic diagram of the reflection equation: the two routes from the initial state $|1, 2, B\rangle$ to the final state $|\bar{1}, \bar{2}, B\rangle$ must be equivalent. B indicates the open boundary

which also satisfies the commuting relation (1.2.6). $G = \text{id}$ corresponds to the periodic boundary condition. If G is a diagonal matrix in the auxiliary space, it corresponds to a twisted boundary condition. If G is a non-diagonal matrix, it then defines an antiperiodic (or topological) boundary condition.

For the open-boundary quantum systems, it was proposed by Sklyanin [22] and Cherednik [80] that, apart from YBE, another equation accounting for the boundaries must also be satisfied to preserve integrability. For a two-body system with open boundaries, there are again two routes from the initial state $|1, 2, B\rangle$ to the final state $|\bar{1}, \bar{2}, B\rangle$ as shown in Fig. 1.2. Because of the uniqueness of the wave function, the following equation must hold for a factorizable system:

$$\begin{aligned} &S_{1,2}(k_1, k_2)K_1(k_1)S_{2,1}(k_2, -k_1)K_2(k_2) \\ &\quad = K_2(k_2)S_{1,2}(k_1, -k_2)K_1(k_1)S_{2,1}(-k_2, -k_1), \end{aligned} \tag{1.2.10}$$

where $K(u)$ denotes the reflection matrix at the boundary. Equation (1.2.10) is called a reflection equation (RE). A general form of RE for the additive systems is written as

$$\begin{aligned} &R_{1,2}(u - v)K_1^-(u)R_{2,1}(u + v)K_2^-(v) \\ &\quad = K_2^-(v)R_{1,2}(u + v)K_1^-(u)R_{2,1}(u - v). \end{aligned} \tag{1.2.11}$$

Generally, the R-matrix satisfies the following unitary and crossing-unitary relations [81, 82]:

$$R_{1,2}(u)\, R_{2,1}(-u) \propto \text{id} \otimes \text{id}, \tag{1.2.12}$$

$$R_{1,2}^{t_1}(u)\, \mathscr{M}_1\, R_{2,1}^{t_1}(-u - h\eta)\, \mathscr{M}_1^{-1} \propto \text{id} \otimes \text{id}, \tag{1.2.13}$$

where t_i denotes transposition in the ith space, h is a model-dependent number (usually called the dual Coxeter number [81]) and η is the crossing parameter, which describes the interaction strength among the quantum objects, and $\mathscr{M}$ is a constant matrix satisfying the relation

$$[\mathscr{M}_1\mathscr{M}_2,\ R_{1,2}(u)] = 0, \quad \mathscr{M}^t = \mathscr{M}. \tag{1.2.14}$$

Given a $\bar{K}^+(u)$ belonging to the solution set $\{K^-(u)\}$ of (1.2.11), let us define

$$K^+(u) = \bar{K}^+(-u - \frac{h}{2}\eta)\mathscr{M}. \tag{1.2.15}$$

Then $K^+(u)$ satisfies the dual RE [22, 80, 83]

$$\begin{aligned} &R_{1,2}(v-u)K_1^+(u)\mathscr{M}_1^{-1}R_{2,1}(-u-v-h\eta)\mathscr{M}_1K_2^+(v) \\ &\quad = K_2^+(v)\mathscr{M}_2^{-1}R_{1,2}(-u-v-h\eta)\mathscr{M}_2K_1^+(u)R_{2,1}(v-u). \end{aligned} \tag{1.2.16}$$

It should be remarked that $\mathscr{M} = \text{id}$ only for some special cases but generally $\mathscr{M} \neq \text{id}$ (see, e.g., Chap. 9).

Corresponding to RE (1.2.10), the general dual RE reads

$$\begin{aligned} &S_{2,1}^{-1}(-k_2,-k_1)K_1^+(k_1)S_{1,2}^{t_2,-1,t_2}(k_1,-k_2)K_2^+(k_2) \\ &\quad = K_2^+(k_2)S_{2,1}^{t_1,-1,t_1}(k_2,-k_1)K_1^+(k_1)S_{1,2}^{-1}(k_1,k_2). \end{aligned} \tag{1.2.17}$$

As for the periodic quantum integrable models, based on YBE and RE, commuting transfer matrix for a quantum integrable model with open boundaries can be constructed. Details will be given in Chap. 2.

1.3 Basic Ingredients of the Coordinate Bethe Ansatz

The idea of the coordinate Bethe Ansatz [13] is clear with the factorizability. For an integrable system with the periodic boundary conditions

$$\Psi_{in}(\ldots, x_j = 0, \ldots) = \Psi_{out}(\ldots, x_j = L, \ldots),$$

we have the equation

$$\begin{aligned} &S_{j,N}(k_j,k_N)\cdots S_{j,j+1}(k_j,k_{j+1}) \\ &\quad \times S_{j,j-1}(k_j,k_{j-1})\cdots S_{j,1}(k_j,k_1)e^{ik_jL}\Psi_{in} = \Psi_{in}. \end{aligned} \tag{1.3.1}$$

For a spinless system, all the S-matrices are c-numbers. The Bethe Ansatz equations (BAEs) determining the Bethe roots $\{k_j\}$ read

$$e^{ik_jL} = \prod_{l\neq j}^{N} S_{j,l}^{-1}(k_j, k_l). \tag{1.3.2}$$

If the particles also possess internal degrees of freedom, $S_{j,l}(k_j, k_l)$ is normally an operator and (1.3.1) becomes another eigenvalue problem

$$t(k_j)\Psi_{in} = e^{-ik_jL}\Psi_{in}, \tag{1.3.3}$$

which can be solved by either the algebraic Bethe Ansatz method introduced in Chap. 2 or the ODBA method introduced in Chap. 3 with the transfer matrix

$$t(k_j) = S_{j,N}(k_j, k_N)\cdots S_{j,j+1}(k_j, k_{j+1})S_{j,j-1}(k_j, k_{j-1})\cdots S_{j,1}(k_j, k_1). \tag{1.3.4}$$

A detailed description is given in Chap. 6.

The coordinate Bethe Ansatz method was first applied to the open integrable systems by Gaudin [84]. For N dynamical particles in an interval $[0, L]$, suppose the initial wave function is

$$\Psi_{in} \sim e^{i\sum_{j=1}^{N} k_jx_j}, \quad x_j = 0 \leq x_1 \ll \cdots \ll x_N \leq L. \tag{1.3.5}$$

We consider the following process: particle j moves from the left boundary to the right boundary by scattering once with all the other particles. Then it is bounced back by the right boundary with a reversed momentum $-k_j$. The reflected particle moves further from the right boundary to the left boundary by scattering once again with all the other particles. Finally it is bounced back by the left boundary and arrives at its initial position. This process is described by the equation

$$\begin{aligned}\bar{K}_j^{+}(k_j)S_{1,j}(k_1, -k_j)\cdots S_{N,j}(k_N, -k_j)K_j^{-}(k_j)\\ \times e^{2ik_jL}S_{j,N}(k_j, k_N)\cdots S_{j,1}(k_j, k_1)\Psi_{in} = \Psi_{in},\end{aligned} \tag{1.3.6}$$

where $\bar{K}_j^{+}(k_j)$ is the reflection matrix to the left boundary (see Chap. 6). Again if the particles have no internal degrees of freedom, all the S-matrices and the K-matrices are c numbers. The BAEs determining the Bethe roots $\{k_j\}$ read

$$\begin{aligned}e^{-2ik_jL} = \bar{K}_j^{+}(k_j)S_{1,j}(k_1, -k_j)\cdots S_{N,j}(k_N, -k_j)\\ \times K_j^{-}(k_j)S_{j,N}(k_j, k_N)\cdots S_{j,1}(k_j, k_1).\end{aligned} \tag{1.3.7}$$

However, if the particles also possess internal degrees of freedom, Eq. (1.3.6) becomes a new eigenvalue equation

$$t(k_j)\Psi_{in} = e^{-2ik_jL}\Psi_{in}, \tag{1.3.8}$$

with the transfer matrix

$$\begin{aligned} t(k_j) = {} & \bar{K}_j^+(k_j)S_{1,j}(k_1, -k_j)\cdots S_{N,j}(k_N, -k_j) \\ & \times K_j^-(k_j)S_{j,N}(k_j, k_N)\cdots S_{j,1}(k_j, k_1), \end{aligned} \tag{1.3.9}$$

which can be solved either by the algebraic Bethe Ansatz method (for constrained K-matrices) or by the ODBA method as introduced in Chap. 6.

1.4 $T - Q$ Relation

The $T - Q$ relation was proposed by Baxter [2] for solving the six-vertex model and the eight-vertex model. By checking the Bethe Ansatz solutions of the six-vertex model, Baxter realized that the eigenvalues $\Lambda(u)$ of the transfer matrix $t(u)$ take a unified form for all the eigenstates

$$\Lambda(u) = a(u)\frac{Q(u-\eta)}{Q(u)} + d(u)\frac{Q(u+\eta)}{Q(u)}. \tag{1.4.1}$$

This allows us to rewrite the above functional relation to a matrix equation, i.e., to replace $\Lambda(u)$ and $Q(u)$ with the corresponding diagonal matrices $t(u)$ and $\hat{Q}(u)$. If the Bethe states span the whole Hilbert space, we may further treat $t(u)$ and $\hat{Q}(u)$ as two mutually commutative operators defined in the corresponding Hilbert space. Based on this observation, Baxter proposed the $T - Q$ relation for the spin-$\frac{1}{2}$ integrable models

$$\hat{Q}(u)t(u) = a(u)\hat{Q}(u-\eta) + d(u)\hat{Q}(u+\eta), \tag{1.4.2}$$

where $a(u)$ and $d(u)$ are two known model-dependent functions. Note that the above operator relation does not depend on the representation basis. Based on this proposal, Baxter successfully obtained the exact solution of the eight-vertex model by constructing the eigenvectors of the $\hat{Q}$-operator. In addition, since $[t(u), t(v)] = 0$, the eigenstates of the transfer matrix do not depend on u. Suppose $|\Psi\rangle$ is a common eigenstate of the transfer matrix and the $\hat{Q}$-operator with

$$\begin{aligned} t(u)|\Psi\rangle &= \Lambda(u)|\Psi\rangle, \\ \hat{Q}(u)|\Psi\rangle &= Q(u)|\Psi\rangle. \end{aligned} \tag{1.4.3}$$

From Eq. (1.4.2) the functional relation Eq. (1.4.1) is recovered.

Indeed, the $T - Q$ relation is universal for most of the integrable models (the single-component ones) possessing a proper reference state, as verified later by the algebraic Bethe Ansatz. Normally $Q(u)$ is a polynomial of some entire function $f(u)$ and can be parameterized as

$$Q(u) = \prod_{j=1}^{M} f(u - \mu_j), \tag{1.4.4}$$

with $f(0) = 0$, where the parameters $\{\mu_j | j = 1, \ldots, M\}$ are usually called Bethe roots. Since $\Lambda(u)$ is also an entire function about u, the regularity of Eq. (1.4.1) requires that all the residues about μ_j must be zero. This induces the BAEs

$$\frac{a(\mu_j)}{d(\mu_j)} = -\frac{Q(\mu_j + \eta)}{Q(\mu_j - \eta)}. \tag{1.4.5}$$

Note that the simplicity of the "poles" μ_j is required in order to derive the above BAEs, which is a common feature of the Bethe roots.

Some remarks are in order here: For most of the integrable models without $U(1)$ symmetry, this relation does not allow polynomial solutions of $Q(u)$ and an extended version, namely, an inhomogeneous $T - Q$ relation [62] should be used. In addition, for the high-rank integrable systems, more Q-functions or operators are included in the corresponding $T - Q$ relations. This generalization will be discussed in Chap. 7.

1.5 Basic Ingredients of the Off-Diagonal Bethe Ansatz

Usually, to derive eigenstates of a Hamiltonian or a transfer matrix, a reference state is definitely needed. However, for the quantum integrable models without $U(1)$ symmetry, some off-diagonal elements of the monodromy matrix enter the expression of the transfer matrix, which induces the absence of an obvious reference state. Nevertheless, a reference state is not necessary for getting the spectrum, if we can find some functional relations to determine the eigenvalues in terms of an extended version of the $T - Q$ relation (1.4.1) or (1.4.2). The eigenstates can thus be naturally retrieved from the eigenvalues (see Chaps. 4 and 5).

Generally, the eigenvalue $\Lambda(u)$ is a degree N polynomial of some entire function $f(u)$ and can be factorized as

$$\Lambda(u) = \Lambda_0 \prod_{j=1}^{N} f(u - z_j), \tag{1.5.1}$$

where Λ_0 is a constant which can be determined by the asymptotic behavior of the transfer matrix. If there are N equations for the N unknowns z_j, $\Lambda(u)$ can be determined completely.

The central idea of the ODBA [62] is to derive the functional $T-Q$ relation based on N operator identities which determine the N unknowns of $\Lambda(u)$. As a matter of fact, for a regular 2×2 matrix, there are two basic invariants, i.e., its trace and its determinant that do not depend on the representation basis or the eigenvectors (for an $n \times n$ regular matrix, there are n invariants.). For the quantum integrable models, the operator-valued monodromy matrix also possesses two basic invariants, i.e., its trace (the transfer matrix) and its quantum determinant. Indeed, for quantum integrable systems, the relationship between the transfer matrix and the quantum determinant at N special points of the spectral parameter can be constructed based on YBE and RE (open boundary systems).

1.5.1 Functional Relations of the XXX Spin-$\frac{1}{2}$ Chain

To show the ODBA procedure clearly, let us consider the periodic XXX spin-$\frac{1}{2}$ chain. The corresponding R-matrix reads

$$R_{0,j}(u) = u + \eta P_{0,j} = u + \frac{1}{2}\eta(1 + \sigma_j \cdot \sigma_0), \tag{1.5.2}$$

where η is the crossing parameter (we put $\eta = 1$ in this case), $\sigma_j = (\sigma_j^x, \sigma_j^y, \sigma_j^z)$ are the Pauli matrices, and $P_{i,j}$ is the permutation operator possessing the properties

$$P_{i,j} O_j = O_i P_{i,j}, \quad P_{i,j}^2 = \mathrm{id}, \quad tr_j P_{i,j} = tr_i P_{i,j} = \mathrm{id}, \tag{1.5.3}$$

for arbitrary operator O defined in the corresponding tensor space. It is easy to show that the R-matrix (1.5.2) also satisfies the following relations which will be used often in the following text

$$\begin{aligned}
&\text{Initial condition}: \ R_{1,2}(0) = P_{1,2}, && (1.5.4)\\
&\text{Unitary relation}: \ R_{1,2}(u)R_{2,1}(-u) = -\varphi(u) \times \mathrm{id}, && (1.5.5)\\
&\qquad\qquad\qquad\quad \varphi(u) = u^2 - 1, &&\\
&\text{Crossing relation}: \ R_{1,2}(u) = -\sigma_1^y R_{1,2}^{t_1}(-u-1)\sigma_1^y, && (1.5.6)\\
&\text{PT-symmetry}: \ R_{1,2}(u) = R_{2,1}(u) = R_{1,2}^{t_1 t_2}(u), && (1.5.7)\\
&Z_2\text{-symmetry}: \ \sigma_1^\alpha \sigma_2^\alpha R_{1,2}(u) = R_{1,2}(u)\sigma_1^\alpha \sigma_2^\alpha, \ \text{ for } \alpha = x, y, z, && (1.5.8)\\
&\text{Fusion condition}: \ R_{1,2}(\pm 1) = \pm 1 + P_{1,2} = \pm 2P_{1,2}^{(\pm)}, && (1.5.9)
\end{aligned}$$

where $P_{1,2}^{(+)}$ ($P_{1,2}^{(-)}$) is the symmetric (antisymmetric) projection operator. With the above basic properties, the following crossing-unitary relation also holds:

$$R_{1,2}^{t_1}(u)R_{1,2}^{t_1}(-u-2) = R_{1,2}^{t_1}(-u-2)R_{1,2}^{t_1}(u) = -\varphi(u+1) \times \mathrm{id}. \tag{1.5.10}$$

The monodromy matrix and the corresponding transfer matrix of the periodic XXX spin-$\frac{1}{2}$ chain are respectively defined as

$$T_0(u) = R_{0,N}(u - \theta_N) \cdots R_{0,1}(u - \theta_1), \tag{1.5.11}$$

$$t(u) = tr_0 T_0(u), \tag{1.5.12}$$

with $\{\theta_j | j = 1, \ldots, N\}$ being some generic site-dependent inhomogeneity parameters. The Hamiltonian of the XXX spin-$\frac{1}{2}$ chain is thus expressed as

$$H = \frac{1}{2}\sum_{j=1}^{N} \sigma_j \cdot \sigma_{j+1} = \left.\frac{\partial \ln t(u)}{\partial u}\right|_{u=0,\{\theta_j=0\}} - \frac{1}{2}N, \tag{1.5.13}$$

with the periodic boundary condition $\sigma_{N+1} \equiv \sigma_1$.

In order to get some functional relations of the transfer matrix, we evaluate the transfer matrix $t(u)$ at the particular points $u = \theta_j$ and $u = \theta_j - 1$. Let us apply the initial condition of the R-matrix to express the transfer matrix $t(\theta_j)$ as

$$\begin{aligned} t(\theta_j) &= tr_0\{R_{0,N}(\theta_j - \theta_N) \cdots R_{0,j+1}(\theta_j - \theta_{j+1}) \\ &\quad \times P_{0,j} R_{0,j-1}(\theta_j - \theta_{j-1}) \cdots R_{0,1}(\theta_j - \theta_1)\} \\ &= R_{j,j-1}(\theta_j - \theta_{j-1}) \cdots R_{j,1}(\theta_j - \theta_1) \\ &\quad \times tr_0\{R_{0,N}(\theta_j - \theta_N) \cdots R_{0,j+1}(\theta_j - \theta_{j+1}) P_{0,j}\} \\ &= R_{j,j-1}(\theta_j - \theta_{j-1}) \cdots R_{j,1}(\theta_j - \theta_1) \\ &\quad \times R_{j,N}(\theta_j - \theta_N) \cdots R_{j,j+1}(\theta_j - \theta_{j+1}). \end{aligned} \tag{1.5.14}$$

In deriving the above equation, the initial condition (1.5.4) of the R-matrix plays a key role, which allows us to rewrite the transfer matrix as a product of R-matrices at the special spectral parameter points θ_j. The transfer matrix $t(\theta_j)$ is a reduced monodromy matrix if the jth quantum space is treated as the auxiliary space.

The crossing relation (1.5.6) makes it possible to express the transfer matrix $t(\theta_j - 1)$ as

$$\begin{aligned} t(\theta_j - 1) &= tr_0\{R_{0,N}(\theta_j - \theta_N - 1) \cdots R_{0,1}(\theta_j - \theta_1 - 1)\} \\ &= (-1)^N \, tr_0\{\sigma_0^y R_{0,N}^{t_0}(-\theta_j + \theta_N) \cdots R_{0,1}^{t_0}(-\theta_j + \theta_1)\sigma_0^y\} \\ &= (-1)^N \, tr_0\{R_{0,1}(-\theta_j + \theta_1) \cdots R_{0,N}(-\theta_j + \theta_N)\} \\ &= (-1)^N R_{j,j+1}(-\theta_j + \theta_{j+1}) \cdots R_{j,N}(-\theta_j + \theta_N) \\ &\quad \times R_{j,1}(-\theta_j + \theta_1) \cdots R_{j,j-1}(-\theta_j + \theta_{j-1}). \end{aligned} \tag{1.5.15}$$

Using the unitary relation (1.5.5), we have

$$t(\theta_j)t(\theta_j - 1) = a(\theta_j)d(\theta_j - 1), \quad j = 1, \ldots, N, \tag{1.5.16}$$

$$a(u) = \prod_{j=1}^{N} (u - \theta_j + 1), \quad d(u) = \prod_{j=1}^{N}(u - \theta_j). \tag{1.5.17}$$

The homogeneous analogue of (1.5.16) reads

$$\frac{\partial^l}{\partial u^l}\{t(u)t(u-1) - a(u)d(u-1)\}|_{u=0,\{\theta_j=0\}} = 0, \quad l = 0, \ldots, N-1. \tag{1.5.18}$$

Applying (1.5.16) to an eigenstate of $t(u)$, the corresponding eigenvalue $\Lambda(u)$ thus satisfies

$$\Lambda(\theta_j)\Lambda(\theta_j - 1) = a(\theta_j)d(\theta_j - 1), \quad j = 1, \ldots, N. \tag{1.5.19}$$

In addition, from the definition of the transfer matrix (1.5.12) it is easy to show that

$$\Lambda(u) \text{ is a degree } N \text{ polynomial of } u, \tag{1.5.20}$$

with the asymptotic behavior

$$\Lambda(u) = 2u^N + \cdots . \tag{1.5.21}$$

1.5.2 Two Theorems on the Complete-Spectrum Characterization

Theorem 1 *Each solution of* (1.5.19)–(1.5.21) *can be parameterized in terms of the* $T - Q$ *relation* (1.4.1) *with a polynomial* Q*-function.*

Proof Given a degree N polynomial $\Lambda(u)$ satisfying (1.5.19)–(1.5.21), we seek the degree N polynomial solution of Q-function satisfying the equation

$$Q(u)\Lambda(u) = a(u)Q(u-1) + d(u)Q(u+1). \tag{1.5.22}$$

We note that the above equation is a polynomial of degree $2N$. If the equation holds at $2N+1$ independent points of u, the equation must also hold for arbitrary u. Obviously, the above equation holds for $u \to \infty$. In addition, as $d(\theta_j) = a(\theta_j - 1) = 0$, we readily obtain that

$$Q(\theta_j)\Lambda(\theta_j) = a(\theta_j)Q(\theta_j - 1), \tag{1.5.23}$$

$$Q(\theta_j - 1)\Lambda(\theta_j - 1) = d(\theta_j - 1)Q(\theta_j). \tag{1.5.24}$$

From (1.5.19) we deduce that only one of (1.5.23) and (1.5.24) is independent. Obviously, (1.5.23) (or equivalently (1.5.24)) allows a degree N polynomial solution of $Q(u)$

$$Q(u) = u^n + \sum_{n=0}^{N-1} \tilde{I}_n u^n = \prod_{j=1}^{N} (u - \mu_j). \tag{1.5.25}$$

Substituting the above Ansatz into (1.5.23) we have N linear equations for the N coefficients $\{\tilde{I}_n | n = 0, \ldots, N-1\}$ which have a unique solution for a given $\Lambda(u)$. We note that when some of $\mu_j \to \infty$ in (1.5.25), the degree of $Q(u)$ can be reduced to M with $0 \leq M \leq N$. Taking $u = \mu_j$ in (1.5.22), we readily have the BAEs

$$a(\mu_j)Q(\mu_j - 1) + d(\mu_j)Q(\mu_j + 1) = 0, \quad j = 0, \ldots, M. \tag{1.5.26}$$

□

Theorem 2 *The functional relations (1.5.19)–(1.5.21) are the sufficient and necessary conditions to completely characterize the spectrum of the transfer matrix (1.5.12) with the R-matrix (1.5.2).*

Proof For each solution $\Lambda(u)$ of the functional relations (1.5.19)–(1.5.21), in terms of the $T-Q$ relation (1.4.1), we can construct an eigenstate of the transfer matrix via algebraic Bethe Ansatz (see Chap. 2) with $\Lambda(u)$ as the eigenvalue. In such sense, each solution of the functional relations (1.5.19)–(1.5.21) corresponds to a right eigenvalue of the transfer matrix. □

As we shall show in the following chapters, a similar conclusion with a generic inhomogeneous $T-Q$ relation also holds for other quantum integrable models.

1.5.3 Inhomogeneous $T - Q$ Relation

Baxter's $T-Q$ relation (1.4.1) gives a convenient parametrization of the eigenvalues of the transfer matrix. It is obvious that this parametrization is not the unique one because there are many ways to characterize a polynomial function, e.g., with its roots or with its coefficients. We can easily demonstrate that for any given parameter ϕ, the following inhomogeneous $T-Q$ relation also satisfies (1.5.19)–(1.5.21) and therefore characterizes the spectrum of the transfer matrix $t(u)$ of the periodic *XXX* spin-$\frac{1}{2}$ chain completely

$$\Lambda(u) = e^{i\phi}a(u)\frac{Q(u-1)}{Q(u)} + e^{-i\phi}d(u)\frac{Q(u+1)}{Q(u)} + 2(1-\cos\phi)\frac{a(u)d(u)}{Q(u)}, \tag{1.5.27}$$

$$Q(u) = \prod_{j=1}^{N}(u - \mu_j), \tag{1.5.28}$$

Table 1.1 The numerical solutions of the BAEs (1.5.29) for $N = 3, \phi = -0.69315i$ and $\{\theta_j = 0\}$

μ_1	μ_2	μ_3	E_n	d
$-2.97259 + 1.15909i$	$-2.51751 - 1.42184i$	$-0.50990 + 0.26274i$	-1.50000	2
$-2.97259 - 1.15909i$	$-2.51751 + 1.42184i$	$-0.50990 - 0.26274i$	-1.50000	2
$-2.88462 + 0.00000i$	$-1.55769 - 2.56650i$	$-1.55769 + 2.56650i$	1.50000	4

The eigenvalues E_n calculated from (1.5.30) are the same as those from the exact diagonalization of the Hamiltonian (1.5.13). The symbol n denotes the number of the energy levels and d indicates the number of degeneracy

provided that the Bethe roots $\{\mu_j | j = 1, \ldots, N\}$ satisfy the BAEs

$$e^{i\phi} a(\mu_j) Q(\mu_j - 1) + e^{-i\phi} d(\mu_j) Q(\mu_j + 1) = 2(\cos\phi - 1) a(\mu_j) d(\mu_j), \tag{1.5.29}$$

and the selection rules $\mu_j \neq \mu_l, \mu_j \neq \theta_l, \theta_l - 1$. The corresponding eigenvalues of the Hamiltonian (1.5.13) read

$$E = \left. \frac{\partial \ln \Lambda(u)}{\partial u} \right|_{u=0, \{\theta_j = 0\}} - \frac{1}{2} N. \tag{1.5.30}$$

The numerical solutions of the BAEs (1.5.29) and the corresponding eigenvalues of the Hamiltonian (1.5.13) for $N = 3$ and $N = 4$ and arbitrarily chosen ϕ are shown in Tables 1.1 and 1.2 respectively, while the calculated $\Lambda(u)$ curves for $N = 4$ are shown in Fig. 1.3. Those numerical simulations imply that the inhomogeneous $T - Q$ relation (1.5.27) and the BAEs (1.5.29) indeed give the correct and complete spectrum

Table 1.2 The numerical solutions of the BAEs (1.5.29) for $N = 4, \phi = -0.69315i$ and $\{\theta_j = 0\}$

μ_1	μ_2	μ_3	μ_4	E_n	d
-3.46085 $- 2.04638i$	-3.46085 $+ 2.04638i$	-0.53915 $- 0.28370i$	-0.53915 $+ 0.28370i$	-4.00000	1
-3.49754 $- 0.00000i$	-2.00000 $+ 2.49853i$	-2.00000 $- 2.49853i$	-0.50246 $- 0.00000i$	-2.00000	3
-3.41695 $- 0.01463i$	-2.20702 $+ 2.20734i$	-1.88461 $- 2.68745i$	-0.49142 $+ 0.49474i$	-0.00000	3
-3.41695 $+ 0.01463i$	-2.20702 $- 2.20734i$	-1.88461 $+ 2.68745i$	-0.49142 $- 0.49474i$	-0.00000	3
-3.38446 $- 2.02080i$	-3.38446 $+ 2.02080i$	-1.11571 $+ 0.00000i$	-0.11537 $+ 0.00000i$	0.00000	1
-3.07558 $+ 1.25638i$	-3.07558 $- 1.25638i$	-0.92442 $+ 3.56865i$	-0.92442 $- 3.56865i$	2.00000	5

The eigenvalues E_n calculated from (1.5.30) are the same as those from the exact diagonalization of the Hamiltonian (1.5.13). The symbol n denotes the number of the energy levels and d indicates the number of degeneracy

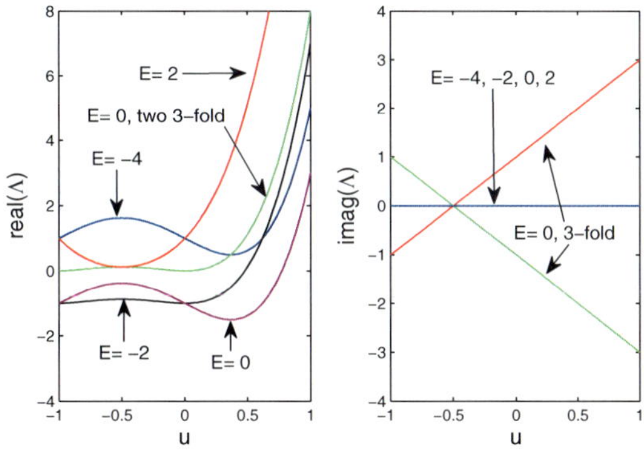

Fig. 1.3 $\Lambda(u)$ vs. u for $N=4, \phi=-0.69315i$ and $\{\theta_j=0\}$. The *curves* calculated from the inhomogeneous $T-Q$ relation (1.5.27) and the BAEs (1.5.29) are the same as those obtained from the exact diagonalization of the transfer matrix $t(u)$

of the periodic *XXX* spin-$\frac{1}{2}$ chain model. Actually, the eigenstates associated with the $T-Q$ relation (1.5.27) and the BAEs (1.5.29) can be constructed via the method that we shall introduce in Chap. 4.

For most of the spin-$\frac{1}{2}$ integrable models, based on YBE and RE, the following operator identities for the transfer matrix $t(u)$ hold:

$$t(\theta_j)t(\theta_j-\eta)=a(\theta_j)d(\theta_j-\eta)\sim \Delta_q(\theta_j), \quad j=1,\ldots,N, \tag{1.5.31}$$

where $a(u)$ and $d(u)$ are two known functions with zeros $\{\theta_j-\eta\}$ and $\{\theta_j\}$, respectively, and $\Delta_q(u)$ is the quantum determinant. We note that the functional version of the above relation (the inversion relation) was first proposed to derive the free energy in the thermodynamic limit [85, 86] and a clue to construct Eq. (1.5.31) was introduced in the framework of analytic Bethe Ansatz [87], where the operator product relations (not really inversion identities for multi-component models) were given in case of homogeneous transfer matrices. Such inversion identities with generic inhomogeneity were also derived in early 1990s in studying the quantum Knizhkov-Zamolodchikov equations (for example, see [88]). A method to construct the operator product identities (1.5.31) used in [64, 65] will be introduced in Chaps. 3–5 and the generalization to high spin cases will be introduced in Chaps. 8 and 9.

Applying Eq. (1.5.31) to an eigenstate $|\Psi\rangle$ of $t(u)$, we obtain

$$\Lambda(\theta_j)\Lambda(\theta_j-\eta)=a(\theta_j)d(\theta_j-\eta), \quad j=1,\ldots,N, \tag{1.5.32}$$

which has also been derived via the SoV method [45–51, 73] for particular models. In fact, for a spin-$\frac{1}{2}$ system, the functional relations together with the analytic properties of the transfer matrix constitute the sufficient and necessary conditions to determine the N unknowns of $\Lambda(u)$ polynomial (as we have proven for the XXX spin chain) and allow us to construct the inhomogeneous functional $T-Q$ relation

$$\Lambda(u) = e^{i\phi(u)}a(u)\frac{Q(u-\eta)Q_1(u-\eta)}{Q(u)Q_2(u)} + e^{-i\phi(u+\eta)}d(u)\frac{Q(u+\eta)Q_2(u+\eta)}{Q(u)Q_1(u)} + c(u)\frac{a(u)d(u)}{Q(u)Q_1(u)Q_2(u)}, \tag{1.5.33}$$

with $Q(u)$ being a degree M polynomial, and $Q_1(u)$ and $Q_2(u)$ being degree M_1 polynomials. $c(u)$ is a polynomial adjust function of degree n which matches the asymptotic behavior of $\Lambda(u)$. The phase $\phi(u)$ can be determined by the initial condition of the transfer matrix. The key point for the inhomogeneous $T-Q$ relation lies in that at the special points $u = \theta_j, \theta_j - \eta$, the last term vanishes. This makes it possible that the generalized functional $T-Q$ relation can satisfy Eq. (1.5.32) automatically for arbitrary Q-functions and $c(u)$. That is, with the help of $d(\theta_j) = a(\theta_j - \eta) = 0$, Eq. (1.5.33) leads to

$$\Lambda(\theta_j) = e^{i\phi(\theta_j)}a(\theta_j)\frac{Q(\theta_j-\eta)Q_1(\theta_j-\eta)}{Q(\theta_j)Q_2(\theta_j)}, \tag{1.5.34}$$

$$\Lambda(\theta_j-\eta) = e^{-i\phi(\theta_j)}d(\theta_j-\eta)\frac{Q(\theta_j)Q_2(\theta_j)}{Q(\theta_j-\eta)Q_1(\theta_j-\eta)}. \tag{1.5.35}$$

The above two equations give Eq. (1.5.32).

To ensure $\Lambda(u)$ to be a polynomial of degree N, we need three conditions: (1) $M + 2M_1 - n = N$; (2) the asymptotic behavior of the right hand side of Eq. (1.5.33) must coincide with that of $t(u)$; (3) the right hand side of Eq. (1.5.33) must be regular which gives the BAEs. Because Eq. (1.5.33) is correct for all the eigenvalues, we naturally conclude that the following inhomogeneous $T-Q$ relation holds:

$$t(u) = e^{i\phi(u)}a(u)\frac{\hat{Q}(u-\eta)\hat{Q}_1(u-\eta)}{\hat{Q}(u)\hat{Q}_2(u)} + e^{-i\phi(u+\eta)}d(u)\frac{\hat{Q}(u+\eta)\hat{Q}_2(u+\eta)}{\hat{Q}(u)\hat{Q}_1(u)} + c(u)\frac{a(u)d(u)}{\hat{Q}(u)\hat{Q}_1(u)\hat{Q}_2(u)}, \tag{1.5.36}$$

with all $\hat{Q}$-operators being mutually commutative.

Equation (1.5.36) is the most general $T-Q$ relation for the spin-$\frac{1}{2}$ quantum integrable systems derived from the operator product identities. Obviously, there is an infinite number of choices for the Q-functions and $c(u)$. As we already demonstrated for the *XXX* spin chain, a minimal $T-Q$ relation with $n = 0$ or $n = 1$ is sufficient to parameterize $\Lambda(u)$ completely (see, e.g., Chap. 4). This was also verified both numerically and analytically [71, 73, 89] for several cases. Usually, we choose either $Q_1(u) = Q_2(u) = 1$ and $n = 0$ or $Q(u) = 1$ and $n = 0, 1$ alternatively, to get a minimal $T-Q$ relation. It should be remarked that for several models a $T-Q$ relation with $M_1 = 0$ can not characterize the eigenvalues correctly (see, e.g., Chap. 3 for the odd N *XYZ* model and Chap. 9 for the IK model), and instead $T-Q$ relations with $M_1 \neq 0$ should be adopted. For the case of $c(u) = 0$, $Q_1(u) = Q_2(u)$ is definitely required and the inhomogeneous $T-Q$ relation is reduced to Baxter's form. In principle, the ODBA scheme makes up for the lack of a reference state. For the multi-component integrable systems, the operator identities (1.5.31) must be generalized to recursive ones [68, 70], which allow us to construct nested $T-Q$ relations. Details will be introduced in Chaps. 7 and 8.

Another important problem in the ODBA scheme is to retrieve the eigenstates, which are crucial to calculate the correlation functions and the dynamical properties. Indeed we can do that based on the spectrum characterized by the inhomogeneous $T-Q$ relation [75]. Details will be introduced in Chaps. 4 and 5.

It should be remarked that the inhomogeneous $T-Q$ relation must imply some unusual algebraic structures and even category structures because of the arbitrariness in choosing the Q-functions and the off-diagonal terms. Indeed, the structure of the nested inhomogeneous $T-Q$ relations (see Chap. 7) already reveals the difference in the algebraic representations [90] because of the non-zero inhomogeneous terms. For many cases, the operator product identities do not allow the $c(u)=0$ solution. Such an irreducible inhomogeneous term must imply a non-trivial topological nature of the system boundaries.

References

1. C.N. Yang, Some exact results for the many-body problem in one dimension with repulsive delta-function interaction. Phys. Rev. Lett. **19**, 1312 (1967)
2. R.J. Baxter, *Exactly Solved Models in Statistical Mechanics* (Academic Press, London, 1982)
3. P.D. Lax, Integrals of nonlinear equations of evolution and solitary waves. Commun. Pure Appl. Math. **21**, 467 (1968)
4. L. Onsager, Crystal statistics I. A two-dimensional model with an order-disorder transition. Phys. Rev. **65**, 117 (1944)
5. E.H. Lieb, F.Y. Wu, Absence of Mott transition in an exact solution of the short-range one-band model in one dimension. Phys. Rev. Lett. **20**, 1445 (1968)
6. L.D. Faddeev, L.A. Takhtajan, What is the spin of a spin wave? Phys. Lett. A **85**, 375 (1981)
7. E.H. Lieb, W. Liniger, Exact analysis of an interacting Bose gas I. The general solution and the ground state. Phys. Rev. **130**, 1605 (1963)
8. E.H. Lieb, Exact analysis of an interacting Bose gas II. The excitation spectrum. Phys. Rev. **130**, 1616 (1963)
9. M. Gaudin, Un systeme à une dimension de fermions en interaction. Phys. Lett. A **24**, 55 (1967)
10. X.-W. Guan, M.T. Batchelor, C. Lee, Fermi gases in one dimension: from Bethe Ansatz to experiments. Rev. Mod. Phys. **85**, 1633 (2013)
11. N. Beisert, C. Ahn, L.F. Alday, Z. Bajnok, J.M. Drummond, L. Freyhult, N. Gromov, R.A. Janik, V. Kazakov, T. Klose, G.P. Korchemsky, C. Kristjansen, M. Magro, T. McLoughlin, J.A. Minahan, R.I. Nepomechie, A. Rej, R. Roiban, S. Schafer-Nameki, C. Sieg, M. Staudacher, A. Torriclli, A.A. Tseytlin, P. Vieira, D. Volin, K. Zoubos, Review of AdS/CFT integrability: an overview. Lett. Math. Phys. **99**, 1 (2012)
12. B. Chen, X.J. Wang, Y.S. Wu, Integrable open spin chain in super Yang-Mills and the plane-wave/SYM duality. J. High Energy Phys. **02**, 029 (2004)
13. H. Bethe, On the theory of metals 1. Eigenvalues and eigenfunction of a linear chain of atoms. Z. Phys. **71**, 205 (1931)
14. R.J. Baxter, 8 vertex model in lattice statistics. Phys. Rev. Lett. **26**, 832 (1971)
15. R.J. Baxter, One-dimensional anisotropic Heisenberg chain. Phys. Rev. Lett. **26**, 834 (1971)

16. E.K. Sklyanin, L.A. Takhtajan, L.D. Faddeev, Qunatum inverse problem method. Theor. Math. Phys. **40**, 688 (1980)
17. L.A. Takhtadzhan, L.D. Faddeev, The quantum method of the inverse problem and the Heisenberg XYZ model. Rush. Math. Surv. **34**, 11 (1979)
18. L.D. Faddeev, Quantum inverse scattering method. Sov. Sci. Rev. Math. C **1**, 107 (1980)
19. E.K. Sklyanin, Quantum version of the method of inverse scattering problem. J. Sov. Math. **19**, 1546 (1982)
20. L.A. Takhtajan, Introduction to Bethe Ansatz. Lect. Notes Phys. **242**, 175 (1985)
21. P.P. Kulish, E.K. Sklyanin, Quantum spectral transform method: recent developments. Lect. Notes Phys. **151**, 61 (1982)
22. E.K. Sklyanin, Boundary conditions for integrable quantum systems. J. Phys. A **21**, 2375 (1988)
23. C.M. Yung, M.T. Batchelor, Exact solution of the spin-s XXZ chain with non-diagonal twists. Nucl. Phys. B **446**, 461 (1995)
24. J. Cao, H.-Q. Lin, K.-J. Shi, Y. Wang, Exact solution of XXZ spin chain with unparallel boundary fields. Nucl. Phys. B **663**, 487 (2003)
25. H. Fan, B.-Y. Hou, K.-J. Shi, Z.-X. Yang, Algebraic Bethe Ansatz for the eight-vertex model with general open boundary conditions. Nucl. Phys. B **478**, 723 (1996)
26. R.I. Nepomechie, Solving the open XXZ spin chain with nondiagonal boundary terms at roots of unity. Nucl. Phys. B **622**, 615 (2002)
27. R.I. Nepomechie, Bethe Ansatz solution of the open XXZ chain with nondiagonal boundary terms. J. Phys. A **37**, 433 (2004)
28. A.M. Povolotsky, Bethe Ansatz solution of zero-range process with nonuniform stationary state. Phys. Rev. E **69**, 061109 (2004)
29. J. de Gier, F.H.L. Essler, Bethe Ansatz solution of the asymmetric exclusion process with open boundaries. Phys. Rev. Lett. **95**, 240601 (2005)
30. J. de Gier, P. Pyatov, Bethe Ansatz for the Temperley-Lieb loop model with open boundaries. J. Stat. Mech. **P03002** (2004)
31. A. Nichols, V. Rittenberg, J. de Gier, One-boundary Temperley-Lieb algebras in the XXZ and loop models. J. Stat. Mech. **P03003** (2005)
32. J. de Gier, A. Nichols, P. Pyatov, V. Rittenberg, Magic in the spectra of the XXZ quantum chain with boundaries at $\Delta = 0$ and $\Delta = -1/2$. Nucl. Phys. B **729**, 387 (2005)
33. A.M. Povolotsky, J.F.F. Mendes, Bethe Ansatz solution of discrete time stochastic processes with fully parallel update. J. Stat. Phys. **123**, 125 (2006)
34. C. Winkelholz, R. Fazio, F.W.J. Hekking, G. Schön, Anomalous density of states of a Luttinger liquid in contact with a superconductor. Phys. Rev. Lett. **77**, 3200 (1996)
35. R.I. Nepomechie, Bethe Ansatz solution of the open XX spin chain with nondiagonal boundary terms. J. Phys. A **34**, 9993 (2001)
36. R.I. Nepomechie, Functional relations and Bethe Ansatz for the XXZ chain. J. Stat. Phys. **111**, 1363 (2003)
37. W.-L. Yang, Y.-Z. Zhang, M. Gould, Exact solution of the XXZ Gaudin model with generic open boundaries. Nucl. Phys. B **698**, 503 (2004)
38. J. de Gier, F.H.L. Essler, Exact spectral gaps of the asymmetric exclusion process with open boundaries. J. Stat. Mech. **P12011** (2006)
39. A. Doikou, P.P. Martins, On quantum group symmetry and Bethe Ansatz for the asymmetric twin spin chain with integrable boundary. J. Stat. Mech. **P06004** (2006)
40. A. Doikou, The open XXZ and associated models at q root of unity. J. Stat. Mech. **P09010** (2006)
41. Z. Bajnok, Equivalences between spin models induced by defects. J. Stat. Mech. **P06010** (2006)
42. W.-L. Yang, R.I. Nepomechie, Y.-Z. Zhang, Q-operator and T-Q relation from the fusion hierarchy. Phys. Lett. B **633**, 664 (2006)
43. W.-L. Yang, Y.-Z. Zhang, T-Q relation and exact solution for the XYZ chain with general nondiagonal boundary terms. Nucl. Phys. B **744**, 312 (2006)
44. W. Galleas, Functional relations from the Yang-Baxter algebra: eigenvalues of the XXZ model with non-diagonal twisted and open boundary conditions. Nucl. Phys. B **790**, 524 (2008)

45. H. Frahm, A. Seel, T. Wirth, Separation of variables in the open XXX chain. Nucl. Phys. B **802**, 351 (2008)
46. H. Frahm, J.H. Grelik, A. Seel, T. Wirth, Functional Bethe Ansatz methods for the open XXX chain. J. Phys. A: Math. Theor. **44**, 015001 (2011)
47. S. Niekamp, T. Wirth, H. Frahm, The XXZ model with anti-periodic twisted boundary conditions. J. Phys. A: Math. Theor. **42**, 195008 (2009)
48. G. Niccoli, Non-diagonal open spin-1/2 XXZ quantum chains by separation of variables: complete spectrum and matrix elements of some quasi-local operators. J. Stat. Mech. **P10025** (2012)
49. G. Niccoli, Antiperiodic spin-1/2 XXZ quantum chains by separation of variables: complete spectrum and form factors. Nucl. Phys. B **870**, 397 (2013)
50. G. Niccoli, Antiperiodic dynamical 6-vertex model I: complete spectrum by SOV, matrix elements of the identity on separate states and connections to the periodic 8-vertex model. J. Phys. A **46**, 075003 (2013)
51. S. Faldella, N. Kitanine, G. Niccoli, The complete spectrum and scalar products for the open spin-1/2 XXZ quantum chains with non-diagonal boundary terms. J. Stat. Mech. **P01011** (2014)
52. S. Belliard, N. Crampé, E. Ragoucy, Algebraic Bethe Ansatz for open XXX model with triangular boundary matrices. Lett. Math. Phys. **103**, 493 (2013)
53. P. Baseilhac, K. Koizumi, Exact spectrum of the XXZ open spin chain from the q-Onsager algebra representation theory. J. Stat. Mech. **P09006** (2007)
54. E.K. Sklyanin, The quantum Toda chain. Lect. Notes Phys. **226**, 196 (1985)
55. E.K. Sklyanin, Separation of variables in the Gaudin model. J. Sov. Math. **47**, 2473 (1989)
56. E.K. Sklyanin, Separation of variables in the classical integrable $SL(3)$ magnetic chain. Comm. Math. Phys. **150**, 181 (1992)
57. E.K. Sklyanin, Separation of variables—new trends. Prog. Theor. Phys. Suppl. **118**, 35 (1995)
58. L. Frappat, R.I. Nepomechie, E. Ragoucy, A complete Bethe Ansatz solution of the open spin-s XXZ chain with general integrable boundary terms. J. Stat. Mech. **P09009** (2007)
59. R. Murgan, Bethe Ansatz of the open spin-s XXZ chain with nondiagonal boundary terms. J. High Energy Phys. **04**, 076 (2009)
60. R. Baiyasi, R. Murgan, Generalized T-Q relations and the open spin-s XXZ chain with nondiagonal boundary terms. J. Stat. Mech. **P10003** (2012)
61. C.S. Melo, G.A.P. Ribeiro, M.J. Martins, Bethe Ansatz for the XXX-S chain with non-diagonal open boundaries. Nucl. Phys. B **711**, 565 (2005)
62. J. Cao, W.-L. Yang, K. Shi, Y. Wang, Off-diagonal Bethe Ansatz and exact solution of a topological spin ring. Phys. Rev. Lett. **111**, 137201 (2013)
63. J. Cao, W.-L. Yang, K. Shi, Y. Wang, Off-diagonal Bethe Ansatz solution of the XXX spin chain with arbitrary boundary conditions. Nucl. Phys. B **875**, 152 (2013)
64. J. Cao, S. Cui, W.-L. Yang, K. Shi, Y. Wang, Spin-1/2 XYZ model revisit: general solutions via off-diagonal Bethe Ansatz. Nucl. Phys. B **886**, 185 (2014)
65. J. Cao, W.-L. Yang, K. Shi, Y. Wang, Off-diagonal Bethe Ansatz solutions of the anisotropic spin-1/2 chains with arbitrary boundary fields. Nucl. Phys. B **877**, 152 (2013)
66. Y.-Y. Li, J. Cao, W.-L. Yang, K. Shi, Y. Wang, Exact solution of the one-dimensional Hubbard model with arbitrary boundary magnetic fields. Nucl. Phys. B **879**, 98 (2014)
67. X. Zhang, J. Cao, W.-L. Yang, K. Shi, Y. Wang, Exact solution of the one-dimensional super-symmetric t-J model with unparallel boundary fields. J. Stat. Mech. **P04031** (2014)
68. J. Cao, W.-L. Yang, K. Shi, Y. Wang, Nested off-diagonal Bethe Ansatz and exact solutions of the $su(n)$ spin chain with generic integrable boundaries. J. High Energy Phys. **04**, 143 (2014)
69. K. Hao, J. Cao, G.-L. Li, W.-L. Yang, K. Shi, Y. Wang, Exact solution of the Izergin-Korepin model with general non-diagonal boundary terms. J. High Energy Phys. **06**, 128 (2014)
70. J. Cao, S. Cui, W.L. Yang, K. Shi, Y. Wang, Exact solution of the spin-s Heisenberg chain with generic non-diagonal boundaries. J. High Energy Phys. **02**, 036 (2015)
71. R.I. Nepomechie, Inhomogeneous T-Q equation for the open XXX chain with general boundary terms: completeness and arbitrary spin. J. Phys. A **46**, 442002 (2013)

72. Y.-Y. Li, J. Cao, W.-L. Yang, K. Shi, Y. Wang, Thermodynamic limit and surface energy of the XXZ spin chain with arbitrary boundary fields. Nucl. Phys. B **884**, 17 (2014)
73. N. Kitanine, J.-M. Maillet, G. Niccoli, Open spin chains with generic integrable boundaries: Baxter equation and Bethe Ansatz completeness from SOV. J. Stat. Mech. **P05015** (2014)
74. S. Belliard, N. Crampé, Heisenberg XXX model with general boundaries: eigenvectors from algebraic Bethe Ansatz. SIGMA **9**, 072 (2013)
75. X. Zhang, Y.-Y. Li, J. Cao, W.-L. Yang, K. Shi, Y. Wang, Retrive the Bethe states of quantum integrable models solved via off-diagonal Bethe Ansatz. arXiv:1407.5294
76. X. Zhang, Y.-Y. Li, J. Cao, W.-L. Yang, K. Shi, Y. Wang, Bethe states of the XXZ spin-1/2 chian with arbitrary boundary fields. Nucl. Phys. B **893**, 70 (2015)
77. J. Liouville, Note sur l'intégration des équations différentielles de la dynamique. Journal de Mathématiques **XX**, 137 (1885)
78. J.B. McGuire, Study of exactly soluble one-dimensional N-body problems. J. Math. Phys. **5**, 622 (1964)
79. C.N. Yang, S matrix for the one-dimensional N-body problem with repulsive or attractive δ-function interaction. Phys. Rev. **168**, 1920 (1968)
80. I. Cherednik, Factorizing particles on a half line and root sytsems. Theor. Math. Phys. **61**, 977 (1984)
81. V. Chari, A. Pressley, *A Guide to Quantum Groups* (Cambridge University Press, Cambridge, 1994)
82. N. Yu Reshetikhin, M.A. Semenov-Tian-Shansky, Central extensions of quantum current groups. Lett. Math. Phys. **19**, 133 (1991)
83. L. Mezincescu, R.I. Nepomechie, Integrability of open spin chains with quantum algebra symmetry. Int. J. Mod. Phys. **A6**, 5231 (1991)
84. M. Gaudin, Boundary energy of a Bose gas in one dimension. Phys. Rev. A **4**, 386 (1971)
85. G. Yu Stroganov, A new calculation method for partition functions in some lattice models. Phys. Lett. **74A**, 116 (1979)
86. C.L. Schultz, Solvable q-state models in lattice statistics and quantum field theory. Phys. Rev. Lett. **46**, 629 (1981)
87. N.Y. Reshetikhin, The functional equation method in the theory of exactly soluble quantum systems. Sov. Phys. JETP **57**, 691 (1983)
88. I. Cherednik, Quantum Knizhnik-Zamolodchikov equations and affine root systems. Comm. Math. Phys. **150**, 109 (1992)
89. Y. Jiang, S. Cui, J. Cao, W.-L. Yang, K. Shi, Y. Wang, Completeness and Bethe root distribution of the spin-$\frac{1}{2}$ Heisenberg chain with arbitrary boundary fields. arXiv:1309.6456
90. E. Frenkel, N. Reshetikhin, Quantum affine algebras and deformations of Virasoro and W-algebras. Commun. Math. Phys. **178**, 237 (1996)

Chapter 2
The Algebraic Bethe Ansatz

The algebraic Bethe Ansatz method for quantum integrable models was proposed by the Leningrad Group [1–7] in the late 1970s, based on YBE. This method was then generalized to open boundary integrable systems by Sklyanin [8] in 1988, through developing an equation accounting for the integrable boundaries. In the past several decades, the algebraic Bethe Ansatz method has become the most popular one for solving quantum integrable models. Particularly, the development of the nested algebraic Bethe Ansatz [9–19] makes it possible to diagonalize multi-component integrable models in a systematic way.

This chapter is devoted to a detailed description of the algebraic Bethe Ansatz method and its nested version, with the isotropic spin-chain models as examples. These approaches are applicable for all the integrable models under the condition that a proper reference state exists, though different tricks may be used to find proper generating operators for the eigenstates according to the properties of R-matrices defined under different algebras [20–24]. Based on the Bethe Ansatz solutions, the methods to construct low-lying elementary excitations and thermodynamics for the spin-$\frac{1}{2}$ chain are introduced. In addition, the fusion procedure and a quantity that is important throughout this book, the quantum determinant, are also introduced. The last section is devoted to a brief introduction of Sklyanin's separation of variables method [25–27].

2.1 The Periodic Heisenberg Spin Chain

2.1.1 The Algebraic Bethe Ansatz

To show the algebraic Bethe Ansatz procedure clearly, let us consider again the spin-$\frac{1}{2}$ Heisenberg chain model. For convenience, we define the homogeneous monodromy matrix and the corresponding transfer matrix as

Y. Wang et al., *Off-Diagonal Bethe Ansatz for Exactly Solvable Models*,
DOI 10.1007/978-3-662-46756-5_2

$$T_0(u) = R_{0,N}(u) \cdots R_{0,1}(u), \tag{2.1.1}$$

and

$$t(u) = tr_0 T_0(u), \tag{2.1.2}$$

respectively with the R-matrix defined in (1.5.2). With the properties of the permutation operator we have

$$\begin{aligned} \left.\frac{\partial \ln t(u)}{\partial u}\right|_{u=0} &= t^{-1}(0) \left.\frac{\partial t(u)}{\partial u}\right|_{u=0} \\ &= \sum_{j=1}^{N} P_{1,2} \cdots P_{1,N} tr_0 [P_{0,N} \cdots P_{0,j+1} P_{0,j-1} \cdots P_{0,1}] = \sum_{j=1}^{N} P_{j,j-1}. \end{aligned} \tag{2.1.3}$$

The Hamiltonian of the Heisenberg spin chain is thus expressed as

$$H = \frac{1}{2} \sum_{j=1}^{N} \sigma_j \cdot \sigma_{j+1} = \left.\frac{\partial \ln t(u)}{\partial u}\right|_{u=0} - \frac{1}{2} N, \tag{2.1.4}$$

with the periodic boundary condition $\sigma_{N+1} \equiv \sigma_1$.

To calculate the commutation relations among the elements of the monodromy matrix, let us write out the explicit forms of the R-matrix and the monodromy matrix in the auxiliary tensor space:

$$R_{0,\bar{0}}(u-v) = \begin{pmatrix} u-v+1 & 0 & 0 & 0 \\ 0 & u-v & 1 & 0 \\ 0 & 1 & u-v & 0 \\ 0 & 0 & 0 & u-v+1 \end{pmatrix}, \tag{2.1.5}$$

$$T_0(u) = \begin{pmatrix} A(u) & B(u) \\ C(u) & D(u) \end{pmatrix} \otimes I_{\bar{0}} = \begin{pmatrix} A(u) & 0 & B(u) & 0 \\ 0 & A(u) & 0 & B(u) \\ C(u) & 0 & D(u) & 0 \\ 0 & C(u) & 0 & D(u) \end{pmatrix}, \tag{2.1.6}$$

$$T_{\bar{0}}(v) = I_0 \otimes \begin{pmatrix} A(v) & B(v) \\ C(v) & D(v) \end{pmatrix} = \begin{pmatrix} A(v) & B(v) & 0 & 0 \\ C(v) & D(v) & 0 & 0 \\ 0 & 0 & A(v) & B(v) \\ 0 & 0 & C(v) & D(v) \end{pmatrix}, \tag{2.1.7}$$

where I_0 and $I_{\bar{0}}$ are the identity operators in the 0th space and $\bar{0}$th space, respectively. With the help of the Yang-Baxter relation (1.2.5), we can easily deduce the following commutation relations:

$$\begin{aligned}
&[A(u), A(v)] = [D(u), D(v)] = 0, \\
&A(u)B(v) = \frac{u-v-1}{u-v}B(v)A(u) + \frac{1}{u-v}B(u)A(v), \\
&D(u)B(v) = \frac{u-v+1}{u-v}B(v)D(u) - \frac{1}{u-v}B(u)D(v), \\
&[B(u), B(v)] = [C(u), C(v)] = 0, \\
&[B(u), C(v)] = \frac{1}{u-v}[D(v)A(u) - D(u)A(v)].
\end{aligned} \tag{2.1.8}$$

Based on the above relations, the following useful formulae can be derived:

$$\begin{aligned}
A(u)B(\mu_1)\cdots B(\mu_M) = &\prod_{j=1}^{M} \frac{u-\mu_j-1}{u-\mu_j} B(\mu_1)\cdots B(\mu_M)A(u) \\
&+ \sum_{j=1}^{M} \frac{1}{u-\mu_j} \prod_{l\neq j}^{M} \frac{\mu_j-\mu_l-1}{\mu_j-\mu_l} B(\mu_1)\cdots B(\mu_{j-1}) \\
&\times B(u)B(\mu_{j+1})\cdots B(\mu_M)A(\mu_j),
\end{aligned} \tag{2.1.9}$$

$$\begin{aligned}
D(u)B(\mu_1)\cdots B(\mu_M) = &\prod_{j=1}^{M} \frac{u-\mu_j+1}{u-\mu_j} B(\mu_1)\cdots B(\mu_M)D(u) \\
&- \sum_{j=1}^{M} \frac{1}{u-\mu_j} \prod_{l\neq j}^{M} \frac{\mu_j-\mu_l+1}{\mu_j-\mu_l} B(\mu_1)\cdots B(\mu_{j-1}) \\
&\times B(u)B(\mu_{j+1})\cdots B(\mu_M)D(\mu_j).
\end{aligned} \tag{2.1.10}$$

Proof From the commutation relations (2.1.8) we know that Eq. (2.1.9) is satisfied for $M = 1$. Assuming that Eq. (2.1.9) is also satisfied for an arbitrary M, we have

$$\begin{aligned}
A(u)B(\mu_{M+1})B(\mu_1)\cdots B(\mu_M) = &\frac{u-\mu_{M+1}-1}{u-\mu_{M+1}} B(\mu_{M+1})A(u)B(\mu_1)\cdots B(\mu_M) \\
&+ \frac{1}{u-\mu_{M+1}} B(u)A(\mu_{M+1})B(\mu_1)\cdots B(\mu_M) \\
= &\prod_{j=1}^{M+1} \frac{u-\mu_j-1}{u-\mu_j} B(\mu_1)\cdots B(\mu_M)B(\mu_{M+1})A(u) \\
&+ \frac{u-\mu_{M+1}-1}{u-\mu_{M+1}} \sum_{j=1}^{M} \frac{1}{u-\mu_j} \prod_{l\neq j}^{M} \frac{\mu_j-\mu_l-1}{\mu_j-\mu_l} \\
&\times B(\mu_1)\cdots B(\mu_{j-1})B(u)B(\mu_{j+1})\cdots B(\mu_{M+1})A(\mu_j) \\
&+ \frac{1}{u-\mu_{M+1}} \prod_{j=1}^{M} \frac{\mu_{M+1}-\mu_j-1}{\mu_{M+1}-\mu_j} B(\mu_1)\cdots B(\mu_M)B(u)A(\mu_{M+1})
\end{aligned}$$

$$+\frac{1}{u-\mu_{M+1}}\sum_{j=1}^{M}\frac{1}{\mu_{M+1}-\mu_j}\prod_{l\neq j}^{M}\frac{\mu_j-\mu_l-1}{\mu_j-\mu_l}$$
$$\times B(\mu_1)\cdots B(\mu_{j-1})B(u)B(\mu_{j+1})\cdots B(\mu_{M+1})A(\mu_j). \tag{2.1.11}$$

Combining the second and the fourth terms in the above equation, we obtain

$$A(u)B(\mu_1)\cdots B(\mu_{M+1}) = \prod_{j=1}^{M+1}\frac{u-\mu_j-1}{u-\mu_j}B(\mu_1)\cdots B(\mu_{M+1})A(u)$$
$$+\sum_{j=1}^{M+1}\frac{1}{u-\mu_j}\prod_{l\neq j}^{M+1}\frac{\mu_j-\mu_l-1}{\mu_j-\mu_l}B(\mu_1)\cdots B(\mu_{j-1})$$
$$\times B(u)B(\mu_{j+1})\cdots B(\mu_{M+1})A(\mu_j). \tag{2.1.12}$$

Therefore, Eq. (2.1.9) is also satisfied for $M+1$. Equation (2.1.10) can be proven similarly. □

Let us define the vacuum state of the system as

$$|0\rangle = |\uparrow\rangle_1 \otimes \cdots \otimes |\uparrow\rangle_N, \tag{2.1.13}$$

where $|\uparrow\rangle_n$ is the local spin-up state of site n (Accordingly, the local spin-down state is denoted as $|\downarrow\rangle_n$.). For convenience, we introduce the notations

$$\sigma_j^{\pm} = \frac{1}{2}(\sigma_j^x \pm i\sigma_j^y). \tag{2.1.14}$$

The Pauli matrices applying on the states $|\uparrow\rangle_j$ and $|\downarrow\rangle_j$ thus behave as

$$\sigma_j^-|\uparrow\rangle_j = |\downarrow\rangle_j, \quad \sigma_j^+|\downarrow\rangle_j = |\uparrow\rangle_j,$$
$$\sigma_j^-|\downarrow\rangle_j = \sigma_j^+|\uparrow\rangle_j = 0,$$
$$\sigma_j^z|\uparrow\rangle_j = |\uparrow\rangle_j, \quad \sigma_j^z|\downarrow\rangle_j = -|\downarrow\rangle_j. \tag{2.1.15}$$

From the definition of the R-matrix we have

$$R_{0,n}(u)|0\rangle = \begin{pmatrix} u+\frac{1}{2}(1+\sigma_n^z) & \sigma_n^- \\ \sigma_n^+ & u+\frac{1}{2}(1-\sigma_n^z)\end{pmatrix}|0\rangle$$
$$= \begin{pmatrix} u+1 & \sigma_n^- \\ 0 & u\end{pmatrix}|0\rangle. \tag{2.1.16}$$

This directly induces

$$\begin{aligned} &A(u)|0\rangle = a(u)|0\rangle = (u+1)^N|0\rangle, \\ &D(u)|0\rangle = d(u)|0\rangle = u^N|0\rangle, \\ &C(u)|0\rangle = 0. \end{aligned} \tag{2.1.17}$$

The operator $B(u)$ can be treated as the spin flipping operator and used to construct the Bethe states

$$|\mu_1,\ldots,\mu_M\rangle = \prod_{j=1}^{M} B(\mu_j)|0\rangle, \tag{2.1.18}$$

where M is the number of flipped spins and $\{\mu_j\}$ is a set of parameters. Note that

$$t(u) = A(u) + D(u). \tag{2.1.19}$$

Applying $t(u)$ to the Bethe state, with the help of the commutation relations (2.1.9) and (2.1.10), we have

$$\begin{aligned} t(u)|\mu_1,\ldots,\mu_M\rangle &= \Lambda(u)|\mu_1,\ldots,\mu_M\rangle \\ &+ \sum_{j=1}^{M} \Lambda_j(u) B(\mu_1)\cdots B(\mu_{j-1})B(u)B(\mu_{j+1})\cdots B(\mu_M)|0\rangle, \end{aligned} \tag{2.1.20}$$

where

$$\Lambda(u) = a(u)\prod_{j=1}^{M}\frac{u-\mu_j-1}{u-\mu_j} + d(u)\prod_{j=1}^{M}\frac{u-\mu_j+1}{u-\mu_j}, \tag{2.1.21}$$

$$\Lambda_j(u) = \frac{1}{u-\mu_j}\left\{a(\mu_j)\prod_{l\neq j}^{M}\frac{\mu_j-\mu_l-1}{\mu_j-\mu_l} - d(\mu_j)\prod_{l\neq j}^{M}\frac{\mu_j-\mu_l+1}{\mu_j-\mu_l}\right\}. \tag{2.1.22}$$

To ensure the Bethe state to be an eigenstate of the transfer matrix, the unwanted terms must vanish, i.e., $\Lambda_j(u) = 0$. This induces the Bethe Ansatz equations (BAEs)

$$\left(1+\frac{1}{\mu_j}\right)^N = \prod_{l\neq j}^{M}\frac{\mu_j-\mu_l+1}{\mu_j-\mu_l-1}, \quad j = 1,\ldots,M. \tag{2.1.23}$$

The solutions $\{\mu_j|j-1,\ldots,M\}$ of the above equations are the Bethe roots.

For convenience, we put $\mu_j = i\lambda_j - \frac{1}{2}$. The BAEs can be rewritten as

$$\left(\frac{\lambda_j - \frac{i}{2}}{\lambda_j + \frac{i}{2}}\right)^N = -\prod_{l=1}^{M} \frac{\lambda_j - \lambda_l - i}{\lambda_j - \lambda_l + i}, \quad j = 1, \ldots, M. \tag{2.1.24}$$

From Eq. (2.1.4) we obtain the eigenvalue of the Hamiltonian in terms of the Bethe roots as

$$E(\lambda_1, \ldots, \lambda_M) = \left.\frac{\partial \ln \Lambda(u)}{\partial u}\right|_{u=0} - \frac{1}{2}N = -\sum_{j=1}^{M} \frac{1}{\lambda_j^2 + \frac{1}{4}} + \frac{1}{2}N. \tag{2.1.25}$$

Obviously, the $T-Q$ relation (1.4.1) holds for this model with the parametrization

$$Q(u) = \prod_{j=1}^{M} (u - \mu_j). \tag{2.1.26}$$

In addition, the unwanted terms $\Lambda_j(u)$ can be expressed in terms of the residue of $\Lambda(u)$ at the point $u = \mu_j$

$$\Lambda_j(u) = \frac{1}{\mu_j - u} \text{res}\, \Lambda(u)|_{u=\mu_j}, \tag{2.1.27}$$

which indicates that the regularity of $\Lambda(u)$ already ensures the "unwanted" terms in Eq. (2.1.20) to be zero [28].

2.1.2 Selection Rules of the Bethe Roots

As we mentioned in Chap. 1, in order to get a self consistent set of BAEs, the poles μ_j must be simple. Indeed by carefully examining the Bethe states we can deduce the Pauli principle for the Bethe roots [29], i.e., the eigenvector is zero as long as $\mu_j = \mu_l$ for $j \neq l$. Such a selection rule can easily be verified by the coordinate Bethe Ansatz. In fact, to preserve the regularity of $\Lambda(u)$, doubly degenerate μ_j (if they exist) must satisfy the condition

$$\text{res}\{(u - \mu_j)\Lambda(u)\}|_{u=\mu_j} = 0, \tag{2.1.28}$$

which gives rise to an additional equation apart from the $M - 1$ Eq. (2.1.23) and makes the $M - 1$ Bethe roots overdetermined.

Moreover, one may find that a pair $\mu_1 = 0$ and $\mu_2 = -1$ satisfy the BAEs (2.1.23). However, this solution also induces a zero Bethe vector. A simple proof is as follows: From the definition we know that

$$T_0(0) = P_{0,N} \cdots P_{0,1} = P_{0,1} P_{1,N} \cdots P_{1,2}. \tag{2.1.29}$$

Therefore,

$$B(0) = \sigma_1^- P_{1,N} \cdots P_{1,2}. \tag{2.1.30}$$

From the crossing symmetry property (1.5.6) we know that

$$\begin{aligned} T_0(-1) &= R_{0,N}(-1) \cdots R_{0,1}(-1) \\ &= (-1)^N \sigma_0^y P_{0,N}^{t_0} \cdots P_{0,1}^{t_0} \sigma_0^y, \end{aligned} \tag{2.1.31}$$

which gives

$$B(-1) = (-1)^{N-1} P_{1,2} \cdots P_{1,N} \sigma_1^-. \tag{2.1.32}$$

Equations (2.1.30) and (2.1.32) imply that

$$B(-1)B(0) = B(0)B(-1) = 0. \tag{2.1.33}$$

2.1.3 Ground State

A remarkable fact is that based on the BAEs, the physical quantities can be derived. Particularly, computing physical quantities becomes simpler in the thermodynamic limit.

Let us first consider the case of all the Bethe roots $\{\lambda_j | j = 1, \ldots, M\}$ being real. Taking the logarithm of (2.1.24), we have

$$\theta_1(\lambda_j) = \frac{2\pi I_j}{N} + \frac{1}{N} \sum_{l=1}^{M} \theta_2(\lambda_j - \lambda_l), \tag{2.1.34}$$

where $\theta_n(x) = 2\arctan(2x/n)$, and $\{I_j\}$ are certain integers (half odd integers) for $N - M$ odd ($N - M$ even). For convenience, we define the counting function

$$Z(\lambda) = \frac{1}{2\pi} \left[\theta_1(\lambda) - \frac{1}{N} \sum_{l=1}^{M} \theta_2(\lambda - \lambda_l) \right]. \tag{2.1.35}$$

Obviously, $Z(\lambda_j) = I_j/N$ corresponds to the Eq. (2.1.34). In principle, each possible I_j may correspond to a λ_j solution of the BAEs. However, those solutions may not be occupied. We treat the occupied solutions as "particles" and the unoccupied solutions as "holes". For any consecutive I_j and $I_{j+1} = I_j + 1$, the following relation holds:

$$\frac{Z(\lambda_{j+1}) - Z(\lambda_j)}{\lambda_{j+1} - \lambda_j} = \frac{1}{N\delta\lambda_j}, \tag{2.1.36}$$

with $\delta\lambda_j = \lambda_{j+1} - \lambda_j$. In the thermodynamic limit $N \to \infty$, Eq. (2.1.36) becomes the density of states $\rho(\lambda) + \rho^h(\lambda)$ in the λ space, where $\rho(\lambda)$ and $\rho^h(\lambda)$ are the densities of the particles and holes, respectively. Taking the derivative of Eq. (2.1.35) with respect to λ, we obtain

$$\rho(\lambda) + \rho^h(\lambda) = \frac{dZ(\lambda)}{d\lambda} = a_1(\lambda) - \int_{-\infty}^{\infty} a_2(\lambda - \mu)\rho(\mu)d\mu, \tag{2.1.37}$$

where

$$a_n(\lambda) = \frac{1}{2\pi}\frac{n}{\lambda^2 + n^2/4}. \tag{2.1.38}$$

From Eq. (2.1.25), we know that each real Bethe root λ_j contributes negative energy. In the ground state, the Bethe roots should fill the whole real axis and leave no hole for an even N, i.e., $\rho^h(\lambda) = 0$. This means that the density of particles in the ground state $\rho_g(\lambda)$ satisfies

$$\rho_g(\lambda) = a_1(\lambda) - \int_{-\infty}^{\infty} a_2(\lambda - \mu)\rho_g(\mu)d\mu. \tag{2.1.39}$$

Equation (2.1.39) can be solved by the Fourier transformation defined for an arbitrary function $F(\lambda)$ as

$$\begin{aligned}\tilde{F}(\omega) &= \int_{-\infty}^{\infty} e^{i\omega\lambda}F(\lambda)d\lambda,\\ F(\lambda) &= \frac{1}{2\pi}\int_{-\infty}^{\infty} e^{-i\omega\lambda}\tilde{F}(\omega)d\omega.\end{aligned} \tag{2.1.40}$$

Taking the Fourier transform of $a_n(\lambda)$, we have

$$\tilde{a}_n(\omega) = e^{-\frac{n|\omega|}{2}}. \tag{2.1.41}$$

Taking the Fourier transform of Eq. (2.1.39), we obtain

$$\tilde{\rho}_g(\omega) = \frac{1}{2\cosh\frac{\omega}{2}}. \tag{2.1.42}$$

Thus the solution of Eq. (2.1.39) is

$$\rho_g(\lambda) = \frac{1}{2\cosh(\pi\lambda)}. \tag{2.1.43}$$

The density of flipped spins relative to the reference state is

$$\frac{M}{N} = \int_{-\infty}^{\infty} \rho_g(\lambda)d\lambda = \frac{1}{2}, \tag{2.1.44}$$

which means that the magnetization of the ground state is zero. The energy density of the ground state reads

$$e_g = -2\pi \int_{-\infty}^{\infty} a_1(\lambda)\rho_g(\lambda)d\lambda + \frac{1}{2} = \frac{1}{2} - 2\ln 2. \tag{2.1.45}$$

For an odd N, there is a hole at $\lambda^h = \pm\infty$ in the real axis which carries zero energy. The energy density of the ground state still takes the form of (2.1.45) but the state is doubly degenerate.

2.1.4 Spinon Excitations

Now let us consider the elementary excitations of the system. We focus on the even N case. The simplest excitation is the case of one less spin flipped, i.e., $M = N/2 - 1$. Such a configuration is described with two holes put at λ_1^h and λ_2^h in the λ sea. In this case, the density of holes is

$$\rho^h(\lambda) = \frac{1}{N}\delta(\lambda - \lambda_1^h) + \frac{1}{N}\delta(\lambda - \lambda_2^h). \tag{2.1.46}$$

The density $\rho(\lambda)$ will deviate from $\rho_g(\lambda)$ by $\delta\rho(\lambda)$ because of the presence of the two holes. From Eqs. (2.1.39) and (2.1.46) we obtain that

$$\delta\rho(\lambda) + \rho^h(\lambda) = -\int_{-\infty}^{\infty} a_2(\lambda - \mu)\delta\rho(\mu)d\mu, \tag{2.1.47}$$

which can be solved by Fourier transformation. After some calculations, we obtain the total spin S_e of this excitation as

$$S_e = -N\int_{-\infty}^{\infty} \delta\rho(\lambda)d\lambda = 1, \tag{2.1.48}$$

and the excitation energy as

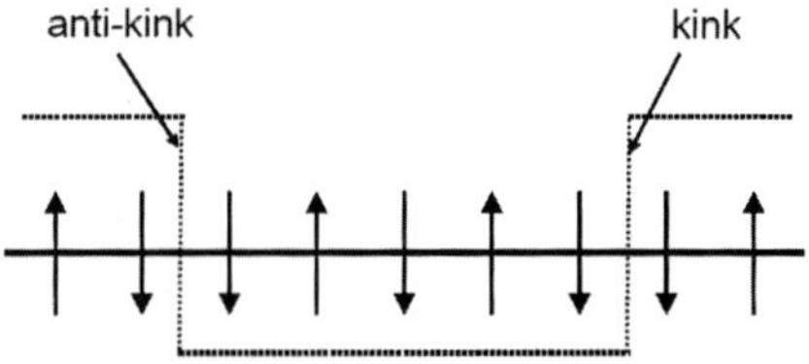

Fig. 2.1 Classical picture of the spin-triplet elementary excitations. Relative to the Neel state, the net spin carried by the flipped domain is one. Each domain boundary (kink or anti-kink) carries a spin of $\frac{1}{2}$

$$\Delta E = -2\pi N \int_{-\infty}^{\infty} a_1(\lambda)\delta\rho(\lambda)d\lambda = \varepsilon(\lambda_1^h) + \varepsilon(\lambda_2^h), \tag{2.1.49}$$

where $\varepsilon(\lambda)$ is the dressed energy with the definition

$$\varepsilon(\lambda) = 2\pi a_1(\lambda) - \int_{-\infty}^{\infty} a_2(\lambda - \mu)\varepsilon(\mu)d\mu. \tag{2.1.50}$$

From Eq. (2.1.49), we see that the energy of such excitations is the summation of the energies of two quasi-holes. Solving Eq. (2.1.50) by Fourier transformation, we obtain

$$\varepsilon(\lambda) = \frac{\pi}{\cosh(\pi\lambda)} = 2\pi\rho_g(\lambda). \tag{2.1.51}$$

Here the two holes together carry spin of one, and each of them may only carry a spin $\frac{1}{2}$. Note that such elementary excitations are unusual compared to those in the higher dimensional magnetic systems where one magnon carries a total spin of one. The classical picture of the spin-triplet excitation is shown in Fig. 2.1. Those excitations are usually called spinons [30], a typical fractional excitation in the one-dimensional quantum systems.

2.1.5 String Solutions

In the above we only considered the real Bethe roots. In fact, the BAEs may have complex solutions. For a complex λ_j with a positive imaginary part, we have

$$\left|\lambda_j - \frac{i}{2}\right| \leq \left|\lambda_j + \frac{i}{2}\right|. \tag{2.1.52}$$

This indicates that the left hand side of BAEs (2.1.24) tends to zero when $N \to \infty$. To keep the equality, the numerator of the right hand side of Eq. (2.1.24) must also tend to zero in this limit, which means that there exists another solution $\lambda_j^* \sim \lambda_j - i$ in the set of solutions. The general complex solutions of the Bethe roots read

$$\lambda_{j,\alpha}^{(n)} = \lambda_\alpha^{(n)} - \frac{i}{2}(n + 1 - 2j) + o(e^{-\delta N}), \quad j = 1, 2, \ldots, n. \tag{2.1.53}$$

This is just the string hypothesis [31]. Here $\lambda_\alpha^{(n)}$ indicates the position of the αth n-string in the real axis, and δ is a small positive number to account for the finite size deviations.

Substituting the string solutions into the BAEs and taking the product for all j in the string, we readily obtain

$$\prod_{j=1}^{n}\left(\frac{\lambda_{j,\alpha}^{(n)} - \frac{i}{2}}{\lambda_{j,\alpha}^{(n)} + \frac{i}{2}}\right)^N = \prod_{j=1}^{n}\prod_{m=1}^{\infty}\prod_{m,l,\beta\neq n,j,\alpha}\frac{\lambda_{j,\alpha}^{(n)} - \lambda_{l,\beta}^{(m)} - i}{\lambda_{j,\alpha}^{(n)} - \lambda_{l,\beta}^{(m)} + i}. \tag{2.1.54}$$

Considering the large N limit and omitting the exponentially small corrections, we reduce the above equation to

$$\begin{aligned}\left(\frac{\lambda_\alpha^{(n)} - \frac{i}{2}n}{\lambda_\alpha^{(n)} + \frac{i}{2}n}\right)^N &= -\prod_{m=1}^{\infty}\prod_{\beta}\frac{\lambda_\alpha^{(n)} \quad \lambda_\beta^{(m)} \quad \frac{i}{2}(m \mid n)}{\lambda_\alpha^{(n)} - \lambda_\beta^{(m)} + \frac{i}{2}(m+n)} \\ &\times\left[\frac{\lambda_\alpha^{(n)} - \lambda_\beta^{(m)} - \frac{i}{2}(m+n-2)}{\lambda_\alpha^{(n)} - \lambda_\beta^{(m)} + \frac{i}{2}(m+n-2)}\right]^2 \times \cdots \\ &\times\left[\frac{\lambda_\alpha^{(n)} - \lambda_\beta^{(m)} - \frac{i}{2}(|m-n|+2)}{\lambda_\alpha^{(n)} - \lambda_\beta^{(m)} + \frac{i}{2}(|m-n|+2)}\right]^2 \frac{\lambda_\alpha^{(n)} - \lambda_\beta^{(m)} - \frac{i}{2}|m-n|}{\lambda_\alpha^{(n)} - \lambda_\beta^{(m)} + \frac{i}{2}|m-n|}.\end{aligned} \tag{2.1.55}$$

Taking the logarithm of the above equation we readily have

$$\theta_n(\lambda_\alpha^{(n)}) = \frac{2\pi I_\alpha^{(n)}}{N} + \frac{1}{N}\sum_{m,\beta}\theta'_{m,n}(\lambda_\alpha^{(n)} - \lambda_\beta^{(m)}), \tag{2.1.56}$$

where $I_\alpha^{(n)}$ are integers or half odd integers depending on the parity of $N - \sum_{n=1}^{\infty} nM_n$ with M_n being the number of n-strings and

$$\begin{aligned}\theta'_{m,n}(\lambda) = &\ \theta_{m+n}(\lambda) + 2\theta_{m+n-2}(\lambda) + \cdots \\ &+ 2\theta_{|m-n|+2}(\lambda) + (1 - \delta_{m,n})\theta_{|m-n|}(\lambda).\end{aligned} \tag{2.1.57}$$

As for the real solution case, we define the counting functions

$$Z_n(\lambda) = \frac{1}{2\pi}\left[\theta_n(\lambda) - \frac{1}{N}\sum_{m,\beta}\theta'_{m,n}(\lambda - \lambda_\beta^{(m)})\right]. \tag{2.1.58}$$

Obviously, $Z_n(\lambda_\alpha^{(n)}) = I_\alpha^{(n)}/N$ corresponds to Eq. (2.1.56). In the thermodynamic limit, we have

$$\frac{dZ_n(\lambda)}{d\lambda} = \rho_n(\lambda) + \rho_n^h(\lambda), \tag{2.1.59}$$

where $\rho_n(\lambda)$ and $\rho_n^h(\lambda)$ are the densities of n-strings and n-string holes, respectively. The density of flipped spins is

$$\frac{M}{N} = \sum_{n=1}^{\infty} n \int_{-\infty}^{\infty} \rho_n(\lambda)d\lambda. \tag{2.1.60}$$

Taking the derivative of (2.1.58), we obtain the relation between $\rho_n^h(\lambda)$ and $\rho_m(\lambda)$ as

$$\rho_n^h(\lambda) = a_n(\lambda) - \sum_{m=1}^{\infty}\int_{-\infty}^{\infty} A_{m,n}(\lambda - \mu)\rho_m(\mu)d\mu, \tag{2.1.61}$$

where

$$A_{m,n}(\lambda) = a_{m+n}(\lambda) + 2a_{m+n-2}(\lambda) + \cdots + 2a_{|m-n|+2}(\lambda) + a_{|m-n|}(\lambda),$$
$$a_0(\lambda) \equiv \delta(\lambda). \tag{2.1.62}$$

Equation (2.1.61) is significant for studying the elementary excitations and thermodynamics.

In order to give the complete picture of the elementary excitations in the system, let us consider another simple type of elementary excitation, i.e., a 2-string at λ_s plus two holes in the real axis. In this case, the corresponding density functions are

$$\rho_1^h(\lambda) = \frac{1}{N}\left[\delta(\lambda - \lambda_1^h) + \delta(\lambda - \lambda_2^h)\right], \tag{2.1.63}$$

$$\rho_2(\lambda) = \frac{1}{N}\delta(\lambda - \lambda_s). \tag{2.1.64}$$

For $n = 1$ in Eq. (2.1.61), we obtain

$$\rho_1(\lambda) + \rho_1^h(\lambda) = a_1(\lambda) - \int_{-\infty}^{\infty} a_2(\lambda - \mu)\rho_1(\mu)d\mu$$
$$- \int_{-\infty}^{\infty}[a_1(\lambda - \mu) + a_3(\lambda - \mu)]\rho_2(\mu)d\mu. \tag{2.1.65}$$

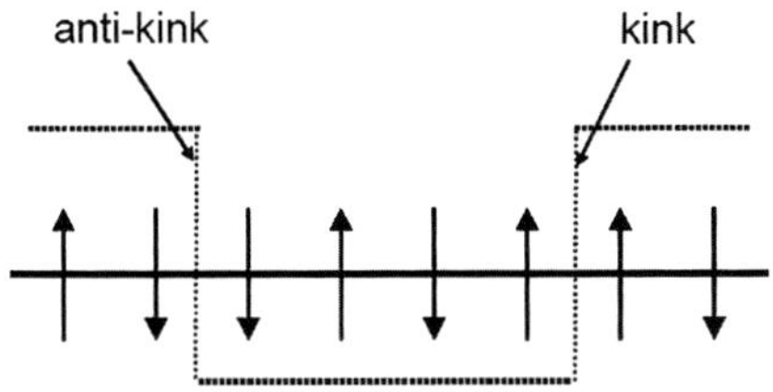

Fig. 2.2 Classical picture of the spin-singlet elementary excitations. Relative to the Neel state, the net spin carried by the flipped domain is zero. One domain boundary carries a spin of $\frac{1}{2}$ and the other carries a spin of -$\frac{1}{2}$

The deviation of the particle density from that of the ground state reads

$$\delta\rho_1(\lambda) = -\rho_1^h(\lambda) - \int_{-\infty}^{\infty} a_2(\lambda-\mu)\delta\rho_1(\mu)d\mu \\ - \int_{-\infty}^{\infty} [a_1(\lambda-\mu) + a_3(\lambda-\mu)]\rho_2(\mu)d\mu. \tag{2.1.66}$$

This allows us to derive the excitation energy as

$$\Delta E = -2\pi N \int_{-\infty}^{\infty} a_1(\lambda)\delta\rho_1(\lambda)d\lambda - 2\pi\left[a_1\left(\lambda_s + \frac{i}{2}\right) + a_1\left(\lambda_s - \frac{i}{2}\right)\right] \\ = \varepsilon(\lambda_1^h) + \varepsilon(\lambda_2^h). \tag{2.1.67}$$

It is easy to check that $M = N/2$ in this case, indicating a spin singlet excitation as shown in Fig. 2.2. Interestingly, the excitation energy takes the same form as that of Eq. (2.1.49). This means that the contribution of the 2-string is completely canceled by that of the $\rho_1(\lambda)$ redistribution induced by the presence of the string. It can be proven that this statement is also valid for the more-holes cases with the presence of arbitrary strings: the n-strings contribute nothing to the energy, and the excitation energy only depends on the positions of the holes. However, the strings do affect the scattering matrix among the holes [32].

2.1.6 Thermodynamics

The thermodynamic Bethe Ansatz was first proposed by Yang and Yang [33] for the Lieb-Liniger model [34] and subsequently generalized to other integrable models by Gaudin [35], Takahashi [36–38] and Johnson and McCoy [39]. The central point lies in deriving the entropy from the distribution of the Bethe roots.

For the present model, the energy of an n-string in the external magnetic field h is

$$\varepsilon_n^0(\lambda) = \sum_{j=1}^{n}\left[\frac{\lambda+\frac{i}{2}(n+1-2j)-\frac{i}{2}}{\lambda+\frac{i}{2}(n+1-2j)+\frac{i}{2}}+\frac{\lambda+\frac{i}{2}(n+1-2j)+\frac{i}{2}}{\lambda+\frac{i}{2}(n+1-2j)-\frac{i}{2}}-2\right]+nh$$
$$= -2\pi a_n(\lambda)+nh. \tag{2.1.68}$$

The density of energy can be calculated by

$$E/N = \sum_{n=1}^{\infty}\int_{-\infty}^{\infty}\varepsilon_n^0(\lambda)\rho_n(\lambda)d\lambda+\frac{1}{2}(1-h). \tag{2.1.69}$$

Let us consider an infinitely small interval $[\lambda, \lambda+d\lambda]$ in the λ space. The number of states allowed to be occupied by an n-string in this interval is

$$N[\rho_n(\lambda)+\rho_n^h(\lambda)]d\lambda.$$

Then the number of the possible physical states in this interval is

$$d\Omega(\lambda) = \prod_{n=1}^{\infty}\frac{[N(\rho_n(\lambda)+\rho_n^h(\lambda))d\lambda]!}{[N\rho_n(\lambda)d\lambda]![N\rho_n^h(\lambda)d\lambda]!}. \tag{2.1.70}$$

With the help of Sterling's formula $\ln N! \approx N\ln N$, we obtain the entropy in the interval

$$dS(\lambda) = \ln d\Omega(\lambda) \approx N\sum_{n=1}^{\infty}\Big\{[\rho_n(\lambda)+\rho_n^h(\lambda)]\ln[\rho_n(\lambda)+\rho_n^h(\lambda)]$$
$$-\rho_n(\lambda)\ln\rho_n(\lambda)-\rho_n^h(\lambda)\ln\rho_n^h(\lambda)\Big\}\,d\lambda. \tag{2.1.71}$$

We define the relative density of the free energy as

$$f = \frac{F}{N}-\frac{1}{2}(1-h)\,, \tag{2.1.72}$$

where $F = E-TS$ is the usual free energy, T is the temperature and S is the entropy. Substituting Eqs. (2.1.69) and (2.1.71) into Eq. (2.1.72), we have

$$f = \sum_{n=1}^{\infty}\int_{-\infty}^{\infty}\varepsilon_n^0(\lambda)\rho_n(\lambda)d\lambda - T\sum_{n=1}^{\infty}\int_{-\infty}^{\infty}\Big\{[\rho_n(\lambda)+\rho_n^h(\lambda)]\ln[\rho_n(\lambda)+\rho_n^h(\lambda)]$$
$$-\rho_n(\lambda)\ln\rho_n(\lambda)-\rho_n^h(\lambda)\ln\rho_n^h(\lambda)\Big\}\,d\lambda. \tag{2.1.73}$$

For a thermal equilibrium state, the free energy should be minimized with the variation taken with respect to $\rho_n(\lambda)$, i.e.,

$$\frac{\delta f}{\delta\rho_n(\lambda)} = 0, \tag{2.1.74}$$

which leads to

$$\sum_{n=1}^{\infty}\int_{-\infty}^{\infty}\Big\{\varepsilon_n^0(\lambda)\delta\rho_n(\lambda) - T\ln[1+\eta_n(\lambda)]\delta\rho_n(\lambda)$$
$$-T\ln[1+\eta_n^{-1}(\lambda)]\delta\rho_n^h(\lambda)\Big\}\,d\lambda = 0, \tag{2.1.75}$$

where

$$\eta_n(\lambda) = \frac{\rho_n^h(\lambda)}{\rho_n(\lambda)}. \tag{2.1.76}$$

Note that $\delta\rho_n(\lambda)$ and $\delta\rho_m^h(\lambda)$ are not independent but are related through the following equation derived from Eq. (2.1.61):

$$\delta\rho_n^h(\lambda) = -\sum_{m=1}^{\infty}\int_{-\infty}^{\infty} A_{m,n}(\lambda-\mu)\delta\rho_m(\mu)d\mu. \tag{2.1.77}$$

Substituting Eq. (2.1.77) into Eq. (2.1.75) and putting the coefficient of $\delta\rho_n(\lambda)$ to zero, we have

$$\ln[1+\eta_n(\lambda)] = \frac{\varepsilon_n^0(\lambda)}{T} + \sum_{m=1}^{\infty}\int_{-\infty}^{\infty} A_{m,n}(\lambda-\mu)\ln[1+\eta_m^{-1}(\mu)]d\mu. \tag{2.1.78}$$

For convenience, we introduce the integral operators $[n]$ as

$$[n]F(\lambda) = \int_{-\infty}^{\infty} a_n(\lambda-\mu)F(\mu)d\mu. \tag{2.1.79}$$

Note that under Fourier transformation, $[n]$ becomes a multiplier $\exp(-n|\omega|/2)$. It can be easily demonstrated that the following relation holds:

$$[m][n] = [m+n]. \tag{2.1.80}$$

Further, we define

$$\hat{A}_{m,n} = [m+n] + 2[m+n-2] + \cdots + 2[|m-n|+2] + [|m-n|],$$
$$\hat{G} = \frac{[1]}{[0]+[2]}, \tag{2.1.81}$$

where the kernel of the operator $\hat{G}$ is

$$\frac{1}{2\pi}\int_{-\infty}^{\infty} e^{-i\lambda\omega}\frac{e^{-\frac{1}{2}|\omega|}}{1+e^{-|\omega|}}d\omega = \frac{1}{2\cosh(\pi\lambda)} = \rho_g(\lambda). \tag{2.1.82}$$

It can be proven that the following operator identities hold:

$$\hat{G}[\hat{A}_{m,n+1} + \hat{A}_{m,n-1}] = -\delta_{m,n} + \hat{A}_{m,n}, \quad n > 1, \tag{2.1.83}$$
$$\hat{G}\hat{A}_{m,2} = -\delta_{1,m} + \hat{A}_{1,m}. \tag{2.1.84}$$

With the help of the above relations, we rewrite Eq. (2.1.78) as

$$\ln(1+\eta_n(\lambda)) = \frac{\varepsilon_n^0(\lambda)}{T} + \sum_{m=1}^{\infty}\hat{A}_{n,m}\ln(1+\eta_m^{-1}(\lambda)). \tag{2.1.85}$$

Applying the integral operator $\hat{G}$ to the summation of Eq. (2.1.85) with $n+1$ and $n-1$, we obtain

$$\begin{aligned}
&\hat{G}[\ln(1+\eta_{n+1}(\lambda)) + \ln(1+\eta_{n-1}(\lambda))] \\
&= \frac{1}{T}\hat{G}(\varepsilon_{n+1}^0(\lambda) + \varepsilon_{n-1}^0(\lambda)) + \sum_{m=1}^{\infty}\hat{G}(\hat{A}_{n+1,m} + \hat{A}_{n-1,m})\ln(1+\eta_m^{-1}(\lambda)) \\
&= \frac{\varepsilon_n^0(\lambda)}{T} - \ln(1+\eta_n^{-1}(\lambda)) + \sum_{m=1}^{\infty}\hat{A}_{n,m}\ln(1+\eta_m^{-1}(\lambda)).
\end{aligned} \tag{2.1.86}$$

Combining Eqs. (2.1.85) and (2.1.86), we arrive at

$$\ln\eta_n(\lambda) = \hat{G}[\ln(1+\eta_{n+1}(\lambda)) + \ln(1+\eta_{n-1}(\lambda))]. \tag{2.1.87}$$

Again, applying the integral operator $\hat{G}$ on Eq. (2.1.85) with $n = 2$, we obtain

$$\ln\eta_1(\lambda) = -\frac{2\pi\rho_g(\lambda)}{T} + \hat{G}\ln(1+\eta_2(\lambda)). \tag{2.1.88}$$

For the case of $n \to \infty$, from Eq. (2.1.85), we learn that

$$\lim_{n\to\infty} \frac{\ln \eta_n}{n} = \frac{h}{T}. \tag{2.1.89}$$

Equations (2.1.87)–(2.1.89) form a closed set of equations for the thermodynamic quantity η_n.

Substituting Eq. (2.1.61) into Eq. (2.1.73) and using Eq. (2.1.75), we obtain

$$f = -T \sum_{n=1}^{\infty} \int_{-\infty}^{\infty} a_n(\lambda) \ln[1 + \eta_n^{-1}(\lambda)] d\lambda. \tag{2.1.90}$$

From (2.1.85), we know that

$$\hat{G} \ln(1 + \eta_1(\lambda)) = \frac{1}{T} \hat{G} \varepsilon_1^0(\lambda) + \sum_{m=1}^{\infty} [m] \ln(1 + \eta_m^{-1}(\lambda)). \tag{2.1.91}$$

Putting $\lambda = 0$ in the above equation we have

$$\begin{aligned} &\sum_{m=1}^{\infty} \int_{-\infty}^{\infty} a_m(\lambda) \ln[1 + \eta_m^{-1}(\lambda)] d\lambda \\ &\quad = \frac{2\ln 2 - \frac{1}{2h}}{T} + \int_{-\infty}^{\infty} \rho_g(\lambda) \ln[1 + \eta_1(\lambda)] d\lambda. \end{aligned} \tag{2.1.92}$$

Substituting (2.1.92) into (2.1.90), we finally get the expression for the free energy

$$F/N = e_g - T \int_{-\infty}^{\infty} \rho_g(\lambda) \ln[1 + \eta_1(\lambda)] d\lambda. \tag{2.1.93}$$

We remark that though initially the string hypothesis is used, the final formula for the free energy is only related to the real root distribution.

Generally, the thermodynamic BAEs cannot be solved exactly. Below, let us consider the low energy limit $T \to 0$ and $h \to 0$. Because $\varepsilon(\lambda) > 0$, the driving term of Eq. (2.1.88) tends to $-\infty$ and $\eta_1(\lambda) \to 0$. This indicates that $\rho_1^h(\lambda) = 0$ at zero temperature, which coincides with the density configuration of the ground state previously derived. In this case, all the $\eta_n(\lambda)$ become constants and the integral Eqs. (2.1.87)–(2.1.89) are reduced to

$$\eta_n^2 = (1 + \eta_{n+1})(1 + \eta_{n-1}), \quad n > 1. \tag{2.1.94}$$

The general solution of the above equations is [31]

$$\eta_n = \left(\frac{bz^n - b^{-1}z^{-n}}{z - z^{-1}}\right)^2 - 1, \tag{2.1.95}$$

where b and z are determined by $\eta_1 = 0$ and (2.1.89), i.e., $b = 1, z = e^{\frac{h}{2T}}$. Therefore, the solution of thermodynamic BAEs in the limit $T \to 0$ is

$$\eta_n = \frac{\sinh^2 \frac{nh}{2T}}{\sinh^2 \frac{h}{2T}} - 1. \tag{2.1.96}$$

In the case of $h = 0$, we have $\eta_n(h = 0) = n^2 - 1$. Putting $\eta_n = \exp[-\varepsilon_n/T]$, then $\varepsilon_1 \sim T^0$. Comparing the T^{-1} terms of Eq. (2.1.85) we have

$$\varepsilon_1(\lambda) = -\varepsilon_1^0(\lambda) - \int_{-\infty}^{\infty} a_2(\lambda - \mu)\varepsilon_1(\mu)d\mu. \tag{2.1.97}$$

From Eqs. (2.1.50) and (2.1.97), we obtain that $\varepsilon_1(\lambda)$ is the dressed energy $\varepsilon(\lambda)$.

Since the quasi momentum is $p(\lambda) = 2\pi Z(\lambda)$, under the Fermi liquid framework [40] we define the density of states as

$$N(\lambda) = \frac{1}{\pi}\left|\frac{dp(\lambda)}{d\varepsilon(\lambda)}\right| = 2\left|\frac{\rho_g(\lambda)}{\varepsilon'(\lambda)}\right|. \tag{2.1.98}$$

In the ground state, the density of states reads $N(\lambda) = \pi^{-2}|\coth(\pi\lambda)|$ and at the Fermi surface it is $N(\infty) = \pi^{-2}$. Up to leading order, the density of free energy reads

$$\begin{aligned} f &= -T\int_{-\infty}^{\infty} N(\lambda)\ln\left[1 + e^{-\frac{|\varepsilon(\lambda)|}{T}}\right]d\varepsilon(\lambda) \\ &\approx -\frac{T^2}{\pi^2}\int_{-\infty}^{\infty} \ln(1 + e^{-|x|})dx = -\frac{1}{6}T^2. \end{aligned} \tag{2.1.99}$$

2.2 The Open Heisenberg Spin Chain

2.2.1 *The Algebraic Bethe Ansatz*

The algebraic Bethe Ansatz for open integrable models can be performed through the combination of YBE and RE. As an example, let us consider the isotropic Heisenberg spin chain with two boundary magnetic fields, a model first exactly solved by Alcaraz et al. via coordinate Bethe Ansatz [41]. The model Hamiltonian is

$$H = \sum_{j=1}^{N-1} \sigma_j \cdot \sigma_{j+1} + h_1 \sigma_1^z + h_N \sigma_N^z, \tag{2.2.1}$$

where h_1 and h_N are the boundary fields.

For the open boundary models, rather than the one-row monodromy matrix, we need to introduce the double-row monodromy matrix

$$\mathscr{U}_0(u) = T_0(u) K_0^-(u) \hat{T}_0(u), \tag{2.2.2}$$

where $K_0^-(u)$ is the solution of RE and

$$\hat{T}_0(u) \equiv (1-u^2)^N T_0^{-1}(-u) = R_{1,0}(u) \cdots R_{N,0}(u). \tag{2.2.3}$$

Here, the R-matrix is defined by (1.5.2) and RE reads

$$\begin{aligned} &R_{1,2}(u-v) K_1^-(u) R_{2,1}(u+v) K_2^-(v) \\ &\quad = K_2^-(v) R_{1,2}(u+v) K_1^-(u) R_{2,1}(u-v). \end{aligned} \tag{2.2.4}$$

It can be demonstrated that $\mathscr{U}_0(u)$ also satisfies the RE

$$\begin{aligned} &R_{1,2}(u-v)\mathscr{U}_1(u)R_{1,2}(u+v)\mathscr{U}_2(v) \\ &\quad = R_{1,2}(u-v)T_1(u)K_1^-(u)\hat{T}_1(u)R_{1,2}(u+v)T_2(v)K_2^-(v)\hat{T}_2(v) \\ &\quad = R_{1,2}(u-v)T_1(u)K_1^-(u)T_2(v)R_{1,2}(u+v)\hat{T}_1(u)K_2^-(v)\hat{T}_2(v) \\ &\quad = R_{1,2}(u-v)T_1(u)T_2(v)K_1^-(u)R_{1,2}(u+v)\hat{T}_1(u)K_2^-(v)\hat{T}_2(v) \\ &\quad = T_2(v)T_1(u)R_{1,2}(u-v)K_1^-(u)R_{1,2}(u+v)K_2^-(v)\hat{T}_1(u)\hat{T}_2(v) \\ &\quad = T_2(v)T_1(u)K_2^-(v)R_{1,2}(u+v)K_1^-(u)R_{1,2}(u-v)\hat{T}_1(u)\hat{T}_2(v) \\ &\quad = T_2(v)K_2^-(v)T_1(u)R_{1,2}(u+v)K_1^-(u)\hat{T}_2(v)\hat{T}_1(u)R_{1,2}(u-v) \\ &\quad = T_2(v)K_2^-(v)T_1(u)R_{1,2}(u+v)\hat{T}_2(v)K_1^-(u)\hat{T}_1(u)R_{12}(u-v) \\ &\quad = T_2(v)K_2^-(v)\hat{T}_2(v)R_{1,2}(u+v)T_1(u)K_1^-(u)\hat{T}_1(u)R_{1,2}(u-v) \\ &\quad = \mathscr{U}_2(v)R_{1,2}(u+v)\mathscr{U}_1(u)R_{1,2}(u-v). \end{aligned} \tag{2.2.5}$$

Note that the following relations are used in deriving the above relation:

$$R_{1,2}(u-v)\hat{T}_1(u)\hat{T}_2(v) = \hat{T}_2(v)\hat{T}_1(u)R_{1,2}(u-v), \tag{2.2.6}$$

$$R_{1,2}^{-1}(u-v) = \frac{1}{1-(u-v)^2} R_{1,2}(-u+v), \tag{2.2.7}$$

$$\hat{T}_1(u)R_{1,2}(u+v)T_2(v) = T_2(v)R_{1,2}(u+v)\hat{T}_1(u). \tag{2.2.8}$$

The transfer matrix of the model with open boundary conditions is constructed by the double-row monodromy matrix as

$$t(u) = tr_0 \left\{ K_0^+(u) \mathcal{U}_0(u) \right\}, \tag{2.2.9}$$

where $K_0^+(u)$ is a solution of the dual RE (1.2.16) which now reads as follows due to $\mathcal{M} = \text{id}$ (see (1.5.10))

$$\begin{aligned} &R_{1,2}(v-u)\, K_1^+(u)\, R_{2,1}(-u-v-2)\, K_2^+(v) \\ &\quad = K_2^+(v) R_{1,2}(-u-v-2)\, K_1^+(u)\, R_{2,1}(v-u). \end{aligned} \tag{2.2.10}$$

With the help of (2.2.5) and the properties (1.5.4)–(1.5.10), we can derive that

$$\begin{aligned}
t(u)\,t(v) &= tr_1 \left\{ K_1^+(u)\mathcal{U}_1(u) \right\} tr_2 \left\{ K_2^+(v)\mathcal{U}_2(v) \right\} \\
&= tr_1 \left\{ K_1^{+t_1}(u)\mathcal{U}_1^{t_1}(u) \right\} tr_2 \left\{ K_2^+(v)\mathcal{U}_2(v) \right\} \\
&= tr_{1,2} \left\{ K_1^{+t_1}(u)\mathcal{U}_1^{t_1}(u)\, K_2^+(v)\mathcal{U}_2(v) \right\} \\
&= tr_{1,2} \left\{ K_1^{+t_1}(u)K_2^+(v)\, \mathcal{U}_1^{t_1}(u)\mathcal{U}_2(v) \right\} \\
&= tr_{1,2} \left\{ K_1^{+t_1}(u)K_2^+(v)\, R_{2,1}^{t_1,-1}(v+u)R_{2,1}^{t_1}(v+u)\, \mathcal{U}_1^{t_1}(u)\mathcal{U}_2(v) \right\} \\
&= tr_{1,2} \left\{ [K_2^+(v)K_1^{+t_1}(u)R_{2,1}^{t_1,-1}(v+u)]^{t_1}\, [R_{2,1}^{t_1}(v+u)\mathcal{U}_1^{t_1}(u)\mathcal{U}_2(v)]^{t_1} \right\} \\
&= tr_{1,2} \left\{ [K_2^+(v)R_{2,1}^{t_1,-1,t_1}(v+u)K_1^+(u)]\, [\mathcal{U}_1(u)R_{2,1}(v+u)\mathcal{U}_2(v)] \right\} \\
&= tr_{1,2} \Big\{ [K_2^+(v)R_{2,1}^{t_1,-1,t_1}(v+u)K_1^+(u)]\, [R_{1,2}^{-1}(u-v)R_{1,2}(u-v)] \\
&\qquad \times [\mathcal{U}_1(u)R_{2,1}(v+u)\mathcal{U}_2(v)] \Big\} \\
&= tr_{1,2} \Big\{ [K_2^+(v)R_{2,1}^{t_1,-1,t_1}(v+u)K_1^+(u)R_{1,2}^{-1}(u-v)] \\
&\qquad \times [R_{1,2}(u-v)\mathcal{U}_1(u)R_{2,1}(v+u)\mathcal{U}_2(v)] \Big\} \\
&\overset{(2.2.5)}{=} tr_{1,2} \Big\{ [K_2^+(v)R_{2,1}^{t_1,-1,t_1}(v+u)K_1^+(u)R_{1,2}^{-1}(u-v)] \\
&\qquad \times [\mathcal{U}_2(v)R_{1,2}(u+v)\mathcal{U}_1(u)R_{2,1}(-v+u)] \Big\} \\
&\overset{(1.2.17)}{=} tr_{1,2} \Big\{ [R_{2,1}^{-1}(-v+u)K_1^+(u)R_{1,2}^{t_2,-1,t_2}(u+v)K_2^+(v)] \\
&\qquad \times [\mathcal{U}_2(v)R_{1,2}(u+v)\mathcal{U}_1(u)R_{2,1}(-v+u)] \Big\} \\
&= tr_{1,2} \Big\{ R_{2,1}^{-1}(-v+u)\, [K_1^+(u)R_{1,2}^{t_2,-1,t_2}(u+v)K_2^+(v) \\
&\qquad \times \mathcal{U}_2(v)R_{1,2}(u+v)\mathcal{U}_1(u)]\, R_{2,1}(-v+u) \Big\} \\
&= tr_{1,2} \left\{ K_1^+(u)R_{1,2}^{t_2,-1,t_2}(u+v)K_2^+(v)\mathcal{U}_2(v)R_{1,2}(u+v)\mathcal{U}_1(u) \right\} \\
&= tr_{1,2} \left\{ [K_1^+(u)R_{1,2}^{t_2,-1,t_2}(u+v)K_2^+(v)]^{t_2}\, [\mathcal{U}_2(v)R_{1,2}(u+v)\mathcal{U}_1(u)]^{t_2} \right\} \\
&= tr_{1,2} \left\{ [K_1^+(u)K_2^{+t_2}(v)R_{1,2}^{t_2,-1}(u+v)]\, [R_{1,2}^{t_2}(u+v)\mathcal{U}_2^{t_2}(v)\mathcal{U}_1(u)] \right\}
\end{aligned}$$

$$
\begin{aligned}
&= tr_{1,2}\left\{[K_1^+(u)K_2^{+t_2}(v)\ \mathcal{U}_2^{t_2}(v)\mathcal{U}_1(u)]\right\} \\
&= tr_{1,2}\left\{[K_2^{+t_2}(v)\mathcal{U}_2^{t_2}(v)]\ [K_1^+(u)\mathcal{U}_1(u)]\right\} \\
&= tr_2\left\{K_2^+(v)\mathcal{U}_2(v)\right\}\ tr_1\left\{K_1^+(u)\mathcal{U}_1(u)\right\} = t(v)\,t(u).
\end{aligned} \tag{2.2.11}
$$

Therefore, the transfer matrices with different spectral parameters are mutually commutative,

$$
[t(u),\ t(v)] = 0. \tag{2.2.12}
$$

The general solutions of $K_0^{\pm}(u)$ were given in [42–45]. Here we choose the diagonal ones

$$
K_0^-(u) = p + u\sigma_0^z, \quad K_0^+(u) = q + (u+1)\sigma_0^z, \tag{2.2.13}
$$

which allow us to perform the algebraic Bethe Ansatz, where p and q are two boundary parameters. Taking the derivative of the logarithm of the transfer matrix, we obtain

$$
\left.\frac{\partial t(u)}{\partial u}\right|_{u=0} = 2pK_N^+(0) + 4pq\sum_{j=1}^{N-1} P_{j,j+1} + 2q\sigma_1^z. \tag{2.2.14}
$$

Therefore, the Hamiltonian (2.2.1) can be constructed by the transfer matrix as

$$
H = \frac{1}{2pq}\left.\frac{\partial t(u)}{\partial u}\right|_{u=0} - N, \tag{2.2.15}
$$

with the boundary parameters p and q determined by h_1 and h_N as

$$
p = \frac{1}{h_1}, \quad q = \frac{1}{h_N}. \tag{2.2.16}
$$

Denote the double-row monodromy matrix as

$$
\mathcal{U}(u) \equiv \begin{pmatrix} \mathcal{A}(u) & \mathcal{B}(u) \\ \mathcal{C}(u) & \mathcal{D}(u) \end{pmatrix} \tag{2.2.17}
$$

By using RE (2.2.5), we can derive the following useful commutation relations:

$$
[\mathcal{B}(u), \mathcal{B}(v)] = [\mathcal{C}(u), \mathcal{C}(v)] = 0, \tag{2.2.18}
$$

$$
\begin{aligned}
\mathcal{A}(u)\mathcal{B}(v) = {} & \frac{(u+v)(u-v-1)}{(u-v)(u+v+1)}\mathcal{B}(v)\mathcal{A}(u) + \frac{u+v}{(u-v)(u+v+1)}\mathcal{B}(u)\mathcal{A}(v) \\
& - \frac{1}{u+v+1}\mathcal{B}(u)\mathcal{D}(v),
\end{aligned} \tag{2.2.19}
$$

$$\mathscr{D}(u)\mathscr{B}(v) = \frac{(u-v+1)(u+v+2)}{(u-v)(u+v+1)}\mathscr{B}(v)\mathscr{D}(u) - \frac{(u+v+2)}{(u-v)(u+v+1)}\mathscr{B}(u)\mathscr{D}(v)$$
$$-\frac{2}{(u-v)(u+v+1)}\mathscr{B}(v)\mathscr{A}(u) + \frac{(u-v+2)}{(u-v)(u+v+1)}\mathscr{B}(u)\mathscr{A}(v). \quad (2.2.20)$$

For convenience, let us introduce

$$\overline{\mathscr{D}}(u) = (2u+1)\mathscr{D}(u) - \mathscr{A}(u). \quad (2.2.21)$$

The transfer matrix thus reads

$$\begin{aligned} t(u) &= (q+u+1)\mathscr{A}(u) + (q-u-1)\mathscr{D}(u) \\ &= \frac{q-u-1}{2u+1}\overline{\mathscr{D}}(u) + \left(\frac{q-u-1}{2u+1} + q + u + 1\right)\mathscr{A}(u), \end{aligned} \quad (2.2.22)$$

and

$$\overline{\mathscr{D}}(u)\mathscr{B}(v) = \frac{(u-v+1)(u+v+2)}{(u-v)(u+v+1)}\mathscr{B}(v)\overline{\mathscr{D}}(u) - \frac{2(u+1)}{(u-v)(2v+1)}\mathscr{B}(u)\overline{\mathscr{D}}(v)$$
$$+\frac{4(u+1)v}{(2v+1)(u+v+1)}\mathscr{B}(u)\mathscr{A}(v), \quad (2.2.23)$$

$$\mathscr{A}(u)\mathscr{B}(v) = \frac{(u+v)(u-v-1)}{(u-v)(u+v+1)}\mathscr{B}(v)\mathscr{A}(u) - \frac{1}{(u+v+1)(2v+1)}\mathscr{B}(u)\overline{\mathscr{D}}(v)$$
$$+\frac{2v}{(u-v)(2v+1)}\mathscr{B}(u)\mathscr{A}(v). \quad (2.2.24)$$

Let us introduce further the notations

$$\begin{aligned} \mathscr{B}_M &= \mathscr{B}(\lambda_1)\cdots\mathscr{B}(\lambda_M), \\ \mathscr{B}_M^j &= \mathscr{B}(\lambda_1)\cdots\mathscr{B}(\lambda_{j-1})\mathscr{B}(u)\mathscr{B}(\lambda_{j+1})\cdots\mathscr{B}(\lambda_M). \end{aligned} \quad (2.2.25)$$

By using the commutation relations (2.2.23) and (2.2.24), we can prove that

$$\mathscr{A}(u)\mathscr{B}_M = \prod_{j=1}^{M}\frac{(u+\lambda_j)(u-\lambda_j-1)}{(u-\lambda_j)(u+\lambda_j+1)}\mathscr{B}_M\mathscr{A}(u)$$
$$-\sum_{j=1}^{M}\frac{1}{(u+\lambda_j+1)(2\lambda_j+1)}\prod_{l\neq j}^{M}\frac{(\lambda_j-\lambda_l+1)(\lambda_j+\lambda_l+2)}{(\lambda_j-\lambda_l)(\lambda_j+\lambda_l+1)}\mathscr{B}_M^j\overline{\mathscr{D}}(\lambda_j)$$

$$+\sum_{j=1}^{M}\frac{2\lambda_j}{(u-\lambda_j)(2\lambda_j+1)}\prod_{l\neq j}^{M}\frac{(\lambda_j+\lambda_l)(\lambda_j-\lambda_l-1)}{(\lambda_j-\lambda_l)(\lambda_j+\lambda_l+1)}\mathscr{B}_M^j\mathscr{A}(\lambda_j), \quad (2.2.26)$$

$$\overline{\mathscr{D}}(u)\mathscr{B}_M=\mathscr{B}_M\overline{\mathscr{D}}(u)\prod_{j=1}^{M}\frac{(u+\lambda_j+2)(u-\lambda_j+1)}{(u+\lambda_j+1)(u-\lambda_j)}$$

$$-\sum_{j=1}^{M}\frac{2(u+1)}{(u-\lambda_j)(2\lambda_j+1)}\prod_{l\neq j}^{M}\frac{(\lambda_j-\lambda_l+1)(\lambda_j+\lambda_l+2)}{(\lambda_j-\lambda_l)(\lambda_j+\lambda_l+1)}\mathscr{B}_M^j\overline{\mathscr{D}}(\lambda_j)$$

$$+\sum_{j=1}^{M}\frac{4\lambda_j(u+1)}{(u+\lambda_j+1)(2\lambda_j+1)}\prod_{l\neq j}^{M}\frac{(\lambda_j+\lambda_l)(\lambda_j-\lambda_l-1)}{(\lambda_j-\lambda_l)(\lambda_j+\lambda_l+1)}\mathscr{B}_M^j\mathscr{A}(\lambda_j). \quad (2.2.27)$$

Proof Obviously, Eq. (2.2.26) is satisfied for $M=1$. Suppose it is also satisfied for an arbitrary M. We have

$$\mathscr{A}(u)\mathscr{B}_{M+1}=\mathscr{A}(u)\mathscr{B}_M\mathscr{B}(\lambda_{M+1})$$

$$=\prod_{j=1}^{M}\frac{(u+\lambda_j)(u-\lambda_j-1)}{(u-\lambda_j)(u+\lambda_j+1)}\mathscr{B}_M\mathscr{A}(u)\mathscr{B}(\lambda_{M+1})$$

$$-\sum_{j=1}^{M}\frac{1}{(u+\lambda_j+1)(2\lambda_j+1)}\prod_{l\neq j}^{M}\frac{(\lambda_j-\lambda_l+1)(\lambda_j+\lambda_l+2)}{(\lambda_j-\lambda_l)(\lambda_j+\lambda_l+1)}\mathscr{B}_M^j\overline{\mathscr{D}}(\lambda_j)\mathscr{B}(\lambda_{M+1})$$

$$+\sum_{j=1}^{M}\frac{2\lambda_j}{(u-\lambda_j)(2\lambda_j+1)}\prod_{l\neq j}^{M}\frac{(\lambda_j+\lambda_l)(\lambda_j-\lambda_l-1)}{(\lambda_j-\lambda_l)(\lambda_j+\lambda_l+1)}\mathscr{B}_M^j\mathscr{A}(\lambda_j)\mathscr{B}(\lambda_{M+1}). \quad (2.2.28)$$

The first term can be calculated as

$$\mathscr{B}_{M+1}\mathscr{A}(u)\prod_{j=1}^{M+1}\frac{(u+\lambda_j)(u-\lambda_j-1)}{(u-\lambda_j)(u+\lambda_j+1)}$$

$$+\prod_{j=1}^{M}\frac{(u+\lambda_j)(u-\lambda_j-1)}{(u-\lambda_j)(u+\lambda_j+1)}\frac{2\lambda_{M+1}}{(u-\lambda_{M+1})(2\lambda_{M+1}+1)}\mathscr{B}_{M+1}^{M+1}\mathscr{A}(\lambda_{M+1})$$

$$-\prod_{j=1}^{M}\frac{(u+\lambda_j)(u-\lambda_j-1)}{(u-\lambda_j)(u+\lambda_j+1)}\frac{\mathscr{B}_{M+1}^{M+1}\overline{\mathscr{D}}(\lambda_{M+1})}{(u-\lambda_{M+1}+1)(2\lambda_{M+1}+1)}. \quad (2.2.29)$$

The second term can be calculated as

$$-\sum_{j=1}^{M}\frac{1}{(u+\lambda_j+1)(2\lambda_j+1)}\prod_{l\neq j}^{M}\frac{(\lambda_j-\lambda_l+1)(\lambda_j+\lambda_l+2)}{(\lambda_j-\lambda_l)(\lambda_j+\lambda_l+1)}$$

$$\times \left[\frac{(\lambda_j - \lambda_{M+1} + 1)(\lambda_j + \lambda_{M+1} + 2)}{(\lambda_j - \lambda_{M+1})(\lambda_j + \lambda_{M+1} + 1)} \mathscr{B}^j_{M+1} \overline{\mathscr{D}}(\lambda_j) \right.$$
$$- \frac{2(\lambda_j + 1)}{(\lambda_j - \lambda_{M+1})(2\lambda_{M+1} + 1)} \mathscr{B}^j_M \mathscr{B}(\lambda_j) \overline{\mathscr{D}}(\lambda_{M+1})$$
$$\left. + \frac{4(\lambda_j + 1)\lambda_{M+1}}{(\lambda_j + \lambda_{M+1} + 1)(2\lambda_{M+1} + 1)} \mathscr{B}^j_M \mathscr{B}(\lambda_j) \mathscr{A}(\lambda_{M+1}) \right]. \tag{2.2.30}$$

The third term can be calculated as

$$\sum_{j=1}^{M} \frac{2\lambda_j}{(u - \lambda_j)(2\lambda_j + 1)} \prod_{l \neq j}^{M} \frac{(\lambda_j + \lambda_l)(\lambda_j - \lambda_l - 1)}{(\lambda_j - \lambda_l)(\lambda_j + \lambda_l + 1)}$$
$$\times \left[\frac{(\lambda_j + \lambda_{M+1})(\lambda_j - \lambda_{M+1} - 1)}{(\lambda_j - \lambda_{M+1})(\lambda_j + \lambda_{M+1} + 1)} \mathscr{B}^j_{M+1} \mathscr{A}(\lambda_j) \right.$$
$$- \frac{1}{(\lambda_j + \lambda_{M+1} + 1)(2\lambda_{M+1} + 1)} \mathscr{B}^j_M \mathscr{B}(\lambda_j) \overline{\mathscr{D}}(\lambda_{M+1})$$
$$\left. + \frac{2\lambda_{M+1}}{(\lambda_j - \lambda_{M+1})(2\lambda_{M+1} + 1)} \mathscr{B}^j_M \mathscr{B}(\lambda_j) \mathscr{A}(\lambda_{M+1}) \right]. \tag{2.2.31}$$

Comparing the coefficients of the terms including $\mathscr{A}(u)$, $\mathscr{A}(\lambda_j)$, $\overline{\mathscr{D}}(\lambda_j)$, $\mathscr{A}(\lambda_{M+1})$, $\overline{\mathscr{D}}(\lambda_{M+1})$ and using the properties

$$\mathscr{B}^{M+1}_{M+1} = \mathscr{B}(u)\mathscr{B}_M, \tag{2.2.32}$$
$$\mathscr{B}_{M+1} = \mathscr{B}(\lambda_{M+1})\mathscr{B}_M, \tag{2.2.33}$$

we arrive at Eq. (2.2.26). Equation (2.2.27) can be proven similarly. □

With the crossing property (1.5.6), we obtain the duality relation between $\hat{T}(u)$ and $T(u)$

$$\begin{aligned} \sigma_0^y [\hat{T}_0(u)]^{t_0} \sigma_0^y &= \sigma_0^y [R_{0,1}(u) \cdots R_{0,N}(u)]^{t_0} \sigma_0^y \\ &= \sigma_0^y R^{t_0}_{0,N}(u) \sigma_0^y \sigma_0^y R^{t_0}_{0,N-1}(u) \sigma_0^y \cdots \sigma_0^y R^{t_0}_{0,1}(u) \sigma_0^y \\ &= (-1)^N R_{0,N}(-u-1) R_{0,N-1}(-u-1) \cdots R_{0,1}(-u-1) \\ &= (-1)^N T_0(-u-1). \end{aligned} \tag{2.2.34}$$

Thus the matrix elements of $\hat{T}(u)$ can be expressed by those of $T(u)$ with a different spectral parameter

$$\hat{T}_0(u) = (-1)^N \begin{pmatrix} D(-u-1) & -B(-u-1) \\ -C(-u-1) & A(-u-1) \end{pmatrix}. \tag{2.2.35}$$

The double-row monodromy matrix is thus factorized as

$$\begin{pmatrix} \mathscr{A}(u) & \mathscr{B}(u) \\ \mathscr{C}(u) & \mathscr{D}(u) \end{pmatrix} = (-1)^N \begin{pmatrix} A(u) & B(u) \\ C(u) & D(u) \end{pmatrix} \\ \times \begin{pmatrix} p+u & 0 \\ 0 & p-u \end{pmatrix} \begin{pmatrix} D(-u-1) & -B(-u-1) \\ -C(-u-1) & A(-u-1) \end{pmatrix}. \tag{2.2.36}$$

With the help of Eq. (2.1.8), we obtain

$$\begin{aligned} \mathscr{A}(u)|0\rangle &= (p+u)(u+1)^{2N}|0\rangle, \\ \overline{\mathscr{D}}(u)|0\rangle &= 2(p-u-1)u^{2N+1}|0\rangle, \\ \mathscr{C}(u)|0\rangle &= 0. \end{aligned} \tag{2.2.37}$$

Therefore, $|0\rangle$ is an eigenstate of $\mathscr{A}(u)$ and $\overline{\mathscr{D}}(u)$. $\mathscr{B}(u)$ can be used as the generating operator for the eigenstates.

Assume that the eigenstates of the transfer matrix take the following form

$$|\,\lambda_1,\ldots,\lambda_M\rangle = \prod_{j=1}^{M} \mathscr{B}(\lambda_j)|0\rangle. \tag{2.2.38}$$

The transfer matrix applied on the state (2.2.38) gives

$$t(u)\,|\,\lambda_1,\ldots,\lambda_M\rangle = \Lambda(u)\,|\,\lambda_1,\ldots,\lambda_M\rangle \\ +\sum_{j=1}^{M} \Lambda_j(u)\mathscr{B}(\lambda_1)\cdots\mathscr{B}(\lambda_{j-1})\mathscr{B}(u)\mathscr{B}(\lambda_{j+1})\cdots\mathscr{B}(\lambda_M)\,|\,0\rangle, \tag{2.2.39}$$

where $\Lambda(u)$ is the eigenvalue term

$$\Lambda(u) = \left(\frac{q-u-1}{2u+1} + q + u + 1\right) \\ \times (p+u)(u+1)^{2N} \prod_{j=1}^{M} \frac{(u+\lambda_j)(u-\lambda_j-1)}{(u-\lambda_j)(u+\lambda_j+1)} \\ +2\frac{q-u-1}{2u+1}(p-u-1)u^{2N+1} \prod_{j=1}^{M} \frac{(u-\lambda_j+1)(u+\lambda_j+2)}{(u-\lambda_j)(u+\lambda_j+1)}, \tag{2.2.40}$$

and the unwanted coefficients $\Lambda_j(u)$ read

$$\Lambda_j(u) = \frac{4(u+1)(q+\lambda_j)(p+\lambda_j)\lambda_j}{(u-\lambda_j)(u+\lambda_j+1)(2\lambda_j+1)}(\lambda_j+1)^{2N} \prod_{l\neq j}^{M} \frac{(\lambda_j+\lambda_l)(\lambda_j-\lambda_l-1)}{(\lambda_j-\lambda_l)(\lambda_j+\lambda_l+1)}$$

$$-\frac{4(u+1)(q-\lambda_j-1)(p-\lambda_j-1)}{(u-\lambda_j)(u+\lambda_j+1)(2\lambda_j+1)}\lambda_j^{2N+1}\prod_{l\neq j}^{M}\frac{(\lambda_j-\lambda_l+1)(\lambda_j+\lambda_l+2)}{(\lambda_j-\lambda_l)(\lambda_j+\lambda_l+1)}.$$

Putting $\Lambda_j(u)=0$, we obtain the BAEs

$$\frac{(\lambda_j+q)(\lambda_j+p)}{(\lambda_j+1-q)(\lambda_j+1-p)}\left(1+\frac{1}{\lambda_j}\right)^{2N} \times\prod_{l\neq j}^{M}\frac{(\lambda_j+\lambda_l)(\lambda_j-\lambda_l-1)}{(\lambda_j-\lambda_l+1)(\lambda_j+\lambda_l+2)}=1,\quad j=1,\ldots,M. \tag{2.2.41}$$

Similarly, the selection rules $\lambda_j\neq\lambda_l$ and $\lambda_j\neq-\lambda_l-1$ are required as those for the periodic boundary case discussed in Sect. 2.1.2.

Assume that $\lambda_j=i\mu_j-\frac{1}{2}$. Equation (2.2.41) can be rewritten as

$$\frac{(\mu_j-i\bar{q})(\mu_j-i\bar{p})}{(\mu_j+i\bar{q})(\mu_j+i\bar{p})}\left(\frac{\mu_j-\frac{i}{2}}{\mu_j+\frac{i}{2}}\right)^{2N}=\prod_{l\neq j}^{M}\frac{(\mu_j-\mu_l-i)(\mu_j+\mu_l-i)}{(\mu_j-\mu_l+i)(\mu_j+\mu_l+i)}, \tag{2.2.42}$$

with $\bar{p}=p-\frac{1}{2},\bar{q}=q-\frac{1}{2}$ and $j=1,\ldots,M$. The eigenvalue of the Hamiltonian is

$$\begin{aligned}E&=\frac{1}{2pq}\frac{\partial\Lambda(u)}{\partial u}\Big|_{u=0}-N\\&=-\sum_{j=1}^{M}\frac{2}{\mu_j^2+\frac{1}{4}}+N-1+\frac{1}{p}+\frac{1}{q}.\end{aligned} \tag{2.2.43}$$

Note that the unwanted terms can also be expressed as

$$\Lambda_j(u)=\frac{(u+1)(2\lambda_j+1)}{(\lambda_j-u)(u+\lambda_j+1)(\lambda_j+1)}\mathrm{res}\Lambda(u)|_{u=\lambda_j}, \tag{2.2.44}$$

which indicates that Baxter's $T-Q$ relation (1.4.1) also holds for this model with

$$\begin{aligned}a(u)&=\frac{2u+2}{2u+1}(u+p)(u+q)(u+1)^{2N},\\d(u)&=\frac{2u}{2u+1}(u-p+1)(u-q+1)u^{2N},\\Q(u)&=\prod_{j=1}^{M}(u-\lambda_j)(u+\lambda_j+1).\end{aligned} \tag{2.2.45}$$

2.2.2 Surface Energy and Boundary Bound States

Both the boundary fields and the open boundary itself contribute finite values to physical quantities such as the ground state energy and the free energy. The surface energy is a typical quantity to describe the boundary effects. It was first studied by Gaudin [46] and subsequently studied by a number of authors [41, 47–54].

Let us consider the case of $\bar{p},\ \bar{q} \geq 0$. In the ground state, all the Bethe roots should take real values. By taking the logarithm of Eq. (2.2.42), we obtain

$$\begin{aligned}\theta_{2\bar{p}}(\mu_j) &+ \theta_{2\bar{q}}(\mu_j) + 2N\theta_1(\mu_j) \\ &= 2\pi I_j + \sum_{l=1}^{M}[\theta_2(\mu_j - \mu_l) + \theta_2(\mu_j + \mu_l)] - \theta_1(\mu_j),\end{aligned} \tag{2.2.46}$$

where the θ_n-functions are defined below (2.1.34) and I_j are integers. Similar to the periodic case, we define

$$\begin{aligned}Z(u) = \frac{1}{2\pi}\Bigg\{&\theta_1(u) + \frac{1}{2N}[\theta_{2\bar{p}}(u) + \theta_{2\bar{q}}(u) + \theta_1(u) \\ &-\sum_{l=1}^{M}(\theta_2(u - \mu_l) + \theta_2(u + \mu_l))]\Bigg\}.\end{aligned} \tag{2.2.47}$$

It is obvious that $Z(\mu_j) = I_j/(2N)$. In the thermodynamic limit, the density distributions are determined by

$$\rho(u) + \rho^h(u) = \frac{dZ(u)}{du}. \tag{2.2.48}$$

Taking the derivative of $Z(u)$, we obtain

$$\begin{aligned}\rho(u) = a_1(u) &+ \frac{1}{2N}\left[a_{2\bar{p}}(u) + a_{2\bar{q}}(u) + a_1(u) - \delta(u)\right] \\ &- \int_{-\infty}^{\infty} a_2(u - v)\rho(v)dv,\end{aligned} \tag{2.2.49}$$

where we put $\rho^h(u) = \frac{1}{2N}\delta(u)$. The existence of $\delta(u)$ in the above equation is due to the hole at $I_j = 0$, which is a solution of the BAEs but leaves a zero wave function.

The density deviation from that of the periodic case satisfies

$$\begin{aligned}\delta\rho(u) = &\frac{1}{2N}\left[a_{2\bar{p}}(u) + a_{2\bar{q}}(u) + a_1(u) - \delta(u)\right] \\ &- \int_{-\infty}^{\infty} a_2(u - v)\delta\rho(v)dv.\end{aligned} \tag{2.2.50}$$

With the help of Fourier transformation, we obtain

$$\delta\tilde{\rho}(\omega) = \frac{1}{2N}\frac{e^{-\bar{p}|\omega|} + e^{-\bar{q}|\omega|} + e^{-\frac{|\omega|}{2}} - 1}{1 + e^{-|\omega|}}. \tag{2.2.51}$$

The surface energy can be calculated as

$$\begin{aligned}
\varepsilon_b &= h_1 + h_N - 1 - 4\pi N \int_{-\infty}^{\infty} a_1(u)\delta\rho(u)du \\
&= h_1 + h_N - 1 - 2N \int_{-\infty}^{\infty} \tilde{a}_1(\omega)\delta\tilde{\rho}(\omega)d\omega \\
&= h_1 + h_N - 1 - 2\int_0^{\infty} \frac{e^{-p\omega} + e^{-q\omega} + e^{-\omega} - e^{-\frac{\omega}{2}}}{1 + e^{-\omega}}d\omega \\
&= h_1 + h_N - 1 + \pi - 2\ln 2 - 2\int_0^{\infty} \frac{e^{-\frac{\omega}{h_1}} + e^{-\frac{\omega}{h_N}}}{1 + e^{-\omega}}d\omega.
\end{aligned} \tag{2.2.52}$$

Note that

$$N\int_{-\infty}^{\infty} \delta\rho(u)du = N\delta\tilde{\rho}(0) = \frac{1}{2}, \tag{2.2.53}$$

indicating that there is a boundary spin in the system. In fact, this boundary spin is carried by a boundary hole at $\lambda^h \to \infty$. If we replace $\rho^h(u)$ with

$$\rho^h(u) = \frac{1}{2N}\left[\delta(u) + \delta(u - \lambda^h) + \delta(u + \lambda^h)\right], \tag{2.2.54}$$

we obtain that $N\int_{-\infty}^{\infty}\delta\rho(u)du = 0$. This boundary hole carries zero energy and corresponds to the Majorana modes at the two boundaries [55].

For $\bar{p} < 0$, an imaginary mode $\mu_j = i\bar{p}$ may exist. This mode contributes a negative bare energy for $-\frac{1}{2} < \bar{p} < 0$ and a positive bare energy for $\bar{p} < -\frac{1}{2}$, indicating that the bound state is only stable in the former case.

In addition to the above solutions, the following boundary string

$$\mu_{b,l}^{m,n} = i\bar{p} + il, \quad l = -n, \ldots, 0, \ldots, m, \tag{2.2.55}$$

may exist, where $n > \bar{p} > 0$ or $m > -\bar{p} > 0$ is needed to preserve the equality of the BAEs. The bare energy of this boundary string reads

$$-2\pi\left[a_{2(m+p)}(0) + a_{2(n-p+1)}(0)\right].$$

The deviation of $\rho(u)$ implied by the boundary string satisfies

$$\delta\rho(u) = -[2]\delta\rho(u) - \frac{1}{2N}\left[a_{2(m+1+\bar{p})}(u)\right.$$
$$\left.+a_{2(m+\bar{p})}(u) + a_{2(n+1-\bar{p})}(u) + a_{2(n-\bar{p})}(u)\right]. \quad (2.2.56)$$

With the help of Fourier transformation we have

$$\delta\tilde{\rho}(\omega) = -\frac{1}{2N}\left[e^{-(m+\bar{p})|\omega|} + e^{-(n-\bar{p})|\omega|}\right]. \quad (2.2.57)$$

Therefore, the contribution of the boundary string to the energy is

$$\begin{aligned}\varepsilon_{bs} = &\int_{-\infty}^{\infty}\left[e^{-(m+\bar{p})|\omega|} + e^{-(n-\bar{p})|\omega|}\right]e^{-\frac{1}{2}|\omega|}d\omega \\ &-2\pi\left[a_{2(m+p)}(0) + a_{2(n-p+1)}(0)\right] \\ = &\ 0. \end{aligned} \quad (2.2.58)$$

The effect of the boundary string is similar to that of the bulk strings, i.e., contributing nothing to the energy.

2.3 Nested Algebraic Bethe Ansatz for $SU(n)$-Invariant Spin Chain

The integrability of the multi-component models was first studied by Sutherland [56] on the basis of Yang's work [57]. Subsequently, Sutherland realized that the corresponding $SU(n)$ spin chain is also exactly solvable [58]. In this section, we introduce the nested algebraic Bethe Ansatz method with the $SU(n)$-invariant quantum spin chain as an example.

The model Hamiltonian reads

$$H = \sum_{j=1}^{N} P_{j,j+1}, \quad (2.3.1)$$

where the permutation operator is defined in the tensor space of n-dimensional linear spaces

$$P_{j,j+1} = \sum_{\mu,\nu=1}^{n} E_j^{\mu,\nu} E_{j+1}^{\nu,\mu}, \quad (2.3.2)$$

and $\mu, \nu = 1, \ldots, n$, $E_j^{\mu,\nu}$ is the Weyl matrix (or the Hubbard operator)

$$E^{\mu,\nu} = |\mu\rangle\langle\nu|. \quad (2.3.3)$$

The R-matrix of the system is

$$R_{i,j}(u) = \alpha(u) + \beta(u) P_{i,j}, \tag{2.3.4}$$

where

$$\alpha(u) = \frac{u}{u+\eta}, \quad \beta(u) = \frac{\eta}{u+\eta}. \tag{2.3.5}$$

One can easily check that the above R-matrix satisfies YBE.

The monodromy matrix of the system is constructed by the R-matrices as

$$T_0(u) = R_{0,N}(u) R_{0,N-1}(u) \cdots R_{0,1}(u). \tag{2.3.6}$$

We can easily deduce the following Yang-Baxter relation:

$$\check{R}_{1,2}(u-v)[T(u) \otimes T(v)] = [T(v) \otimes T(u)]\check{R}_{1,2}(u-v), \tag{2.3.7}$$

where $\check{R}$ is the braided R-matrix with the definition

$$\check{R}_{1,2}(u) = P_{1,2} R_{1,2}(u). \tag{2.3.8}$$

The braided R-matrices satisfy the braided YBE

$$\check{R}_{1,2}(u-v)\check{R}_{2,3}(u)\check{R}_{1,2}(v) = \check{R}_{2,3}(v)\check{R}_{1,2}(u)\check{R}_{2,3}(u-v). \tag{2.3.9}$$

We write out the explicit form of the monodromy matrix in the auxiliary space:

$$T(u) = \begin{pmatrix} A_{1,1}(u) & \cdots & A_{1,n-1}(u) & B_1(u) \\ \cdots & \cdots & \cdots & \cdots \\ A_{n-1,1}(u) & \cdots & A_{n-1,n-1}(u) & B_{n-1}(u) \\ C_1(u) & \cdots & C_{n-1}(u) & D(u) \end{pmatrix}. \tag{2.3.10}$$

The transfer matrix of the system is the trace of the monodromy matrix in the auxiliary space

$$t(u) = tr_0 T_0(u) = A_{1,1}(u) + A_{2,2}(u) + \cdots + D(u). \tag{2.3.11}$$

From the Yang-Baxter relation (2.3.7) one can easily check that the transfer matrices with different spectral parameters are mutually commutative,

$$[t(u), \ t(v)] = 0. \tag{2.3.12}$$

Thus the system is integrable and the Hamiltonian (2.3.1) can be derived from the transfer matrix $t(u)$ as

$$H = \eta \left.\frac{\partial \ln t(u)}{\partial u}\right|_{u=0} + N. \tag{2.3.13}$$

From (2.3.7), we can derive the useful commutation relations:

$$D(u)C_{b_1}(\lambda) = \frac{1}{\alpha(\lambda-u)}C_{b_1}(\lambda)D(u) - \frac{\beta(\lambda-u)}{\alpha(\lambda-u)}C_{b_1}(u)D(\lambda), \tag{2.3.14}$$

$$\begin{aligned} A_{b_1,b_2}(u)C_{b_3}(\lambda) = {} & \frac{\check{R}^{(1)}(u-\lambda)^{b_3,b_2}_{b_4,b_5}}{\alpha(u-\lambda)}C_{b_5}(\lambda)A_{b_1,b_4}(u) \\ & - \frac{\beta(u-\lambda)}{\alpha(u-\lambda)}C_{b_2}(u)A_{b_1,b_3}(\lambda), \end{aligned} \tag{2.3.15}$$

$$C_{b_1}(u)C_{b_2}(\lambda) = \check{R}^{(1)}(u-\lambda)^{b_2,b_1}_{c_1,c_2}C_{c_2}(\lambda)C_{c_1}(u), \tag{2.3.16}$$

where all the subscripts and superscripts take values of $1, \ldots, n-1$, repeated indices mean summation and

$$\begin{aligned} \check{R}^{(1)}_{i,j}(u) &= \sum_{b_1,b_2=1}^{n-1} \beta(u)E_i^{b_1,b_1} \otimes E_j^{b_2,b_2} + \sum_{b_1,b_2=1}^{n-1} \alpha(u)E_i^{b_1,b_2} \otimes E_j^{b_2,b_1} \\ &\equiv \beta(u) + \alpha(u)P^{(1)}_{i,j}, \end{aligned} \tag{2.3.17}$$

where $P^{(1)}_{i,j}$ is the permutation operator defined in the $SU(n-1)$ algebra. The $\check{R}^{(1)}(u)$ is the braided R-matrix of the $SU(n-1)$-invariant spin chain.

To construct the eigenstate, we choose the local vacuum as $|0\rangle_j = (0, 0, \ldots, 1)^t$, where t means transposition. The global vacuum state is the direct product of the local vacuum, $|0\rangle = \otimes_{j=1}^N |0\rangle_j$. Obviously,

$$R_{0,j}(u)|0\rangle_j = \begin{pmatrix} \alpha(u) & 0 & \cdots & 0 \\ 0 & \alpha(u) & \cdots & 0 \\ \cdots & \cdots & \cdots & 0 \\ \beta(u)E_j^{1,n} & \beta(u)E_j^{2,n} & \cdots & 1 \end{pmatrix} |0\rangle_j. \tag{2.3.18}$$

The above relation allows us to arrive at

$$T(u)|0\rangle = \begin{pmatrix} \alpha^N(u) & 0 & \cdots & 0 \\ 0 & \alpha^N(u) & \cdots & 0 \\ \cdots & \cdots & \cdots & \cdots \\ C_1(u) & C_2(u) & \cdots & 1 \end{pmatrix} |0\rangle. \tag{2.3.19}$$

Suppose that the Bethe states take the form:

$$|\lambda_1^{(1)}, \ldots, \lambda_{L_1}^{(1)}|F_1\rangle = C_{a_1}(\lambda_1^{(1)}) \cdots C_{a_{L_1}}(\lambda_{L_1}^{(1)})|0\rangle F_1^{a_l \cdots a_1}, \tag{2.3.20}$$

where $F_1^{a_{L_1}\cdots a_1}$ is a function of the Bethe roots $\lambda_j^{(1)}$ and L_1 is the number of the first set of Bethe roots. Applying the transfer matrix to the Bethe state (2.3.20) and using the commutation relations (2.3.14)–(2.3.16) repeatedly, we readily have

$$
\begin{aligned}
&t(u)C_{a_1}(\lambda_1^{(1)})\cdots C_{a_{L_1}}(\lambda_{L_1}^{(1)})|0\rangle F_1^{a_{L_1}\cdots a_1}\\
&=\left\{\alpha^N(u)\Lambda^{(1)}(u)\prod_{j=1}^{L_1}\frac{1}{\alpha(u-\lambda_j^{(1)})}+\prod_{j=1}^{L_1}\frac{1}{\alpha(\lambda_j^{(1)}-u)}\right\}|\lambda_1^{(1)},\ldots,\lambda_{L_1}^{(1)}|F_1\rangle\\
&\quad-\sum_{j=1}^{L_1}\frac{\beta(u-\lambda_j^{(1)})}{\alpha(u-\lambda_j^{(1)})}\left\{\prod_{k=1,\neq j}^{L_1}\frac{1}{\alpha(\lambda_j^{(1)}-\lambda_k^{(1)})}\alpha^N(\lambda_j^{(1)})t^{(1)}(\lambda_j^{(1)})\right.\\
&\quad\left.-\prod_{k=1,\neq j}^{L_1}\frac{1}{\alpha(\lambda_k^{(1)}-\lambda_j^{(1)})}\right\}|\ldots,\lambda_{j-1}^{(1)},u,\lambda_{j+1}^{(1)},\ldots|F_1\rangle,
\end{aligned}
\tag{2.3.21}
$$

where $\Lambda^{(1)}(u)$ is the eigenvalue of the next transfer matrix $t^{(1)}(u)$ defined below and

$$
|\ldots,\lambda_{j-1}^{(1)},u,\lambda_{j+1}^{(1)},\ldots|F_1\rangle\equiv\cdots C_{a_{j-1}}(\lambda_{j-1}^{(1)})C_{a_j}(u)C_{a_{j+1}}(\lambda_{j+1}^{(1)})\cdots|0\rangle F_1^{a_{L_1}\cdots a_1},
$$

indicates the unwanted terms. If the Bethe state is an eigenstate of the transfer matrix, the unwanted terms must be canceled. This leads to the first set of BAEs:

$$
\begin{aligned}
&\prod_{k\neq j}^{L_1}\frac{\alpha(\lambda_j^{(1)}-\lambda_k^{(1)})}{\alpha(\lambda_k^{(1)}-\lambda_j^{(1)})}\frac{1}{\alpha^N(\lambda_j^{(1)})}F_1^{b_{L_1}\cdots b_1}=t^{(1)}(\lambda_j^{(1)})_{a_1\cdots a_{L_1}}^{b_1\cdots b_{N_1}}F_1^{a_{L_1}\cdots a_1},\\
&\qquad j=1,2,\ldots,L_1.
\end{aligned}
\tag{2.3.22}
$$

The corresponding eigenvalue of the transfer matrix thus reads

$$
\Lambda(u)=\prod_{j=1}^{L_1}\frac{1}{\alpha(u-\lambda_j^{(1)})}\alpha^N(u)\Lambda^{(1)}(u)+\prod_{j=1}^{L_1}\frac{1}{\alpha(\lambda_j^{(1)}-u)}.
\tag{2.3.23}
$$

We note that the first set of BAEs is in fact a new eigenvalue problem.

Let us define the nested monodromy matrix as

$$
T_0^{(1)}(u)=R_{0,L_1}^{(1)}(u-\lambda_{L_1}^{(1)})R_{0,L_1-1}^{(1)}(u-\lambda_{L_1-1}^{(1)})\cdots R_{0,1}^{(1)}(u-\lambda_1^{(1)}),
\tag{2.3.24}
$$

where $R_{0,j}^{(1)}(u)=P_{0,j}^{(1)}\check{R}_{0,j}^{(1)}(u)$. Then the nested transfer matrix $t^{(1)}(u)$ is

$$
t^{(1)}(u)=tr_0T_0^{(1)}(u).
\tag{2.3.25}
$$

It can be checked that the following Yang-Baxter relation holds

$$\check{R}^{(1)}_{1,2}(u-v)[T^{(1)}(u)\otimes T^{(1)}(v)] = [T^{(1)}(v)\otimes T^{(1)}(u)]\check{R}^{(1)}_{1,2}(u-v). \quad (2.3.26)$$

Note that now some inhomogeneous parameters $\{\ldots,\lambda_j^{(1)},\ldots\}$ enter the nested monodromy matrix and the nested transfer matrix. The above process reduces the eigenvalue problem to the $SU(n-1)$ level. Repeating the process, we obtain

$$\begin{aligned}\Lambda^{(r)}(u) &= \prod_{j=1}^{L_{r+1}}\frac{1}{\alpha(u-\lambda_j^{(r+1)})}\prod_{l=1}^{L_r}\alpha(u-\lambda_l^{(r)})\Lambda^{(r+1)}(u)\\ &+\prod_{j=1}^{L_{r+1}}\frac{1}{\alpha(\lambda_j^{(r+1)}-u)}, \quad r=1,\ldots,n-1, \end{aligned} \quad (2.3.27)$$

with the boundary condition $\Lambda^{(n)}(u)=1$, where L_r is the number of the rth set of Bethe roots. The BAEs are given by

$$\begin{aligned}\Lambda^{(r)}(\lambda_j^{(r)}) &= \prod_{k\neq j}^{L_r}\frac{\alpha(\lambda_j^{(r)}-\lambda_k^{(r)})}{\alpha(\lambda_k^{(r)}-\lambda_j^{(r)})}\frac{1}{\alpha^N(\lambda_j^{(r)})},\\ &j=1,\ldots,L_r, \quad r=1,\ldots,n-1. \end{aligned} \quad (2.3.28)$$

Substituting Eq. (2.3.27) into the above equation we readily have

$$\begin{aligned}\prod_{k\neq j}^{L_r}\frac{\lambda_j^{(r)}-\lambda_k^{(r)}-1}{\lambda_j^{(r)}-\lambda_k^{(r)}+1} &= \prod_{l=1}^{L_{r-1}}\frac{\lambda_j^{(r)}-\lambda_l^{(r-1)}}{\lambda_j^{(r)}-\lambda_l^{(r-1)}+1}\prod_{m=1}^{L_{r+1}}\frac{\lambda_j^{(r)}-\lambda_m^{(r+1)}-1}{\lambda_j^{(r)}-\lambda_m^{(r+1)}},\\ j=1,\ldots,L_r, &\quad r=1,2,\ldots,n-1. \end{aligned} \quad (2.3.29)$$

For convenience, we put $\lambda_j^{(r)}\to i\mu_j^{(r)}-r/2$. The BAEs are thus transformed to more symmetric form

$$\begin{aligned}\prod_{k\neq j}^{L_r}\frac{\mu_j^{(r)}-\mu_k^{(r)}-i}{\mu_j^{(r)}-\mu_k^{(r)}+i} &= \prod_{l=1}^{L_{r-1}}\frac{\mu_j^{(r)}-\mu_l^{(r-1)}-\frac{i}{2}}{\mu_j^{(r)}-\mu_l^{(r-1)}+\frac{i}{2}}\prod_{m=1}^{L_{r+1}}\frac{\mu_j^{(r)}-\mu_m^{(r+1)}-\frac{i}{2}}{\mu_j^{(r)}-\mu_m^{(r+1)}+\frac{i}{2}},\\ j=1,\ldots,L_r, &\quad r=1,2,\ldots,n-1. \end{aligned} \quad (2.3.30)$$

Note that $L_0=N$, $L_N=0$ and $\lambda_j^{(0)}=0$ are assumed. The eigenvalue of the transfer matrix is

$$\Lambda(u)=\alpha^N(u)\sum_{r=1}^{n-1}\prod_{j=1}^{L_r}\frac{1}{\alpha(u-\lambda_j^{(r)})}\prod_{l=1}^{L_{r+1}}\frac{1}{\alpha(\lambda_j^{(r+1)}-u)}$$

$$+\prod_{j=1}^{L_1}\frac{1}{\alpha(\lambda_j^{(1)}-u)}, \tag{2.3.31}$$

which allows us to derive the eigenvalue of the Hamiltonian in terms of the Bethe roots:

$$E(\mu_1^{(1)},\ldots,\mu_{L_1}^{(1)})=\frac{\partial\ln\Lambda(u)}{\partial u}\bigg|_{u=0}=-\sum_{j=1}^{L_1}\frac{1}{\mu_j^{(1)2}+\frac{1}{4}}+N. \tag{2.3.32}$$

The physical properties including the ground state energy, the elementary excitations and the thermodynamics of this model can also be derived by a similar process for the spin-$\frac{1}{2}$ model. For these topics we direct the readers' attention to some excellent reviews [59–62]. Finally, we note that for the $SU(n)$-invariant quantum spin chain (2.3.1), the nested $T-Q$ relation can also be constructed. Details will be given in Chap. 7.

2.4 Quantum Determinant, Projectors and Fusion

An important quantity throughout this book is the quantum determinant, which is related to the inverse monodromy matrix. To show how the quantum determinant is defined, let us first check the inverse R-matrix for the XXX spin-$\frac{1}{2}$ chain. The crossing relation (1.5.6) indicates that

$$R_{0,j}(-u)=-\sigma_0^y R_{0,j}^{t_0}(u-\eta)\sigma_0^y. \tag{2.4.1}$$

The unitary relation (1.5.5) indicates that

$$R_{0,j}^{-1}(u)=\varphi^{-1}(u)\sigma_0^y R_{0,j}^{t_0}(u-\eta)\sigma_0^y. \tag{2.4.2}$$

This allows us to define the inverse monodromy matrix

$$\begin{aligned}T_0^{-1}(u)&=[R_{0,N}(u)\cdots R_{0,1}(u)]^{-1}=R_{0,1}^{-1}(u)\cdots R_{0,N}^{-1}(u)\\&=a^{-1}(u)d^{-1}(u-\eta)\sigma_0^y R_{0,1}^{t_0}(u-\eta)\cdots R_{0,N}^{t_0}(u-\eta)\sigma_0^y\\&=a^{-1}(u)d^{-1}(u-\eta)\sigma_0^y T_0^{t_0}(u-\eta)\sigma_0^y,\end{aligned} \tag{2.4.3}$$

where $a(u)$ and $d(u)$ are two R-matrix dependent functions given in (2.1.17) for the R-matrix (1.5.2). The quantum determinant is thus defined as

$$\mathrm{Det}_q\{T(u)\}=T_0(u)\sigma_0^y T_0^{t_0}(u-\eta)\sigma_0^y=a(u)d(u-\eta). \tag{2.4.4}$$

Since

$$\sigma_0^y T_0^{t_0}(u-\eta)\sigma_0^y = \begin{pmatrix} D(u-\eta) & -B(u-\eta) \\ -C(u-\eta) & A(u-\eta) \end{pmatrix}, \tag{2.4.5}$$

we have

$$\begin{aligned} \mathrm{Det}_q\{T(u)\} &= A(u)D(u-\eta) - B(u)C(u-\eta) \\ &= D(u)A(u-\eta) - C(u)B(u-\eta). \end{aligned} \tag{2.4.6}$$

In addition, the following operator identities also hold:

$$\begin{aligned} &A(u)B(u-\eta) = B(u)A(u-\eta), \quad C(u)D(u-\eta) = D(u)C(u-\eta), \\ &D(u-\eta)B(u) = B(u-\eta)D(u), \quad A(u-\eta)C(u) = C(u-\eta)A(u). \end{aligned} \tag{2.4.7}$$

The exact definition of the quantum determinant is given by the fusion procedure [7, 63–66]. Given a tensor space $\mathbf{V} \otimes \mathbf{V}$ spanned by an orthonormal basis $\{|\Phi_{j,\alpha}\rangle | j, \alpha = 0, 1, \ldots\}$, a one-dimensional projection operator, which projects all vectors onto a one-dimensional subspace $\mathbf{V}^{(j,\alpha)}$, is defined as

$$P_{1,2}^{(j,\alpha)} = |\Phi_{j,\alpha}\rangle\langle\Phi_{j,\alpha}|, \tag{2.4.8}$$

which possesses the properties

$$P_{1,2}^{(j,\alpha)} P_{1,2}^{(l,\beta)} = \delta_{j,l}\delta_{\alpha,\beta} P_{1,2}^{(j,\alpha)}. \tag{2.4.9}$$

Since all operators defined in the tensor space can be expressed as linear combinations of $|\Phi_{j,\alpha}\rangle\langle\Phi_{k,\beta}|$, for a given operator $A_{1,2}(u)$, the following relation holds:

$$P_{1,2}^{(j,\alpha)} A_{1,2}(u) P_{1,2}^{(j,\alpha)} = \mathbb{A}_{(j,\alpha)}(u) P_{1,2}^{(j,\alpha)}, \tag{2.4.10}$$

with $\mathbb{A}_{(j,\alpha)}(u)$ being a scalar function.

For any R-matrix possessing the properties (1.5.4)–(1.5.9), we define the quantum determinant of the one-row monodromy matrices as

$$\begin{aligned} \mathrm{Det}_q\{T(u)\} &= tr_{1,2}\left\{P_{1,2}^{(-)} T_1(u-\eta) T_2(u) P_{1,2}^{(-)}\right\}, \\ \mathrm{Det}_q\{\hat{T}(u)\} &= tr_{1,2}\left\{P_{1,2}^{(-)} \hat{T}_1(u-\eta) \hat{T}_2(u) P_{1,2}^{(-)}\right\}. \end{aligned} \tag{2.4.11}$$

Since $tr_{1,2}P_{1,2}^{(-)} = 1$ and $P_{1,2}^{(-)}$ is a one-dimensional projector, $\mathrm{Det}_q\{T(u)\}$ must be a scalar function. With YBE and the fusion condition (1.5.9) we have

$$P_{1,2}^{(-)} R_{1,j}(u-\eta) R_{2,j}(u) = R_{2,j}(u) R_{1,j}(u-\eta) P_{1,2}^{(-)}$$

$$= P_{1,2}^{(-)} R_{1,j}(u-\eta) R_{2,j}(u) P_{1,2}^{(-)} = \mathrm{Det}_q\{R(u)\} P_{1,2}^{(-)}, \qquad (2.4.12)$$

with

$$\mathrm{Det}_q\{R(u)\} = tr_{1,2}\left\{P_{1,2}^{(-)} R_{1,j}(u-\eta) R_{2,j}(u) P_{1,2}^{(-)}\right\}. \qquad (2.4.13)$$

The above relation leads to

$$\mathrm{Det}_q\{T(u)\} = \prod_{j=1}^{N} \mathrm{Det}_q\{R(u)\}. \qquad (2.4.14)$$

Accordingly, the quantum determinants of the reflection matrices, which are useful to compute the quantum determinant of the double-row monodromy matrix, are defined as

$$\mathrm{Det}_q\{K^-(u)\} = tr_{1,2}\left\{P_{1,2}^{(-)} K_1^-(u-\eta) R_{1,2}(2u-\eta) K_2^-(u)\right\}, \qquad (2.4.15)$$

$$\mathrm{Det}_q\{K^+(u)\} = tr_{1,2}\left\{P_{1,2}^{(-)} K_2^+(u) R_{1,2}(-2u-\eta) K_1^+(u-\eta)\right\}. \qquad (2.4.16)$$

We note that the quantities $\mathrm{Det}_q\{K^{\pm}(u)\}$ and $\mathrm{Det}_q\{R(u)\}$ can easily be derived with the explicit expressions for the R-matrix, K-matrices and $P_{1,2}^{(-)}$. The quantum determinant for high-rank R-matrices can be defined similarly with singlet projectors. Details will be given in Chap. 7.

In fact, the quantum determinant is only a special case of fusion with a singlet projector. The fusion procedure can be generalized to cases with high-dimensional projectors in the associated algebras. For a given j, the states $\{|\Phi_{j,\alpha}\rangle | \alpha = 1, \ldots, n_j\}$ span an n_j-dimensional subspace. The corresponding projection operator onto this subspace is thus defined as

$$P_{1,2}^{(j)} = \sum_{\alpha=1}^{n_j} P_{1,2}^{(j,\alpha)}, \qquad (2.4.17)$$

which possesses the property

$$P_{1,2}^{(j)} P_{1,2}^{(l)} = \delta_{j,l} P_{1,2}^{(j)}. \qquad (2.4.18)$$

Given an R-matrix defined in the tensor space $\mathbf{V}_1 \otimes \mathbf{V}_2$ (the dimensions of the two vector spaces $\mathbf{V}_1$ and $\mathbf{V}_2$ may be different), at some special point $u = u_0$ (e.g., $u_0 = \pm\eta$ in (1.5.9)) the corresponding R-matrix becomes decorated projector, namely,

$$R_{1,2}(u_0) = P_{1,2}^{(j)} \times \gamma_{1,2}^{(j)} = P_{1,2}^{(j)} R_{1,2}(u_0), \qquad (2.4.19)$$

where $P_{1,2}^{(j)}$ is a projector with the same rank of $R_{1,2}(u_0)$ and $\gamma_{1,2}^{(j)}$ is some non-degenerate matrix. For some particular R-matrices, the corresponding value of u_0, $P_{1,2}^{(j)}$ and $\gamma_{1,2}^{(j)}$ will be seen in the following chapters. From YBE we obtain

$$\begin{aligned} R_{2,m}(u)R_{1,m}(u+u_0)R_{1,2}(u_0) &= P_{1,2}^{(j)}R_{1,2}(u_0)R_{1,m}(u+u_0)R_{2,m}(u) \\ &= P_{1,2}^{(j)}R_{2,m}(u)R_{1,m}(u+u_0)R_{1,2}(u_0). \end{aligned} \tag{2.4.20}$$

Multiplying (2.4.20) by the inversion of $\gamma_{1,2}^{(j)}$ from the right side, we have

$$P_{1,2}^{(j)}R_{2,m}(u)R_{1,m}(u+u_0)P_{1,2}^{(j)} = R_{2,m}(u)R_{1,m}(u+u_0)P_{1,2}^{(j)}. \tag{2.4.21}$$

Similarly, we can derive

$$P_{1,2}^{(j)}T_2(u)T_1(u+u_0)P_{1,2}^{(j)} = T_2(u)T_1(u+u_0)P_{1,2}^{(j)}. \tag{2.4.22}$$

The relations (2.4.19)–(2.4.22) are useful to construct operator product identities of high-rank and high-spin integrable models. Details will be given in Chaps. 7–9.

2.5 Sklyanin's Separation of Variables

According to Liouville's theorem, a remarkable feature of classical integrable systems is that their variables are completely separable. Sklyanin realized that quantum integrable models also possess such a feature and the separation of variables of quantum integrable models can be performed in the framework of the algebraic Bethe Ansatz [25–27]. We use a simple example, i.e., the periodic spin-$\frac{1}{2}$ Heisenberg chain to explain the main idea of the quantum SoV method.

2.5.1 SoV Basis

Let us start from the monodromy matrix like (1.5.11) denoted by

$$T(u) = \begin{pmatrix} A(u) & B(u) \\ C(u) & D(u) \end{pmatrix}. \tag{2.5.1}$$

From the definition of the R-matrix we know that $D(u)$ is an operator valued polynomial of u with a degree N and can be expressed as

$$D(u) = (u - \mathbf{d}_1)(u - \mathbf{d}_2)\cdots(u - \mathbf{d}_N), \tag{2.5.2}$$

where $\{\mathbf{d}_j | j = 1, \ldots, N\}$ are certain u-independent operators. From the commutative property $[D(u), D(v)] = 0$ we readily have

$$[\mathbf{d}_j, \ \mathbf{d}_l] = 0, \quad j, l = 1, \ldots, N, \tag{2.5.3}$$

which indicate that $\{\mathbf{d}_j | j = 1, \ldots, N\}$ form a mutually commutative operator family and have common eigenstates. These operators thus serve as the quantum counterpart of action variables or the canonical momenta in the Liouville theory.

Given a common eigenstate $|\Omega\rangle$ of $\{\mathbf{d}_j | j = 1, \ldots, N\}$, let us assume

$$\mathbf{d}_j |\Omega\rangle = d_j |\Omega\rangle, \quad j = 1, \ldots, N, \tag{2.5.4}$$

where $\{d_j | j = 1, \ldots, N\}$ are the corresponding eigenvalues. Let us also assume that the operator $D(u)$ is simple, i.e., there should be 2^N (which is equal to the dimension of the Hilbert space) possible sets of $\{d_j | j = 1, \ldots, N\}$. Such a condition can always be realized with generic inhomogeneity included in the monodromy matrix, which allows us to choose one set of them and assume that $d_j \neq d_l \neq d_j \pm \eta$ for $j \neq l$.

Obviously,

$$D(d_j)|\Omega\rangle = 0, \quad j = 1, \ldots, N. \tag{2.5.5}$$

This relation allows us to define other non-null eigenstates of $D(u)$ such as

$$|d_{p_1}, \ldots, d_{p_n}\rangle = \prod_{j=1}^{n} B(d_{p_j})|\Omega\rangle, \tag{2.5.6}$$

with $p_j \in \{1, \ldots, N\}$, $p_1 < p_2 < \cdots < p_n$ and $0 \leq n \leq N$ and

$$D(u)|d_{p_1}, \ldots, d_{p_n}\rangle = \prod_{j=1}^{n}(u - d_{p_j} + \eta) \prod_{l \neq \{p_1, \ldots, p_n\}}^{N} (u - d_l)|d_{p_1}, \ldots, d_{p_n}\rangle. \tag{2.5.7}$$

With the inhomogeneous parameters $\{\theta_j | j = 1, \ldots, N\}$, a natural choice of the initial state is $|\Omega\rangle = |0\rangle$. In this case, $d_j = \theta_j$ and the eigenstates of $D(u)$ read

$$|\theta_{p_1}, \cdots, \theta_{p_n}\rangle = \prod_{j=1}^{n} B(\theta_{p_j})|0\rangle, \quad n = 0, \cdots, N, \tag{2.5.8}$$

$$\langle\theta_{p_1}, \cdots, \theta_{p_n}| = \prod_{j=1}^{n} \langle 0|C(\theta_{p_j}), \quad n = 0, \cdots, N. \tag{2.5.9}$$

Let us consider the quantity $\langle\theta_{q_1},\ldots,\theta_{q_m}|D(u)|\theta_{p_1},\ldots,\theta_{p_n}\rangle$. Acting $D(u)$ to the left and to the right alternatively, we readily have

$$\langle\theta_{q_1},\ldots,\theta_{q_m}|\theta_{p_1},\ldots,\theta_{p_n}\rangle = f_n(\theta_{p_1},\ldots,\theta_{p_n})\delta_{m,n}\prod_{j=1}^{n}\delta_{p_j,q_j}, \quad (2.5.10)$$

with

$$f_n(\theta_{p_1},\ldots,\theta_{p_n}) = \langle\theta_{p_1},\ldots,\theta_{p_n}|\theta_{p_1},\ldots,\theta_{p_n}\rangle. \quad (2.5.11)$$

The total number of states defined in (2.5.8) from $n = 0$ to $n = N$ is exactly 2^N by a simple counting. Therefore, the left eigenstates (2.5.9) and the right eigenstates (2.5.8) are orthogonal and respectively form a left basis and a right basis of the Hilbert space [67].

2.5.2 Functional Relations

Let $\langle\Psi|$ denote a left eigenstate of the transfer matrix $t(u) = A(u) + D(u)$ with the eigenvalue $\Lambda(u)$. In addition, we define the scalar product $F_n(u_1,\cdots,u_n)$ as

$$F_n(u_1,\ldots,u_n) = \langle\Psi|\prod_{j=1}^{n}B(u_j)|0\rangle, \quad n = 0,\ldots,N. \quad (2.5.12)$$

Note the fact that $B(\theta_j)B(\theta_j-\eta) = 0$, which can be proven with a similar procedure in Sect. 2.1.2. With the help of the commutation relations (2.1.9)–(2.1.10) and by computing the quantities

$$\langle\Psi|t(\theta_j-\eta)|\theta_1,\ldots,\theta_n\rangle, \quad \langle\Psi|t(\theta_j)|\theta_1,\ldots,\theta_j-\eta,\ldots,\theta_n\rangle,$$

we obtain

$$\Lambda(\theta_j-\eta)F_n(\theta_1,\ldots,\theta_n) = -\prod_{l\neq j}^{n}\frac{\theta_j-\theta_l-\eta}{\theta_j-\theta_l}a(\theta_j)F_n(\ldots,\theta_j-\eta,\ldots), \quad (2.5.13)$$

$$\Lambda(\theta_j)F_n(\ldots,\theta_j-\eta,\ldots) = -\prod_{l\neq j}^{n}\frac{\theta_j-\theta_l}{\theta_j-\theta_l-\eta}d(\theta_j-\eta)F_n(\theta_1,\ldots,\theta_n),$$
$$j = 1,\ldots,n, \quad (2.5.14)$$

which readily give the functional relations (1.5.19), where $a(u) = d(u+\eta) = \prod_{j=1}^{N}(u-\theta_j+\eta)$. Provided that $\Lambda(u)$ is parameterized by the homogeneous $T-Q$

relation (1.5.22), the associated eigenstates are the usual Bethe states and the solution of the above Eqs. (2.5.13) and (2.5.14) can be given in terms of a certain determinant such as (4.6.18). Detailed derivation of $F_n(\theta_{p_1}, \ldots, \theta_{p_n})$ for antiperiodic and open boundary conditions will be introduced in Chaps. 4 and 5 respectively. The derivation of this quantity for the periodic boundary case can be found in [29, 67].

2.5.3 Operator Decompositions

From the definition of the monodromy matrix we know that $B(u)$ and $C(u)$ are operator valued polynomials of u with a degree $N-1$. From the commutation relations (2.1.8), we can see that the coefficients of $B(u)$ [or $C(u)$] are mutually commutative. Accordingly, we can make the following useful operator decompositions

$$B(u)=\sum_{j=1}^{N}\prod_{l\neq j}^{N}\frac{u-b_l}{b_j-b_l}B(b_j),\quad C(u)=\sum_{j=1}^{N}\prod_{l\neq j}^{N}\frac{u-b_l}{b_j-b_l}C(b_j),\tag{2.5.15}$$

where $\{b_j|j=1,\ldots,N\}$ are arbitrary complex numbers with $b_j\neq b_l\neq b_j\pm\eta$.

The above operator decompositions are convenient to compute inner products and scalar products. Put $b_j=\theta_j-\eta$ for $j=1,\ldots,n$ and $b_j=\theta_j$ for $j=n+1,\ldots,N$. We have

$$\begin{aligned}B(\theta_n)=&\sum_{l=1}^{n}\prod_{k\neq l}^{n}\frac{\theta_n-\theta_k+\eta}{\theta_l-\theta_k}\prod_{k=n+1}^{N}\frac{\theta_n-\theta_k}{\theta_l-\theta_k-\eta}B(\theta_l-\eta)\\&+\sum_{l=n+1}^{N}\prod_{k=1}^{n}\frac{\theta_n-\theta_k+\eta}{\theta_l-\theta_k+\eta}\prod_{k=n+1,\neq l}^{N}\frac{\theta_n-\theta_k}{\theta_l-\theta_k}B(\theta_l).\end{aligned}\tag{2.5.16}$$

With the above relation and the fact $B(\theta_j)B(\theta_j-\eta)=0$, we readily obtain

$$\begin{aligned}f_n(\theta_1,\ldots,\theta_n)=&\prod_{k=1}^{n-1}\frac{\theta_n-\theta_k+\eta}{\theta_n-\theta_k}\prod_{k=n+1}^{N}\frac{\theta_n-\theta_k}{\theta_n-\theta_k-\eta}\\&\times\langle\theta_1,\ldots,\theta_{n-1}|C(\theta_n)B(\theta_n-\eta)|\theta_1,\ldots,\theta_{n-1}\rangle.\end{aligned}\tag{2.5.17}$$

The expression (2.4.6) of the quantum determinant implies

$$\begin{aligned}f_n(\theta_1,\ldots,\theta_n)=&-a(\theta_n)d(\theta_n-\eta)\prod_{k=1}^{n-1}\frac{\theta_n-\theta_k+\eta}{\theta_n-\theta_k}\\&\times\prod_{k=n+1}^{N}\frac{\theta_n-\theta_k}{\theta_n-\theta_k-\eta}f_{n-1}(\theta_1,\ldots,\theta_{n-1}),\end{aligned}\tag{2.5.18}$$

which directly gives the solution

$$f_n(\theta_1, \ldots, \theta_n) = \prod_{j=1}^{n} \left\{ a(\theta_j) d_j(\theta_j) \prod_{k \neq j}^{n} \frac{\theta_j - \theta_k + \eta}{\theta_j - \theta_k} \right\}, \tag{2.5.19}$$

with $d_j(\theta_j) = \eta \prod_{l \neq j}^{N} (\theta_j - \theta_l)$.

The eigenstate $\langle \Psi |$ can be expressed as

$$\langle \Psi | = \sum_{\{p_j\}} \frac{F_n(\theta_{p_1}, \ldots, \theta_{p_n})}{f_n(\theta_{p_1}, \ldots, \theta_{p_n})} \langle \theta_{p_1}, \ldots, \theta_{p_n} |. \tag{2.5.20}$$

Similarly, the right eigenstate $|\Psi\rangle$ can be derived as

$$|\Psi\rangle = \sum_{\{p_j\}} \frac{F_n(\theta_{p_1}, \ldots, \theta_{p_n})}{f_n(\theta_{p_1}, \ldots, \theta_{p_n})} |\theta_{p_1}, \ldots, \theta_{p_n}\rangle. \tag{2.5.21}$$

We remark that for each matrix element of a monodromy matrix satisfying the Yang-Baxter relation or Sklyanin's reflection relation, its eigenstates can be constructed in a similar way, as long as the elements with different spectral parameters are mutually commutative. In such a sense, the SoV scheme gives a precise definition of quantum integrability or Yang-Baxter integrability. As we shall show in Chaps. 4 and 5, depending on the boundary conditions, either diagonal or off-diagonal elements of the monodromy matrices can be used to construct a convenient basis. The key point is that the number of independent eigenstates must be the same as the dimension of the Hilbert space. Obviously, the eigenstates of $C(u)$ [or $B(u)$] for the periodic spin chain can not form a complete basis.

References

1. E.K. Sklyanin, L.D. Faddeev, Quantum mechanical approach to completely integrable field theory models. Sov. Phys. Dokl. **23**, 902 (1978)
2. L.A. Takhtadzhan, L.D. Faddeev, The quantum method of the inverse problem and the Heisenberg XYZ model. Rush. Math. Surv. **34**, 11 (1979)
3. E.K. Sklyanin, L.A. Takhtajan, L.D. Faddeev, Qunatum inverse problem method. Theor. Math. Phys. **40**, 688 (1980)
4. L.D. Faddeev, Quantum inverse scattering method. Sov. Sci. Rev. Math. C **1**, 107 (1980)
5. E.K. Sklyanin, Quantum version of the method of inverse scattering problem. J. Sov. Math. **19**, 1546 (1982)
6. L.A. Takhtajan, Introduction to Bethe Ansatz. Lect. Notes Phys. **242**, 175 (1985)
7. P.P. Kulish, E.K. Sklyanin, Quantum spectral transform method: recent developments. Lect. Notes Phys. **151**, 61 (1982)
8. E.K. Sklyanin, Boundary conditions for integrable quantum systems. J. Phys. A **21**, 2375 (1988)
9. P.P. Kulish, Classical and quantum inverse problem method and generalized Bethe Ansatz. Physica **3D**, 246 (1981)

10. P.P. Kulish, N. Reshetikhin, Generalized Heisenberg ferromagnet and the Gross-Neveu model. Sov. Phys. JETP **53**, 108 (1981)
11. H.J. de Vega, M. Karowski, Exact Bethe Ansatz solution of $O(2N)$ symmetric theories. Nucl. Phys. B **280**, 225 (1987)
12. H.J. de Vega, E. Lopes, Exact solution of the Perk-Shultz model. Phys. Rev. Lett. **67**, 489 (1991)
13. E. Lopes, Exact solution of the multi-component generalized six-vertex model. Nucl. Phys. B **370**, 636 (1992)
14. F.H.L. Essler, V.E. Korepin, Higher conservation laws and algebraic Bethe Ansatz for the supersymmetric t-J model. Phys. Rev. B **46**, 9147 (1992)
15. A. Foerster, M. Karowski, Algebraic properties of the Bethe Ansatz for an $spl(2,1)$-supersymmetric t-J model. Nucl. Phys. B **396**, 611 (1993)
16. A. Foerster, M. Karowski, The supersymmetric t-J model with quantum group invariance. Nucl. Phys. B **408**, 512 (1993)
17. H.J. de Vega, A. González-Ruiz, Exact solution of the $SU_q(n)$-invariant quantum spin chains. Nucl. Phys. **B 417**, 553 (1994)
18. H.J. de Vega, A. González-Ruiz, Exact Bethe Ansatz solution for A_{n-1} chains with non-$SU_q(n)$ invariant open boundary conditions. Mod. Phys. Lett. **A 9**, 2207 (1994)
19. W. Galleas, M.J. Martins, Solution of the $SU(N)$ vertex model with non-diagonal open boundaries. Phys. Lett. A **335**, 167 (2005)
20. M.J. Martins, The exact solution and the finite-size behaviour of the $Osp(1|2)$-invariant spin chain. Nucl. Phys. B **450**, 768 (1995)
21. M.J. Martins, P.B. Ramos, The algebraic Bethe Ansatz for rational braid-monoid lattice models. Nucl. Phys. B **500**, 579 (1997)
22. M.J. Martins, The thermodynamic limit and the finite-size behaviour of the fundamental $Sp(2N)$ spin chain. Nucl. Phys. B **636**, 583 (2002)
23. R.-H. Yue, H. Fan, B.-Y. Hou, Exact diagonalization of the quantum supersymmetric $SU_q(n|m)$ model. Nucl. Phys. B **462**, 167 (1996)
24. G.-L. Li, R.-H. Yue, B.-Y. Hou, Nested Bethe Ansatz for Perk-Schultz model with open boundary conditions. Nucl. Phys. B **586**, 711 (2000)
25. E.K. Sklyanin, The quantum Toda chain. Lect. Notes Phys. **226**, 196 (1985)
26. E.K. Sklyanin, Separation of variables in the Gaudin model. J. Sov. Math. **47**, 2473 (1989)
27. E.K. Sklyanin, Separation of variables—new trends. Prog. Theor. Phys. Suppl. **118**, 35 (1995)
28. R.J. Baxter, *Exactly Solved Models in Statistical Mechanics* (Academic Press, London, 1982)
29. V.E. Korepin, N.M. Bogoliubov, A.G. Izergin, *Quantum Inverse Scattering Method and Correlation Functions* (Cambridge University Press, Cambridge, 1993)
30. L.D. Faddeev, L.A. Takhtajan, What is the spin of a spin wave? Phys. Lett. A **85**, 375 (1981)
31. M. Takahashi, *Thermodynamics of One-Dimensional Solvable Models* (Cambridge University Press, Cambridge, 1999)
32. N. Andrei, K. Furuya, J.H. Lowenstein, Solution of the Kondo problem. Rev. Mod. Phys. **55**, 331 (1983)
33. C.N. Yang, C.P. Yang, Thermodynamics of a one-dimensional system of bosons with repulsive delta-function interaction. J. Math. Phys. **10**, 1115 (1969)
34. E.H. Lieb, W. Liniger, Exact analysis of an interacting Bose gas I. The general solution and the ground state. Phys. Rev. **130**, 1605 (1963)
35. M. Gaudin, Thermodynamics of the Heisenberg-Ising ring for $\Delta > 1$. Phys. Rev. Lett. **26**, 1301 (1971)
36. M. Takahashi, One-dimensional Heisenberg model at finite temperature. Progress Theoret. Phys. **46**, 401 (1971)
37. M. Takahashi, One-dimensional Hubbard model at finite temperature. Progress Theoret. Phys. **47**, 69 (1972)
38. M. Takahashi, Low-temperature specific-heat of one-dimensional Hubbard model. Progress Theoret. Phys. **52**, 103 (1974)

39. J.D. Johnson, B. McCoy, Low-temperature thermodynamics of the $|\Delta| \geq 1$ Heisenberg-Ising ring. Phys. Rev. A **6**, 1613 (1972)
40. Y. Wang, Fermi-liquid features of the one-dimensional Luttinger liquid. Int. J. Mod. Phys. B **12**, 3465 (1998)
41. F.C. Alcaraz, M.N. Barber, M.T. Batchelor, R.J. Baxter, G.R.W. Quispel, Surface exponents of the quantum XXZ, Ashkin-Teller and Potts models. J. Phys. A **20**, 6397 (1987)
42. H.J. de Vega, A. González-Ruiz, Boundary K-matrices for the six vertex and the $n(2n-1)$ A_{n-1} vertex models. J. Phys. A **26**, L519 (1993)
43. H.J. de Vega, A. González-Ruiz, Boundary K-matrices for the XYZ, XXZ and XXX spin chains. J. Phys. A **27**, 6129 (1994)
44. A. Lima-Santos, Reflection K-matrices for 19-vertex models. Nucl. Phys. B **558**, 637 (1999)
45. R. Malara, A. Lima-Santos, On $A_{n-1}^{(1)}$, $B_n^{(1)}$, $C_n^{(1)}$, $D_n^{(1)}$, $A_{2n}^{(2)}$, $A_{2n-1}^{(2)}$ and $D_{n+1}^{(2)}$ reflection K-matrices. J. Stat. Mech. P09013 (2006)
46. M. Gaudin, Boundary energy of a Bose gas in one dimension. Phys. Rev. A **4**, 386 (1971)
47. A.L. Owuarek, R.J. Baxter, Surface free energy of the critical six-vertex model with free boundaries. J. Phys. A **22**, 1141 (1989)
48. C.J. Hamer, G.R.W. Quispel, M.T. Batchelor, Conformal anomaly and surface energy for Potts and Ashkin-Teller quantum chains. J. Phys. A **20**, 5677 (1987)
49. M.T. Batchelor, C.J. Hamer, Surface energy of integrable quantum spin chains. J. Phys. A **23**, 761 (1990)
50. P.A. Bares, G. Blatter, M. Ogata, Exact solution of the t-J model in one dimension at $2t = \pm J$: ground state and excitation spectrum. Phys. Rev. B **44**, 130 (1991)
51. M.T. Grisaru, L. Mezincescu, R.I. Nepomechie, Direct calculation of the boundary S-matrix for the open Heisenberg chain. J. Phys. A **28**, 1027 (1995)
52. F.H.L. Essler, The supersymmetric t-J model with a boundary. J. Phys. A **29**, 6183 (1996)
53. A. Kapustin, S. Skorik, Surface excitations and surface energy of the antiferromagnetic XXZ chain by the Bethe Ansatz approach. J. Phys. A **29**, 1629 (1996)
54. R. Murgan, R.I. Nepomechie, C. Shi, Boundary energy of the general open XXZ chain at roots of unity. J. High Energy Phys. **01**, 038 (2007)
55. A. Kitaev, Anyons in an exactly solved model and beyond. Ann. Phys. **(NY) 321**, 2 (2006)
56. B. Sutherland, Further results for the many-body problem in one dimension. Phys. Rev. Lett. **20**, 98 (1968)
57. C.N. Yang, Some exact results for the many-body problem in one dimension with repulsive delta-function interaction. Phys. Rev. Lett. **19**, 1312 (1967)
58. B. Sutherland, Model for a multicomponent quantum system. Phys. Rev. B **12**, 3795 (1975)
59. A.M. Tsvelik, P.B. Wiegmann, Exact results in the theory of magnetic alloys. Adv. Phys. **32**, 453 (1983)
60. P. Schlottmann, Exact results for highly correlated electron systems in one dimension. Int. J. Mod. Phys. B **11**, 415 (1997)
61. T. Deguchi, F.H.L. Essler, F. Göhmann, A. Klümper, V.E. Korepin, K. Kusakabe, Thermodynamics and excitations of the one-dimensional Hubbard model. Phys. Rep. **331**, 197 (2000)
62. B. Sutherland, *Beautiful Models: 70 Years of Exactly Solved Quantum Many-Body Problems* (World Scientify Publishing, Singapore, 2004)
63. P.P. Kulish, N.Y. Reshetikhin, E.K. Sklyanin, Yang-Baxter equation and representation theory. I. Lett. Math. Phys. **5**, 393 (1981)
64. P.P. Kulish, N.Y. Reshetikhin, Quantum linear problem for the sine-Gordon equation and higher representation. J. Sov. Math. **23**, 2435 (1983)
65. A.N. Kirillov, N.Y. Reshetikhin, Exact solution of the Heisenberg XXZ model of spin s. J. Sov. Math. **35**, 2627 (1986)
66. A.N. Kirillov, N.Y. Reshetikhin, Exact solution of the integrable XXZ Heisenberg model with arbitrary spin I. The ground state and the excitation spectrum. J. Phys. A **20**, 1565 (1987)
67. N. Kitanine, J.M. Maillet, V. Terras, Form factors of the XXZ Heisenberg spin-1/2 finite chain. Nucl. Phys. B **554**, 647 (1999)

Chapter 3
The Periodic Anisotropic Spin-$\frac{1}{2}$ Chains

Based on the pioneering work of Bethe [1] in which the coordinate Bethe Ansatz method was invented and the exact solution of the spin-$\frac{1}{2}$ Heisenberg chain model was obtained [2], several authors continued the study of the physical properties of this model. For example, Hulthén introduced the fundamental integral equations [3] to derive the ground state energy in the large size limit, which is essential in the development of the systematic Bethe Ansatz approaches. Nevertheless, it was not until 1958 that the anisotropic spin-$\frac{1}{2}$ chain model (*XXZ* case) was first solved by Orbach [4] (see also [5, 6]) via the coordinate Bethe Ansatz. Subsequently, Yang and Yang [7–9] revisited the *XXZ* model with quite a few new results. McCoy and Wu [10] realized the intrinsic relationship between the *XXZ* spin chain and the classical six-vertex model, which was studied extensively by Lieb [11–14] and Sutherland [15], while Sutherland noted the relationship between the *XYZ* spin chain and the eight-vertex model [16]. Baxter [17–19] derived the Hamiltonian from the transfer matrix of the classical two-dimensional vertex model and demonstrated its integrability with the exact correspondence between the elements of the R-matrix of the quantum spin-chain model and the Boltzmann weights of the classical vertex model. With this observation, Baxter successfully diagonalized both the quantum *XXZ* model and the *XYZ* model for an even number of sites. Since Baxter's famous works on the eight-vertex model [17, 20] and the quantum XYZ model [18, 19], many papers have been devoted to this topic in the literature [21–36]. Especially, based on the R-matrices obtained from the vertex models and the classical integrable systems, the algebraic Bethe Ansatz method was developed. This allowed Faddeev and Takhatajan [37] to apply the method to the anisotropic spin chain models, where Baxter's local gauge transformation was used to find a pseudo vacuum for the *XYZ* model with an even number of sites. The physical properties including the ground state, the elementary excitations, the thermodynamics and the correlation functions [38–40] have been extensively studied by many authors. For these topics we direct the readers' attention to some well-written books [38, 41–44] and the references therein.

Y. Wang et al., *Off-Diagonal Bethe Ansatz for Exactly Solvable Models*,
DOI 10.1007/978-3-662-46756-5_3

This chapter introduces the application of ODBA in the anisotropic spin-chain models with periodic boundary conditions. By constructing the operator product identities and the asymptotic behavior of the transfer matrices, we re-derive the known solutions of the periodic *XXZ* model and the *XYZ* model. A remarkable fact is that ODBA greatly simplifies the derivation process for the *XYZ* model [45]. In particular, the exact solution of the *XYZ* model with general coupling constants and an odd number of sites, which had been a longstanding problem for over 40 years, can be constructed via this method.

3.1 The *XXZ* Model

3.1.1 The Hamiltonian

The *XXZ* spin-$\frac{1}{2}$ chain describes N magnetic objectives coupled by anisotropic Heisenberg exchanges between the nearest neighbor sites. The model Hamiltonian reads

$$H = \frac{1}{2}\sum_{j=1}^{N}[\sigma_j^x\sigma_{j+1}^x + \sigma_j^y\sigma_{j+1}^y + \cos\eta\sigma_j^z\sigma_{j+1}^z], \tag{3.1.1}$$

where η is the crossing parameter describing the anisotropy of the couplings. Here the periodic boundary condition $\sigma_{N+1} = \sigma_1$ is assumed. The associated R-matrix of this model is

$$\begin{aligned} R_{0,j}(u) = &\frac{1}{2}\left[\frac{\sin(u+\eta)}{\sin\eta}(1+\sigma_j^z\sigma_0^z) + \frac{\sin u}{\sin\eta}(1-\sigma_j^z\sigma_0^z)\right] \\ &+ \frac{1}{2}(\sigma_j^x\sigma_0^x + \sigma_j^y\sigma_0^y), \end{aligned} \tag{3.1.2}$$

which satisfies YBE. We define the inhomogeneous monodromy matrix $T_0(u)$ and the transfer matrix $t(u)$ as

$$T_0(u) = R_{0,N}(u-\theta_N)\cdots R_{0,1}(u-\theta_1), \tag{3.1.3}$$

$$t(u) = tr_0 T_0(u), \tag{3.1.4}$$

where θ_j are the site-dependent inhomogeneous parameters. By the same procedure used in Chap. 1, we can demonstrate that

$$[t(u), t(v)] = 0. \tag{3.1.5}$$

With the help of the following properties of $t(u)$:

$$t(0)|_{\{\theta_j=0\}} = tr_0\{P_{0,N}\cdots P_{0,1}\} = P_{1,N}\cdots P_{1,2},$$
$$\left.\frac{\partial t(u)}{\partial u}\right|_{u=0,\{\theta_j=0\}} = \frac{1}{2\sin\eta}\sum_{j=1}^{N} tr_0\left\{P_{0,N}\cdots P_{0,j+1}\right.$$
$$\left.\times[\cos\eta(1+\sigma_0^z\sigma_j^z)+1-\sigma_0^z\sigma_j^z]P_{0,j-1}\cdots P_{0,1}\right\}, \quad (3.1.6)$$

the Hamiltonian (3.1.1) can be expressed as

$$H = \sin\eta\left.\frac{\partial \ln t(u)}{\partial u}\right|_{u=0,\{\theta_j=0\}} - \frac{N}{2}\cos\eta. \quad (3.1.7)$$

Therefore, the eigenvalues of the Hamiltonian can be observed from $\Lambda(u)$, the eigenvalue of the transfer matrix

$$E = \sin\eta\left.\frac{\partial \ln \Lambda(u)}{\partial u}\right|_{u=0,\{\theta_j=0\}} - \frac{N}{2}\cos\eta. \quad (3.1.8)$$

3.1.2 Operator Product Identities of the Transfer Matrix

We note that the R-matrix (3.1.2) possesses properties similar to (1.5.4)–(1.5.9):

$$\text{Initial condition: } R_{1,2}(0) = P_{1,2}, \quad (3.1.9)$$
$$\text{Unitarity: } R_{1,2}(u)R_{2,1}(-u) = -\frac{\sin(u+\eta)\sin(u-\eta)}{\sin^2\eta}\times \text{id}, \quad (3.1.10)$$
$$\text{Crossing relation: } R_{1,2}(u) = -\sigma_1^y R_{1,2}^{t_1}(-u-\eta)\sigma_1^y, \quad (3.1.11)$$
$$\text{PT-symmetry: } R_{1,2}(u) = R_{2,1}(u) = R_{1,2}^{t_1 t_2}(u), \quad (3.1.12)$$
$$Z_2\text{-symmetry: } \sigma_1^\alpha\sigma_2^\alpha R_{1,2}(u) = R_{1,2}(u)\sigma_1^\alpha\sigma_2^\alpha, \quad \text{for } \alpha = x, y, z, \quad (3.1.13)$$
$$\text{Fusion condition: } R_{1,2}(-\eta) = -2P_{1,2}^{(-)}. \quad (3.1.14)$$

With the same procedure [45, 46] we introduced in Chap. 1, let us apply the initial condition of the R-matrix to express the transfer matrix $t(\theta_j)$ as

$$t(\theta_j) = R_{j,j-1}(\theta_j-\theta_{j-1})\cdots R_{j,1}(\theta_j-\theta_1)$$
$$\times R_{j,N}(\theta_j-\theta_N)\cdots R_{j,j+1}(\theta_j-\theta_{j+1}). \quad (3.1.15)$$

The crossing relation (3.1.11) makes it possible to express the transfer matrix $t(\theta_j - \eta)$ as

$$t(\theta_j - \eta) = (-1)^N R_{j,j+1}(-\theta_j + \theta_{j+1}) \cdots R_{j,N}(-\theta_j + \theta_N) \\ \times R_{j,1}(-\theta_j + \theta_1) \cdots R_{j,j-1}(-\theta_j + \theta_{j-1}). \quad (3.1.16)$$

Using the unitary relation (3.1.10), we have

$$t(\theta_j)t(\theta_j - \eta) = a(\theta_j)d(\theta_j - \eta) = \mathrm{Det}_q \left\{T(\theta_j)\right\}, \quad (3.1.17)$$

with

$$a(u) = \prod_{j=1}^{N} \frac{\sin(u - \theta_j + \eta)}{\sin \eta}, \quad (3.1.18)$$

$$d(u) = \prod_{j=1}^{N} \frac{\sin(u - \theta_j)}{\sin \eta}. \quad (3.1.19)$$

The commutativity of the transfer matrices with different spectral parameters implies that they have common eigenstates. Let us assume that $|\Psi\rangle$ is an eigenstate of $t(u)$, that does not depend upon u, with the eigenvalue $\Lambda(u)$, i.e.,

$$t(u)|\Psi\rangle = \Lambda(u)|\Psi\rangle. \quad (3.1.20)$$

The operator identity (3.1.17) implies that the corresponding eigenvalue $\Lambda(u)$ satisfies the relation

$$\Lambda(\theta_j)\Lambda(\theta_j - \eta) = a(\theta_j)d(\theta_j - \eta), \quad j = 1, \ldots, N. \quad (3.1.21)$$

Such functional relations were also derived in [47–52] via the separation of variables method for some models related to the $SU(2)$ algebra. We remark that using the inhomogeneity parameters $\{\theta_j\}$ makes it convenient to get N independent relations (3.1.21).

3.1.3 $\Lambda(u)$ as a Trigonometric Polynomial

In order to determine the dependence of the function $\Lambda(u)$ on the spectral parameter u, let us introduce the following notation for the R-matrix

$$R_{0,j}(u - \theta_j) = \begin{pmatrix} \bar{a}_j(u) & \bar{b}_j(u) \\ \bar{c}_j(u) & \bar{d}_j(u) \end{pmatrix}. \quad (3.1.22)$$

From the definition of the R-matrix we know that $\bar{a}_j(u)$ and $\bar{d}_j(u)$ are two operator-valued Laurent polynomials of $x = e^{iu}$:

$$\begin{aligned} \bar{a}_j(u) &= a_j^{(+)} x + a_j^{(-)} x^{-1}, \\ \bar{d}_j(u) &= d_j^{(+)} x + d_j^{(-)} x^{-1}, \end{aligned} \tag{3.1.23}$$

with $a_j^{(\pm)}$ and $d_j^{(\pm)}$ being some operator-valued coefficients, while $\bar{b}_j(u)$ and $\bar{c}_j(u)$ being u-independent operators.

For convenience, let us denote the monodromy matrix $T(u)$ spanned in the auxiliary space as

$$T(u) = \begin{pmatrix} A(u) & B(u) \\ C(u) & D(u) \end{pmatrix}.$$

With a simple recursive process we can demonstrate that

$$\begin{aligned} &A(u) \text{ and } D(u) \text{ are degree } N \text{ Laurent polynomials in } e^{iu}; \\ &B(u) \text{ and } C(u) \text{ are degree } N-1 \text{ Laurent polynomials in } e^{iu}. \end{aligned} \tag{3.1.24}$$

Obviously,

$$t(u+\pi) = (-1)^N t(u). \tag{3.1.25}$$

This allows us to parameterize $\Lambda(u)$ as

$$\Lambda(u) = \Lambda_0 \prod_{j=1}^{N} \sin(u - z_j), \tag{3.1.26}$$

with $N+1$ unknowns Λ_0 and $\{z_j | j = 1, \ldots, N\}$.

Equation (3.1.21) readily gives N relations for the unknowns. To determine the $\Lambda(u)$ polynomial, we need one more condition. For this purpose, let us examine further the asymptotic behavior of the elements of the monodromy matrix for $x \to \pm\infty$. As

$$\bar{a}_j(u) = \frac{xe^{-i\theta_j}}{4i \sin\eta} \left[e^{i\eta}(1+\sigma_j^z) + 1 - \sigma_j^z \right] - \frac{x^{-1} e^{i\theta_j}}{4i \sin\eta} \left[e^{-i\eta}(1+\sigma_j^z) + 1 - \sigma_j^z \right],$$

$$\bar{d}_j(u) = \frac{xe^{-i\theta_j}}{4i \sin\eta} \left[e^{i\eta}(1-\sigma_j^z) + 1 + \sigma_j^z \right] - \frac{x^{-1} e^{i\theta_j}}{4i \sin\eta} \left[e^{-i\eta}(1-\sigma_j^z) + 1 + \sigma_j^z \right],$$

we have

$$\lim_{iu\to+\infty} A(u) = (2i\sin\eta)^{-N} e^{iNu - i\sum_{j=1}^{N}\theta_j + i(N-\hat{M})\eta} + \cdots,$$
$$\lim_{iu\to+\infty} D(u) = (2i\sin\eta)^{-N} e^{iNu - i\sum_{j=1}^{N}\theta_j + i\hat{M}\eta} + \cdots, \tag{3.1.27}$$

where

$$\hat{M} = \frac{1}{2}\left(N - \sum_{j=1}^{N}\sigma_j^z\right), \tag{3.1.28}$$

is a conserved quantity with eigenvalues $M = 0, 1, \ldots, N$. Therefore, the asymptotic behavior of $\Lambda(u)$ reads

$$\lim_{iu\to+\infty} \Lambda(u) = \frac{1}{(2i\sin\eta)^N}\left\{e^{i\left[Nu-\sum_{j=1}^{N}\theta_j+(N-M)\eta\right]} + e^{i\left[Nu-\sum_{j=1}^{N}\theta_j+M\eta\right]}\right\} + \cdots. \tag{3.1.29}$$

Together with Eq. (3.1.21), it is sufficient to determine $\Lambda(u)$ completely.

3.1.4 Functional $T-Q$ Relation and Bethe Ansatz Equations

Adopting the same procedure used in proving Theorem 1 in Chap. 1, we can demonstrate that the eigenvalues of the transfer matrix can be parameterized in terms of the $T-Q$ relation

$$\Lambda(u) = a(u)\frac{Q(u-\eta)}{Q(u)} + d(u)\frac{Q(u+\eta)}{Q(u)}, \tag{3.1.30}$$

with

$$Q(u) = \prod_{j=1}^{M}\sin(u-\mu_j), \tag{3.1.31}$$

where M is an arbitrary non-negative integer. Note that $d(\theta_j) = a(\theta_j - \eta) = 0$. The above expression for $\Lambda(u)$ satisfies Eq. (3.1.21) automatically for arbitrary polynomial $Q(u)$. The form of $Q(u)$ is confined by the quasi-periodicity of Eq. (3.1.25) and the asymptotic behavior of Eq. (3.1.29).

In fact, $\Lambda(u)$ must be a trigonometric polynomial of degree N. The residues at the Bethe roots μ_j in Eq. (3.1.30) must be zero. This regular property of $\Lambda(u)$ leads to the following BAEs:

$$\frac{a(\mu_j)}{d(\mu_j)} = -\frac{Q(\mu_j+\eta)}{Q(\mu_j-\eta)}, \quad j=1,\ldots,M. \tag{3.1.32}$$

For convenience, we put $\mu_j = i\frac{\lambda_j}{2} - \frac{\eta}{2}$ for real η and $\mu_j = \frac{\lambda_j}{2} - \frac{\eta}{2}$ for imaginary $\eta = i\gamma$. In the homogeneous limit $\theta_j \to 0$, the BAEs are rewritten as

$$\frac{\sinh^N \frac{1}{2}(\lambda_j + i\eta)}{\sinh^N \frac{1}{2}(\lambda_j - i\eta)} = -\prod_{l=1}^{M} \frac{\sinh \frac{1}{2}(\lambda_j - \lambda_l + 2i\eta)}{\sinh \frac{1}{2}(\lambda_j - \lambda_l - 2i\eta)}, \tag{3.1.33}$$

for real η and

$$\frac{\sin^N \frac{1}{2}(\lambda_j + i\gamma)}{\sin^N \frac{1}{2}(\lambda_j - i\gamma)} = -\prod_{l=1}^{M} \frac{\sin \frac{1}{2}(\lambda_j - \lambda_l + 2i\gamma)}{\sin \frac{1}{2}(\lambda_j - \lambda_l - 2i\gamma)}, \tag{3.1.34}$$

for imaginary $\eta = i\gamma$.

The eigenvalue of the Hamiltonian (3.1.1) is thus parameterized as

$$E(\{\lambda_j\}) = -\sum_{j=1}^{M} \frac{\sin^2 \eta}{\cosh^2 \frac{\lambda_j}{2} - \cos^2 \frac{\eta}{2}} + \frac{1}{2N} \cos \eta, \tag{3.1.35}$$

for real η and

$$E(\{\lambda_j\}) = -\sum_{j=1}^{M} \frac{\sinh^2 \gamma}{\cosh^2 \frac{\gamma}{2} - \cos^2 \frac{\lambda_j}{2}} + \frac{1}{2} N \cosh \gamma, \tag{3.1.36}$$

for imaginary η. The BAEs and the eigenvalues are exactly the same as those derived from the conventional Bethe Ansatz methods.

3.1.5 Ground States and Elementary Excitations

For real η, the system is in the gapless easy-plane region. The ground state is still described by real Bethe roots filling the whole real axis. As for the XXX case, we introduce

$$a_n(\lambda) = \frac{1}{2\pi} \frac{\sin(n\eta)}{\cosh \lambda - \cos(n\eta)}. \tag{3.1.37}$$

The Bethe root distribution in the ground state satisfies

$$a_1(\lambda) = \rho_g(\lambda) + \int_{-\infty}^{\infty} a_2(\mu)\rho_g(\lambda - \mu)d\lambda. \tag{3.1.38}$$

Without losing generality, we consider the case $\eta \in (0, \pi)$. Under Fourier transformation it is easy to deduce that

$$\tilde{a}_n(\omega) = \int_{-\infty}^{\infty} e^{i\omega\lambda} a_n(\lambda)d\lambda = \frac{\sinh(\pi\omega - 2\delta_n\pi\omega)}{\sinh(\pi\omega)}, \tag{3.1.39}$$

$$\tilde{\rho}_g(\omega) = \frac{\tilde{a}_1(\omega)}{1 + \tilde{a}_2(\omega)} = \frac{1}{2\cosh(\eta\omega)}, \tag{3.1.40}$$

with $\delta_n \equiv \frac{n\eta}{2\pi} - \lfloor\frac{n\eta}{2\pi}\rfloor$ denoting the fractional part of $\frac{n\eta}{2\pi}$. The solution of Eq. (3.1.38) is

$$\rho_g(\lambda) = \frac{1}{2\pi}\int_{-\infty}^{\infty} e^{-i\lambda\omega}\tilde{\rho}_g(\omega)d\omega = \frac{1}{4\eta\cosh(\frac{\pi\lambda}{2\eta})}. \tag{3.1.41}$$

The total magnetization is thus

$$\frac{N}{2} - N\int_{-\infty}^{\infty} \rho_g(\lambda)d\lambda = \frac{N}{2} - N\tilde{\rho}_g(0) = 0, \tag{3.1.42}$$

indicating a spin singlet ground state for an even N (for an odd N there is a hole at infinity and the ground state is doubly degenerate.). The density of the ground state energy reads

$$\begin{aligned} e_g &= -2\pi\sin\eta\int_{-\infty}^{\infty} a_1(\lambda)\rho_g(\lambda)d\lambda + \frac{1}{2}\cos\eta \\ &= -\frac{1}{2}\sin\eta\int_{-\infty}^{\infty}\frac{\sinh(\pi\omega - \eta\omega)}{\sinh(\pi\omega)\cosh(\eta\omega)}d\omega + \frac{1}{2}\cos\eta. \end{aligned} \tag{3.1.43}$$

The dressed energy function in this case reads

$$\varepsilon(\lambda) = 2\pi\sin\eta a_1(\lambda) - \int_{-\infty}^{\infty} a_2(\mu)\varepsilon(\lambda - \mu)d\mu. \tag{3.1.44}$$

Using Fourier transformation, we obtain

$$\varepsilon(\lambda) = \frac{\pi\sin\eta}{2\eta\cosh\left(\frac{\pi\lambda}{2\eta}\right)}. \tag{3.1.45}$$

Obviously, the lowest energy is at $\lambda = \pm\infty$. With the same procedure used in Chap. 2, we can deduce that the energy of a spinon excitation can also be described by $\varepsilon(\lambda^h)$ for a hole at λ^h.

For imaginary η, the system is in the gapped easy-axis region. Without losing generality, we assume $\gamma > 0$. Similarly, let us introduce

$$a_n(\lambda) = \frac{1}{2\pi}\frac{\sinh(n\gamma)}{\cosh(n\gamma) - \cos\lambda}. \tag{3.1.46}$$

The Fourier transform of $a_n(\lambda)$ is

$$\widetilde{a}_n(\omega) = \frac{1}{2\pi}\int_{-\pi}^{\pi} e^{i\omega\lambda}\frac{\sinh(n\gamma)}{\cosh(n\gamma) - \cos\lambda}d\lambda = e^{-n\gamma|\omega|}. \tag{3.1.47}$$

Note that ω takes values of integers. For the ground state, the Bethe roots still take real values and fill the region $(-\pi, \pi]$. The density of Bethe roots in the ground state satisfies

$$\rho_g(\lambda) = a_1(\lambda) - \int_{-\pi}^{\pi} a_2(\mu)\rho_g(\lambda - \mu)d\mu. \tag{3.1.48}$$

Using Fourier transformation we have

$$\widetilde{\rho}_g(\omega) = \frac{1}{2\cosh(\gamma\omega)}, \tag{3.1.49}$$

$$\rho_g(\lambda) = \frac{1}{2\pi}\sum_{\omega=-\infty}^{\infty} e^{-i\lambda\omega}\frac{1}{2\cosh(\gamma\omega)}. \tag{3.1.50}$$

The total magnetization in the ground state for an even N is still zero. Similarly, the dressed energy in this case is given by

$$\varepsilon(\lambda) = 2\pi\sinh\gamma a_1(\lambda) - \int_{-\pi}^{\pi} a_2(\lambda - \mu)\varepsilon(\lambda - \mu)d\mu. \tag{3.1.51}$$

Using Fourier transformation, we obtain

$$\varepsilon(\lambda) = \sinh\gamma\sum_{\omega=-\infty}^{\infty}\frac{e^{-i\omega\lambda}}{2\cosh(\omega\gamma)}. \tag{3.1.52}$$

Because λ_j are confined in a finite region, the elementary excitations possess a finite gap

$$\Delta = 2\varepsilon(\pi) = \sinh\gamma\sum_{\omega=-\infty}^{\infty}\frac{(-1)^{\omega}}{\cosh(\omega\gamma)}. \tag{3.1.53}$$

3.2 The XYZ Model

3.2.1 The Hamiltonian

The spin-$\frac{1}{2}$ XYZ model has a variety of applications in statistical physics (associated with the eight-vertex model), in condensed matter physics (one-dimensional magnetism) and in quantum information. The model with an even N (number of sites) was solved by Baxter via the $T-Q$ method and by Takhatajan and Faddeev via the algebraic Bethe Ansatz method. In their approaches, a reference state plays an important role. However, a proper reference state has been found only for an even N but not for an odd N. Here we introduce the ODBA method to approach this model.

The model Hamiltonian we shall consider reads

$$H=\frac{1}{2}\sum_{j=1}^{N}(J_x\sigma_j^x\sigma_{j+1}^x+J_y\sigma_j^y\sigma_{j+1}^y+J_z\sigma_j^z\sigma_{j+1}^z). \tag{3.2.1}$$

For convenience, let us introduce the following elliptic functions for rational numbers b_1, b_2 and generic complex number τ with $\mathrm{Im}(\tau)>0$:

$$\theta\begin{bmatrix} b_1 \\ b_2 \end{bmatrix}(u,\tau)=\sum_{m=-\infty}^{\infty}\exp\left\{i\pi\left[(m+b_1)^2\tau+2(m+b_1)(u+b_2)\right]\right\}, \tag{3.2.2}$$

$$\sigma(u)=\theta\begin{bmatrix} \frac{1}{2} \\ \frac{1}{2} \end{bmatrix}(u,\tau),\quad \zeta(u)=\frac{\partial}{\partial u}\{\ln\sigma(u)\}. \tag{3.2.3}$$

These elliptic functions satisfy the following identities:

$$\begin{aligned}\sigma(u+x)\sigma(u-x)\sigma(v+y)\sigma(v-y)-\sigma(u+y)\sigma(u-y)\sigma(v+x)\sigma(v-x)\\ =\sigma(u+v)\sigma(u-v)\sigma(x+y)\sigma(x-y),\end{aligned} \tag{3.2.4}$$

$$\sigma(2u)=\frac{2\sigma(u)\sigma(u+\frac{1}{2})\sigma(u+\frac{\tau}{2})\sigma(u-\frac{1}{2}-\frac{\tau}{2})}{\sigma(\frac{1}{2})\sigma(\frac{\tau}{2})\sigma(-\frac{1}{2}-\frac{\tau}{2})}, \tag{3.2.5}$$

$$\frac{\sigma(u)}{\sigma(\frac{\tau}{2})}=\frac{\theta\begin{bmatrix} 0 \\ \frac{1}{2} \end{bmatrix}(u,2\tau)\,\theta\begin{bmatrix} \frac{1}{2} \\ \frac{1}{2} \end{bmatrix}(u,2\tau)}{\theta\begin{bmatrix} 0 \\ \frac{1}{2} \end{bmatrix}(\frac{\tau}{2},2\tau)\,\theta\begin{bmatrix} \frac{1}{2} \\ \frac{1}{2} \end{bmatrix}(\frac{\tau}{2},2\tau)}, \tag{3.2.6}$$

$$\theta\begin{bmatrix} \frac{1}{2} \\ \frac{1}{2} \end{bmatrix}(2u,2\tau)=\theta\begin{bmatrix} \frac{1}{2} \\ \frac{1}{2} \end{bmatrix}(\tau,2\tau)\times\frac{\sigma(u)\sigma(u+\frac{1}{2})}{\sigma(\frac{\tau}{2})\sigma(\frac{1}{2}+\frac{\tau}{2})}, \tag{3.2.7}$$

$$\theta\begin{bmatrix} 0 \\ \frac{1}{2} \end{bmatrix}(2u,2\tau)=\theta\begin{bmatrix} 0 \\ \frac{1}{2} \end{bmatrix}(0,2\tau)\times\frac{\sigma(u-\frac{\tau}{2})\sigma(u+\frac{1}{2}+\frac{\tau}{2})}{\sigma(-\frac{\tau}{2})\sigma(\frac{1}{2}+\frac{\tau}{2})}. \tag{3.2.8}$$

The coupling constants are thus parameterized in terms of the above elliptic functions as

$$J_x = e^{i\pi\eta}\frac{\sigma(\eta+\frac{\tau}{2})}{\sigma(\frac{\tau}{2})}, \quad J_y = e^{i\pi\eta}\frac{\sigma(\eta+\frac{1+\tau}{2})}{\sigma(\frac{1+\tau}{2})}, \quad J_z = \frac{\sigma(\eta+\frac{1}{2})}{\sigma(\frac{1}{2})}, \tag{3.2.9}$$

where the crossing parameter η is a generic complex number.

The R-matrix $R(u) \in \mathrm{End}(\mathbf{V}\otimes\mathbf{V})$ associated with the present model is given by

$$R(u) = \begin{pmatrix} \alpha_1(u) & & & \alpha_4(u) \\ & \alpha_2(u) & \alpha_3(u) & \\ & \alpha_3(u) & \alpha_2(u) & \\ \alpha_4(u) & & & \alpha_1(u) \end{pmatrix}, \tag{3.2.10}$$

where the non-vanishing entries read [53]

$$\alpha_1(u) = \frac{\theta\begin{bmatrix}0\\ \frac{1}{2}\end{bmatrix}(u,2\tau)\,\theta\begin{bmatrix}\frac{1}{2}\\ \frac{1}{2}\end{bmatrix}(u+\eta,2\tau)}{\theta\begin{bmatrix}0\\ \frac{1}{2}\end{bmatrix}(0,2\tau)\,\theta\begin{bmatrix}\frac{1}{2}\\ \frac{1}{2}\end{bmatrix}(\eta,2\tau)}, \tag{3.2.11}$$

$$\alpha_2(u) = \frac{\theta\begin{bmatrix}\frac{1}{2}\\ \frac{1}{2}\end{bmatrix}(u,2\tau)\,\theta\begin{bmatrix}0\\ \frac{1}{2}\end{bmatrix}(u+\eta,2\tau)}{\theta\begin{bmatrix}0\\ \frac{1}{2}\end{bmatrix}(0,2\tau)\,\theta\begin{bmatrix}\frac{1}{2}\\ \frac{1}{2}\end{bmatrix}(\eta,2\tau)}, \tag{3.2.12}$$

$$\alpha_3(u) = \frac{\theta\begin{bmatrix}0\\ \frac{1}{2}\end{bmatrix}(u,2\tau)\,\theta\begin{bmatrix}0\\ \frac{1}{2}\end{bmatrix}(u+\eta,2\tau)}{\theta\begin{bmatrix}0\\ \frac{1}{2}\end{bmatrix}(0,2\tau)\,\theta\begin{bmatrix}0\\ \frac{1}{2}\end{bmatrix}(\eta,2\tau)}, \tag{3.2.13}$$

$$\alpha_4(u) = \frac{\theta\begin{bmatrix}\frac{1}{2}\\ \frac{1}{2}\end{bmatrix}(u,2\tau)\,\theta\begin{bmatrix}\frac{1}{2}\\ \frac{1}{2}\end{bmatrix}(u+\eta,2\tau)}{\theta\begin{bmatrix}0\\ \frac{1}{2}\end{bmatrix}(0,2\tau)\,\theta\begin{bmatrix}0\\ \frac{1}{2}\end{bmatrix}(\eta,2\tau)}. \tag{3.2.14}$$

It can be easily checked that the R-matrix defined by (3.2.10) satisfies YBE and possesses the following properties:

$$\text{Initial condition}: \; R_{1,2}(0) = P_{1,2}, \tag{3.2.15}$$

$$\text{Unitary relation}: \; R_{1,2}(u)R_{2,1}(-u) = -\xi(u)\times \mathrm{id}, \tag{3.2.16}$$

$$\xi(u) = \frac{\sigma(u-\eta)\sigma(u+\eta)}{\sigma(\eta)\sigma(\eta)},$$

$$\text{Crossing relation}: \; R_{1,2}(u) = V_1 R_{1,2}^{t_2}(-u-\eta)V_1, \quad V = -i\sigma^y, \tag{3.2.17}$$

$$\text{PT-symmetry}: \; R_{1,2}(u) = R_{2,1}(u) = R^{t_1 t_2}_{1,2}(u), \tag{3.2.18}$$

$$Z_2\text{-symmetry}: \; \sigma^i_1\sigma^i_2 R_{1,2}(u) = R_{1,2}(u)\sigma^i_1\sigma^i_2, \quad \text{for} \quad i = x, y, z, \tag{3.2.19}$$

$$\text{Fusion condition}: \; R_{1,2}(-\eta) = -(1 - P_{1,2}) = -2P^{(-)}_{1,2}. \tag{3.2.20}$$

The periodic boundary Hamiltonian (3.2.1) is thus given in terms of the transfer matrix $t(u)$ defined by (3.1.3)–(3.1.4) with the R-matrix (3.2.10) as

$$H = \frac{\sigma(\eta)}{\sigma'(0)}\left\{ \frac{\partial \ln t(u)}{\partial u}\Big|_{u=0,\{\theta_j=0\}} - \frac{1}{2}N\zeta(\eta)\right\}, \tag{3.2.21}$$

with $\sigma'(0) = \frac{\partial}{\partial u}\sigma(u)\big|_{u=0}$ and the function $\zeta(u)$ given by (3.2.3).

3.2.2 Operator Product Identities

With a procedure similar to that introduced for the XXX model, we find that the following operator identities hold:

$$t(\theta_j)t(\theta_j - \eta) = a(\theta_j)d(\theta_j - \eta) \equiv \text{Det}_q\left\{T(\theta_j)\right\}, \; j = 1, \ldots, N, \tag{3.2.22}$$

with

$$a(u) = \prod_{l=1}^{N} \frac{\sigma(u - \theta_l + \eta)}{\sigma(\eta)},$$

$$d(u) = a(u - \eta) = \prod_{l=1}^{N} \frac{\sigma(u - \theta_l)}{\sigma(\eta)}. \tag{3.2.23}$$

The quasi-periodicity of the elliptic function $\sigma(u)$

$$\sigma(u + \tau) = -e^{-2i\pi(u+\frac{\tau}{2})}\sigma(u), \quad \sigma(u + 1) = -\sigma(u), \tag{3.2.24}$$

directly induces the quasi-periodic properties of the R-matrix:

$$R_{1,2}(u + 1) = -\sigma^z_1 R_{1,2}(u)\sigma^z_1, \tag{3.2.25}$$

$$R_{1,2}(u + \tau) = -e^{-2i\pi(u+\frac{\eta}{2}+\frac{\tau}{2})}\sigma^x_1 R_{1,2}(u)\sigma^x_1. \tag{3.2.26}$$

With the above relations we can easily deduce that the transfer matrix $t(u)$ possesses the following quasi-periodic properties:

$$t(u + \tau) = (-1)^N e^{-2\pi i\left\{Nu+N(\frac{\eta+\tau}{2})-\sum_{j=1}^{N}\theta_j\right\}}t(u), \tag{3.2.27}$$

$$t(u + 1) = (-1)^N t(u). \tag{3.2.28}$$

In addition, from the unitary relation (3.2.16) and the definition of the transfer matrix we may derive the following operator identity [45]:

$$\prod_{j=1}^{N} t(\theta_j) = \prod_{j=1}^{N} a(\theta_j) \times \text{id}. \tag{3.2.29}$$

The quasi-periodic properties (3.2.27) and (3.2.28), the relations (3.2.22) and (3.2.29) constitute the sufficient conditions to determine the spectrum of $t(u)$.

3.2.3 The Inhomogeneous $T - Q$ Relation

Applying $t(u)$ to an arbitrary eigenstate $|\Psi\rangle$, we can deduce that the corresponding eigenvalue $\Lambda(u)$ possesses relations similar to those of $t(u)$. From (3.2.27) and (3.2.28) we can derive

$$\Lambda(u+1) = (-1)^N \Lambda(u), \tag{3.2.30}$$

$$\Lambda(u+\tau) = (-1)^N e^{-2\pi i\left\{Nu+N\left(\frac{\eta+\tau}{2}\right)-\sum_{j=1}^{N}\theta_j\right\}} \Lambda(u). \tag{3.2.31}$$

In addition, the analytic properties of the R-matrix indicate that

$$\Lambda(u) \text{ is an entire function of } u. \tag{3.2.32}$$

The above property together with the quasi-periodic properties (3.2.30)–(3.2.31) implies that $\Lambda(u)$, as a function of u, is an elliptic polynomial of degree N. Similarly, (3.2.22) and (3.2.29) lead to

$$\Lambda(\theta_j)\Lambda(\theta_j-\eta) = a(\theta_j)d(\theta_j-\eta), \quad j=1,\ldots,N, \tag{3.2.33}$$

$$\prod_{j=1}^{N} \Lambda(\theta_j) = \prod_{j=1}^{N} a(\theta_j). \tag{3.2.34}$$

The Eqs. (3.2.30)–(3.2.34) provide sufficient conditions to determine the function $\Lambda(u)$ and allow us to construct the inhomogeneous $T - Q$ relation

$$\Lambda(u) = e^{2i\pi l_1 u + i\phi} a(u) \frac{Q_1(u-\eta)}{Q_2(u)} + e^{-2i\pi l_1(u+\eta)-i\phi} d(u) \frac{Q_2(u+\eta)}{Q_1(u)} + c\,\frac{\sigma^m\left(u+\frac{\eta}{2}\right) a(u)d(u)}{\sigma^m(\eta)Q_1(u)Q_2(u)}, \tag{3.2.35}$$

where l_1 is an integer, ϕ and c are two constants, and M and m are two non-negative integers which satisfy the relation

$$N+m=2M. \tag{3.2.36}$$

Note that m is even if N is even, while m is odd if N is odd. The functions $Q_1(u)$ and $Q_2(u)$ are parameterized by $2M$ Bethe roots $\{\mu_j|j=1,\ldots,M\}$ and $\{\nu_j|j=1,\ldots,M\}$ as follows

$$Q_1(u)=\prod_{j=1}^{M}\frac{\sigma(u-\mu_j)}{\sigma(\eta)},\quad Q_2(u)=\prod_{j=1}^{M}\frac{\sigma(u-\nu_j)}{\sigma(\eta)}. \tag{3.2.37}$$

One can easily check that $\Lambda(u)$ given by (3.2.35) is a solution of Eqs. (3.2.30)–(3.2.33), provided that the $2M+2$ parameters satisfy the BAEs:

$$\left(\frac{N}{2}-M\right)\eta-\sum_{j=1}^{M}(\mu_j-\nu_j)=l_1\tau+m_1,\quad l_1,\,m_1\in Z, \tag{3.2.38}$$

$$M\eta-\sum_{l=1}^{N}\theta_l+\sum_{j=1}^{M}(\mu_j+\nu_j)=m_2,\quad m_2\in Z, \tag{3.2.39}$$

$$\frac{c\,e^{2i\pi(l_1\mu_j+l_1\eta)+i\phi}\sigma^m\left(\mu_j+\frac{\eta}{2}\right)}{\sigma^m(\eta)}a(\mu_j)=-Q_2(\mu_j)Q_2(\mu_j+\eta), \tag{3.2.40}$$

$$\frac{c\,e^{-2i\pi l_1\nu_j-i\phi}\sigma^m\left(\nu_j+\frac{\eta}{2}\right)}{\sigma^m(\eta)}d(\nu_j)=-Q_1(\nu_j)Q_1(\nu_j-\eta). \tag{3.2.41}$$

It should be noted that (3.2.40) and (3.2.41) are required by the analyticity of $\Lambda(u)$ while (3.2.38) and (3.2.39) are required by the quasi-periodicity of $\Lambda(u)$.

In the homogeneous limit $\theta_j\to 0$, the inhomogeneous $T-Q$ relation is reduced to

$$\begin{aligned}\Lambda(u)=&\,e^{2i\pi l_1 u+i\phi}\frac{\sigma^N(u+\eta)}{\sigma^N(\eta)}\frac{Q_1(u-\eta)}{Q_2(u)}+\frac{e^{-2i\pi l_1(u+\eta)-i\phi}\sigma^N(u)}{\sigma^N(\eta)}\frac{Q_2(u+\eta)}{Q_1(u)}\\&+\frac{c\,\sigma^m\left(u+\frac{\eta}{2}\right)}{\sigma^m(\eta)Q_1(u)Q_2(u)}\frac{\sigma^N(u+\eta)\sigma^N(u)}{\sigma^N(\eta)\sigma^N(\eta)}.\end{aligned} \tag{3.2.42}$$

Here the $2M+2$ parameters c, ϕ, $\{\mu_j\}$ and $\{\nu_j\}$ satisfy the following BAEs

$$\left(\frac{N}{2}-M\right)\eta-\sum_{j=1}^{M}(\mu_j-\nu_j)=l_1\tau+m_1,\quad l_1,\,m_1\in Z, \tag{3.2.43}$$

$$M\eta+\sum_{j=1}^{M}(\mu_j+\nu_j)=m_2,\quad m_2\in Z, \tag{3.2.44}$$

$$\frac{ce^{2i\pi l_1(\mu_j+\eta)+i\phi}\sigma^m\left(\mu_j+\frac{\eta}{2}\right)}{\sigma^m(\eta)}\frac{\sigma^N(\mu_j+\eta)}{\sigma^N(\eta)}=-Q_2(\mu_j)Q_2(\mu_j+\eta), \tag{3.2.45}$$

$$\frac{ce^{-2i\pi l_1\nu_j-i\phi}\sigma^m\left(\nu_j+\frac{\eta}{2}\right)}{\sigma^m(\eta)}\frac{\sigma^N(\nu_j)}{\sigma^N(\eta)}=-Q_1(\nu_j)Q_1(\nu_j-\eta). \tag{3.2.46}$$

The resulting selection rule (3.2.34) becomes

$$\Lambda(0)=e^{i\phi}\prod_{j=1}^{M}\frac{\sigma(\mu_j+\eta)}{\sigma(\nu_j)}=e^{\frac{2i\pi k}{N}},\quad k=1,\ldots,N. \tag{3.2.47}$$

The eigenvalue of the Hamiltonian (3.2.1) with periodic boundary conditions is given by

$$E=\frac{\sigma(\eta)}{\sigma'(0)}\left\{\sum_{j=1}^{M}\left[\zeta(\nu_j)-\zeta(\mu_j+\eta)\right]+\frac{1}{2}N\zeta(\eta)+2i\pi l_1\right\}. \tag{3.2.48}$$

3.2.4 Even N Case

When N is an even number, it follows from the Eqs. (3.2.45) and (3.2.46) that either $\mu_j=\nu_k$ or $\mu_j=\nu_k-\eta$ leads to $c=0$ and hence induces a one-to-one correspondence between $\{\mu_j\}$ and $\{\nu_k\}$, i.e., either $\mu_j=\nu_k$ or $\mu_j=\nu_k-\eta$. With Eq. (3.2.43) we conclude that for a generic η and $c=0$, the parameters have to obey the relations

$$l_1=0,\quad N=2M,\quad \{\mu_j\}=\{\nu_j\}\equiv\{\lambda_j\}. \tag{3.2.49}$$

The resulting $T-Q$ relation (3.2.42) is thus reduced to a conventional one:

$$\Lambda(u)=e^{i\phi}\frac{\sigma^N(u+\eta)}{\sigma^N(\eta)}\frac{Q(u-\eta)}{Q(u)}+e^{-i\phi}\frac{\sigma^N(u)}{\sigma^N(\eta)}\frac{Q(u+\eta)}{Q(u)}, \tag{3.2.50}$$

$$Q(u)=\prod_{l=1}^{M}\frac{\sigma(u-\lambda_l)}{\sigma(\eta)}. \tag{3.2.51}$$

Accordingly, the $M+1$ parameters ϕ and $\{\lambda_j\}$ satisfy the following BAEs and selection rule:

$$\frac{\sigma^N(\lambda_j+\eta)}{\sigma^N(\lambda_j)}=-e^{-2i\phi}\frac{Q(\lambda_j+\eta)}{Q(\lambda_j-\eta)},\quad j=1,\ldots,M,\tag{3.2.52}$$

$$e^{i\phi}\prod_{j=1}^{M}\frac{\sigma(\lambda_j+\eta)}{\sigma(\lambda_j)}=e^{\frac{2i\pi k}{N}},\quad k=1,\ldots,N.\tag{3.2.53}$$

The eigenvalue of the Hamiltonian reads

$$E=\frac{\sigma(\eta)}{\sigma'(0)}\left\{\sum_{j=1}^{M}[\zeta(\lambda_j)-\zeta(\lambda_j+\eta)]+\frac{1}{2}N\zeta(\eta)\right\}.\tag{3.2.54}$$

We note that the BAEs (3.2.52) coincide exactly with those obtained in Refs. [18, 37, 53]. The selection rule (3.2.53) in fact determines the amplitude of $\Lambda(u)$ and the parameter ϕ. Baxter discussed that $M=N/2$ gives a complete set of solutions of the Hamiltonian for even N [54]. Our numerical solutions for $N=4$ and randomly chosen η and τ (listed in Table 3.1) also support this conjecture. In such a sense, $c\neq 0$ must not lead to new solutions but, instead, to different parameterizations of the eigenvalues.

A similar phenomenon also appears in the XXZ spin chain with unparallel boundary fields [55–59], where the number M in the BAEs is also fixed. The numerical simulation [60] indicates that the BAEs with a fixed M indeed give complete solutions of the model.

3.2.5 Odd N Case

When N is an odd number, it follows from (3.2.45) and (3.2.46) that $c=0$ leads to the parameters $\{\mu_j\}$ and $\{\nu_j\}$ having to form pairs with either $\mu_j=\nu_k$ or $\mu_j=\nu_k-\eta$. However, for a generic η neither $\mu_j=\nu_k$ nor $\mu_j=\nu_k-\eta$ can satisfy Eq. (3.2.43). This contradiction indicates that the solutions of the BAEs (3.2.43)–(3.2.46) with $c=0$ do not actually exist for an odd N. Therefore, for the XYZ chain with an odd number of sites and generic η, the generalized $T-Q$ relation (3.2.42) cannot be reduced to a conventional one, as shown in (3.2.50). But when η takes some discrete values, the corresponding $T-Q$ relation (3.2.42) can indeed be reduced to a conventional one no matter if N is even or odd. For some special values of η, solutions with $c=0$ may exist. In this case, the Bethe roots $\{\mu_j\}$ and $\{\nu_j\}$ have to satisfy the following relations:

$$\mu_j=\nu_j\equiv\lambda_j,\quad j=1,\ldots,M_1,$$
$$\mu_{j+M_1}=\nu_{j+M_1}-\eta,\quad j=1,\ldots,M-M_1.$$

Table 3.1 Numerical solutions of the BAEs (3.2.52) and (3.2.53) for $N = 4$, $\eta = 0.4$ and $\tau = i$

λ_1	λ_2	k	E_n	n
$0.80000 + 0.11349i$	$0.80000 + 0.88651i$	2	−3.21353	1
$0.80000 + 0.00000i$	$0.80000 + 0.50000i$	1	−2.34227	2
$0.80000 + 0.00000i$	$0.30000 + 0.50000i$	1	−1.71217	3
$0.30000 + 0.00000i$	$0.80000 + 0.00000i$	0	−0.61387	4
$0.30000 + 0.70000i$	$0.80000 + 0.80000i$	3	0.00000	5
$0.30000 + 0.30000i$	$0.80000 + 0.20000i$	1	0.00000	5
$0.30000 + 0.86676i$	$0.80000 + 0.13324i$	2	0.00000	5
$0.30000 + 0.13324i$	$0.80000 + 0.86676i$	2	0.00000	5
$0.62340 + 0.25000i$	$0.97660 + 0.25000i$	1	0.00000	5
$0.62340 + 0.75000i$	$0.97660 + 0.75000i$	3	0.00000	5
− − −	− − −	−	0.00000	5
$0.03367 + 0.50000i$	$0.56633 + 0.50000i$	2	0.58230	6
$0.30000 + 0.50000i$	$0.80000 + 0.50000i$	2	0.61387	7
$0.30000 + 0.00000i$	$0.80000 + 0.50000i$	1	1.71217	8
$0.30000 + 0.00000i$	$0.30000 + 0.50000i$	1	2.34227	9
$0.30000 + 0.16022i$	$0.30000 + 0.83978i$	2	2.63122	10

The eigenvalues E_n calculated from (3.2.54) are the same as those from the exact diagonalization of the Hamiltonian. The symbol n denotes the number of the energy levels. Here "- - -" indicates a degenerate set of Bethe roots (reproduced from [45])

Note that (3.2.44) is not necessary any more, since $c = 0$. The relation (3.2.43) now reads

$$\left(\frac{N}{2} - M_1\right)\eta = l_1\tau + m_1, \quad l_1, m_1 \in Z. \tag{3.2.55}$$

This implies that if the crossing parameter η takes some discrete values

$$\eta = \frac{2l_1}{N - 2M_1}\tau + \frac{2m_1}{N - 2M_1}, \tag{3.2.56}$$

for any given non-negative integer M_1 and integers l_1 and m_1, the inhomogeneous $T - Q$ relation (3.2.42) is reduced to a conventional homogeneous one:

$$\begin{aligned}\Lambda(u) = & e^{2i\pi l_1 u + i\phi}\frac{\sigma^N(u+\eta)}{\sigma^N(\eta)}\frac{Q(u-\eta)}{Q(u)} \\ & + e^{-[2i\pi l_1(u+\eta)+i\phi]}\frac{\sigma^N(u)}{\sigma^N(\eta)}\frac{Q(u+\eta)}{Q(u)},\end{aligned} \tag{3.2.57}$$

$$Q(u) = \prod_{l=1}^{M_1}\frac{\sigma(u-\lambda_l)}{\sigma(\eta)}. \tag{3.2.58}$$

The $M_1 + 1$ parameters ϕ and $\{\lambda_j\}$ satisfy the associated BAEs:

$$e^{2i\pi(2l_1\lambda_j+l_1\eta)+2i\phi}\frac{\sigma^N(\lambda_j+\eta)}{\sigma^N(\lambda_j)} = -\frac{Q(\lambda_j+\eta)}{Q(\lambda_j-\eta)}, \quad j = 1, \ldots, M_1, \tag{3.2.59}$$

$$e^{i\phi}\prod_{j=1}^{M_1}\frac{\sigma(\lambda_j+\eta)}{\sigma(\lambda_j)} = e^{\frac{2i\pi k}{N}}, \quad k = 1, \ldots, N. \tag{3.2.60}$$

It should be emphasized that $M_1 = N$ may give a complete set of solutions and that the degenerate points given by Eq. (3.2.56) become dense in the whole complex η-plane in the thermodynamic limit ($N \to \infty$). This makes it possible to obtain the thermodynamic properties (up to the order of $O(N^{-2})$) [61] of the XYZ model for generic values of η via the conventional thermodynamic Bethe Ansatz method [42].

3.2.6 An Alternative Inhomogeneous $T - Q$ Relation

Based on the functional relations (3.2.33), we can construct another inhomogeneous $T - Q$ relation

$$\begin{aligned}\Lambda(u) = e^{2i\pi l_1 u+i\phi}a(u)\frac{Q_1(u-\eta)Q(u-\eta)}{Q_2(u)Q(u)} + e^{-[2i\pi l_1(u+\eta)+i\phi]} \\ \times d(u)\frac{Q_2(u+\eta)Q(u+\eta)}{Q_1(u)Q(u)} + c\frac{a(u)d(u)}{Q_1(u)Q_2(u)Q(u)},\end{aligned} \tag{3.2.61}$$

where c is a constant, $Q_1(u)$, $Q_2(u)$ and $Q(u)$ are given by (3.2.37) and (3.2.58), respectively, with $2M + M_1 = N$. For an even N, M_1 must be even and M takes the possible values of

$$M = 0, 1, \ldots, \frac{N}{2}. \tag{3.2.62}$$

$M = 0$ indicates that $\Lambda(u)$ can be parameterized by one function $Q(u)$. For odd N however, M_1 must be odd and M takes the possible values of

$$M = 1, \ldots, \frac{N-1}{2}. \tag{3.2.63}$$

The advantage of the present Ansatz is that only N Bethe roots are needed, which is the minimum number of parameters to determine the spectrum of the transfer matrix for odd N and generic η.

From the quasi-periodic properties (3.2.30)–(3.2.31) and the regularity of $\Lambda(u)$, we obtain the following BAEs:

Table 3.2 Numerical solutions of the BAEs (3.2.64)–(3.2.69) for $N = 3$, $M = M_1 = 1$, $\eta = 0.20$, $\tau = i$ and $l_1 = m_1 = m_2 = 0$

μ_1	ν_1	λ_1
$0.35000 + 0.02632i$	$0.45000 + 0.02632i$	$-1.10000 - 0.05263i$
$0.35000 - 0.02632i$	$0.45000 - 0.02632i$	$-1.10000 + 0.05263i$
$-0.15000 + 0.08693i$	$-0.05000 + 0.08693i$	$-0.10000 - 0.17387i$
$-0.15000 - 0.08693i$	$-0.05000 - 0.08693i$	$-0.10000 + 0.17387i$
$-0.65000 - 0.27875i$	$-0.55000 - 0.27875i$	$0.90000 + 0.55749i$
$-0.28066 + 0.31196i$	$-0.18066 + 0.31196i$	$0.16133 - 0.62392i$
$0.15828 + 0.12139i$	$0.25828 + 0.12139i$	$-0.71655 - 0.24279i$
$-0.42198 + 0.50000i$	$-0.32198 + 0.50000i$	$0.44397 - 1.00000i$

c	ϕ	k	E_n	n
$-0.08948 + 0.00000i$	$-0.08501 - 0.00000i$	1	-1.40865	1
$-0.08948 + 0.00000i$	$0.08501 - 0.00000i$	2	-1.40865	1
$3.04065 + 0.00000i$	$4.10893 - 0.00000i$	2	-1.40865	1
$3.04065 - 0.00000i$	$-4.10893 - 0.00000i$	1	-1.40865	1
$-0.28951 - 0.00000i$	$0.35925 - 0.00000i$	0	1.18468	2
$-0.61188 + 0.36729i$	$-0.27657 + 0.04967i$	0	1.18468	2
$-0.09303 - 0.16695i$	$-0.29190 + 0.31832i$	0	1.63263	3
$3.33371 - 7.57925i$	$-0.94248 - 0.14392i$	0	1.63263	3

The eigenvalues E_n calculated from (3.2.70) are the same as those from the exact diagonalization of the Hamiltonian. The symbol n denotes the number of the energy levels (reproduced from [45])

$$\left(\frac{N}{2} - M - M_1\right)\eta - \sum_{j=1}^{M}(\mu_j - \nu_j) = l_1\tau + m_1, \quad l_1, m_1 \in Z, \tag{3.2.64}$$

$$\frac{N}{2}\eta + \sum_{j=1}^{M}(\mu_j + \nu_j) + \sum_{j=1}^{M_1}\lambda_j = m_2, \quad m_2 \in Z, \tag{3.2.65}$$

$$ca(\mu_j) + e^{-2i\pi l_1(\mu_j+\eta)-i\phi}Q_2(\mu_j + \eta)Q_2(\mu_j)Q(\mu_j + \eta) = 0, \tag{3.2.66}$$

$$cd(\nu_j) + e^{2i\pi l_1\nu_j+i\phi}Q_1(\nu_j - \eta)Q_1(\nu_j)Q(\nu_j - \eta) = 0, \tag{3.2.67}$$

$$\begin{aligned} &ca(\lambda_j)d(\lambda_j) + e^{2i\pi l_1\lambda_j+i\phi}a(\lambda_j)Q_1(\lambda_j - \eta)Q_1(\lambda_j)Q(\lambda_j - \eta) \\ &+e^{-2i\pi l_1(\lambda_j+\eta)-i\phi}d(\lambda_j)Q_2(\lambda_j + \eta)Q_2(\lambda_j)Q(\lambda_j + \eta) = 0, \end{aligned} \tag{3.2.68}$$

$$e^{i\phi}\prod_{j=1}^{M}\frac{\sigma(\mu_j + \eta)}{\sigma(\nu_j)}\prod_{l=1}^{M_1}\frac{\sigma(\lambda_j + \eta)}{\sigma(\lambda_j)} = e^{\frac{2i\pi k}{N}}, \quad k = 1, \ldots, N. \tag{3.2.69}$$

Note that $M = 0$ is forbidden for generic η and odd N case to ensure the validity of Eq. (3.2.64). The eigenvalue of the Hamiltonian (3.2.1) is given by

Table 3.3 Numerical solutions of the BAEs (3.2.64)–(3.2.69) for $N = 5$, $\eta = 0.20$, $M = 1$, $\tau = i$ and $l_1 = m_1 = m_2 = 0$

μ_1	ν_1	λ_1	λ_2	λ_3	c	ϕ	k	E_n	n
$-0.55827 - 0.02265i$	$-0.25827 - 0.02265i$	$-0.10018 + 0.08190i$	$-0.10011 - 0.01038i$	$0.51684 - 0.02622i$	$-0.08617 - 0.00699i$	$0.10696 - 0.03689i$	1	-3.51343	1
$-0.55827 + 0.02265i$	$-0.25827 + 0.02265i$	$-0.10018 - 0.08190i$	$-0.10011 + 0.01038i$	$0.51684 + 0.02622i$	$-0.08617 + 0.00699i$	$-0.10696 - 0.03689i$	4	-3.51343	1
$-0.35211 + 0.07575i$	$-0.05211 + 0.07575i$	$-2.07865 + 0.00704i$	$0.90992 - 0.07849i$	$1.07296 - 0.08006i$	$-0.26973 + 6.70848i$	$1.26914 + 1.97113i$	1	-3.51343	1
$-0.35211 - 0.07575i$	$-0.05211 - 0.07575i$	$-2.07865 - 0.00704i$	$0.90992 + 0.07849i$	$1.07296 + 0.08006i$	$-0.26973 - 6.70848i$	$-1.26914 + 1.97113i$	4	-3.51343	1
$-1.42021 - 0.00000i$	$-1.12021 - 0.00000i$	$-0.06738 + 0.09747i$	$2.17518 + 0.00000i$	$-0.06738 - 0.09747i$	$-3.95699 + 0.00000i$	$0.00000 + 1.74063i$	0	-1.42192	2
$0.44197 - 0.00000i$	$0.74197 - 0.00000i$	$-4.09984 + 0.09712i$	$-1.09984 - 0.09712i$	$3.51573 - 0.00000i$	$-0.09625 - 0.00000i$	$-0.00000 - 0.02711i$	0	-1.42192	2
$-0.37881 - 0.02198i$	$-0.07881 - 0.02198i$	$-0.08454 + 0.09207i$	$-0.02749 - 0.11816i$	$0.06967 + 0.07005i$	$10.06245 - 5.20522i$	$-2.60644 + 2.33291i$	2	-1.25055	3
$-0.37881 + 0.02198i$	$-0.07881 + 0.02198i$	$-0.08454 - 0.09207i$	$-0.02749 + 0.11816i$	$0.06967 - 0.07005i$	$10.06245 + 5.20522i$	$2.60644 + 2.33291i$	3	-1.25055	3
$0.25000 - 0.04330i$	$0.55000 - 0.04330i$	$-0.77352 - 0.43943i$	$-0.42648 + 0.56057i$	$-0.10000 - 0.03454i$	$-0.60291 + 1.15659i$	$0.09819 - 0.00000i$	2	-1.25055	3
$0.25000 + 0.04330i$	$0.55000 + 0.04330i$	$-0.77352 + 0.43943i$	$0.57352 - 0.56057i$	$-1.10000 + 0.03454i$	$-0.60291 - 1.15659i$	$-0.09819 - 0.00000i$	3	-1.25055	3
$-0.45598 - 0.05509i$	$-0.15598 - 0.05509i$	$-0.87080 - 0.05853i$	$1.09356 + 0.20136i$	$-0.11080 - 0.03264i$	$-1.36773 + 1.60135i$	$0.51434 + 1.26882i$	2	-0.86239	4
$-0.45598 + 0.05509i$	$-0.15598 + 0.05509i$	$-0.87080 + 0.05853i$	$1.09356 - 0.20136i$	$-0.11080 + 0.03264i$	$-1.36773 - 1.60135i$	$-0.51434 + 1.26882i$	3	-0.86239	4
$0.25000 + 0.00965i$	$0.55000 + 0.00965i$	$-2.10000 - 0.03122i$	$0.40000 - 0.16249i$	$0.40000 + 0.17440i$	$-0.08524 - 0.00000i$	$-0.03396 - 0.00000i$	2	-0.86239	4

(continued)

Table 3.3 (continued)

μ_1	ν_1	λ_1	λ_2	λ_3	c	ϕ	k	E_n	n
0.25000 − 0.00965i	0.55000 − 0.00965i	−2.10000 + 0.03122i	0.40000 + 0.16249i	0.40000 − 0.17440i	−0.08524 + 0.00000i	0.03396 − 0.00000i	3	−0.86239	4
−0.75000 + 0.11275i	−0.45000 + 0.11275i	−1.90029 − 0.08059i	−0.29971 − 0.08059i	2.90000 − 0.06432i	0.23171 − 0.00000i	−0.58316 + 0.00000i	2	0.70428	5
−0.75000 − 0.11275i	−0.45000 − 0.11275i	−1.90029 + 0.08059i	−0.29971 + 0.08059i	2.90000 + 0.06432i	0.23171 + 0.00000i	0.58316 + 0.00000i	3	0.70428	5
−1.55828 + 0.03094i	−1.25828 + 0.03094i	−0.20115 − 0.04899i	1.51859 + 0.03613i	0.99912 − 0.04902i	−0.07119 + 0.01370i	−0.16566 − 0.04516i	2	0.70428	5
−1.55828 − 0.03094i	−1.25828 − 0.03094i	−0.20115 + 0.04899i	1.51859 − 0.03613i	0.99912 + 0.04902i	−0.07119 − 0.01370i	0.16566 − 0.04516i	3	0.70428	5
0.05720 + 0.13634i	0.35720 + 0.13634i	−2.06490 − 0.15055i	1.37567 + 0.08642i	−0.22517 − 0.20856i	−0.23061 + 0.32198i	−0.81379 − 0.24504i	1	1.02350	6
0.05720 − 0.13634i	0.35720 − 0.13634i	−2.06490 + 0.15055i	1.37567 − 0.08642i	−0.22517 + 0.20856i	−0.23061 − 0.32198i	0.81379 − 0.24504i	4	1.02350	6
0.25000 − 0.11861i	0.55000 − 0.11861i	−1.10000 − 0.14588i	−0.42425 + 0.69155i	0.22425 − 0.30845i	−0.38487 + 0.76419i	0.23725 − 0.00000i	1	1.02350	6
0.25000 + 0.11861i	0.55000 + 0.11861i	−1.10000 + 0.14588i	−0.42425 − 0.69155i	0.22425 + 0.30845i	−0.38487 − 0.76419i	−0.23725 − 0.00000i	4	1.02350	6
0.90991 − 0.30523i	1.20991 − 0.30523i	−3.10036 − 0.12527i	0.11226 + 0.64836i	0.36829 + 0.08736i	0.51259 + 0.05215i	7.25665 + 0.11304i	1	1.08128	7
0.90991 + 0.30523i	1.20991 + 0.30523i	−3.10036 + 0.12527i	0.11226 − 0.64836i	0.36829 − 0.08736i	0.51259 − 0.05215i	−7.25665 + 0.11304i	4	1.08128	7
0.25000 + 0.30649i	0.55000 + 0.30649i	−1.10000 − 0.12487i	−0.60000 − 0.55900i	0.40000 + 0.07089i	0.25098 − 0.00000i	5.34851 + 0.00000i	1	1.08128	7
0.25000 − 0.30649i	0.55000 − 0.30649i	−1.10000 + 0.12487i	−0.60000 + 0.55900i	0.40000 − 0.07089i	0.25098 − 0.00000i	−5.34851 − 0.00000i	4	1.08128	7

(continued)

Table 3.3 (continued)

μ_1	ν_1	λ_1	λ_2	λ_3	c	ϕ	k	E_n	n
0.04768 $-0.36108i$	0.34768 $-0.36108i$	−0.70037 $+0.60978i$	0.07267 $-0.43720i$	−0.26765 $+0.54957i$	−0.45482 $+0.61880i$	0.94953 $+0.00473i$	0	2.00622	8
−0.75000 $+0.35671i$	−0.45000 $+0.35671i$	−1.46212 $-0.58320i$	−0.10000 $-0.54702i$	2.26212 $+0.41680i$	−0.43735 $-0.51456i$	−0.93370 $-0.00000i$	0	2.00622	8
0.25000 $+0.31675i$	0.55000 $+0.31675i$	−1.60000 $+0.08844i$	−0.10000 $-0.43539i$	0.40000 $-0.28655i$	0.42608 $-0.00000i$	−1.21388 $+0.00000i$	0	2.35931	9
−1.08828 $-0.00000i$	−0.78828 $-0.00000i$	0.11378 $-0.00000i$	0.13139 $+0.44094i$	1.13139 $-0.44094i$	−1.85262 $+0.00000i$	−0.00000 $+0.50365i$	0	2.35931	9
0.39331 $-0.09117i$	0.69331 $-0.09117i$	0.46344 $-0.00267i$	0.48984 $-0.16274i$	−2.53991 $+0.34774i$	−0.01886 $-0.06929i$	0.30512 $-0.18507i$	0	2.69100	10
−1.36802 $+0.13801i$	−1.06802 $+0.13801i$	−0.26550 $+0.14338i$	1.63970 $-0.35821i$	0.56183 $-0.06119i$	−0.14181 $+0.59662i$	−0.50305 $-0.73047i$	0	2.69100	10

The eigenvalues E_n calculated from (3.2.70) are the same as those from the exact diagonalization of the Hamiltonian. The symbol n denotes the number of the energy levels (reproduced from [45])

$$E = \frac{\sigma(\eta)}{\sigma'(0)} \left\{ \sum_{j=1}^{M} \left[\zeta(\nu_j) - \zeta(\mu_j + \eta) \right] \right.$$

$$\left. + \sum_{j=1}^{M_1} \left[\zeta(\lambda_j) - \zeta(\lambda_j + \eta) \right] + \frac{1}{2} N \zeta(\eta) + 2i\pi l_1 \right\}. \qquad (3.2.70)$$

It seems that there is some arbitrariness for the choice of l_1, m_1 and m_2 due to the periodicity of the model. Chosen sets of numerical results for $N = 3$ and $N = 5$ are respectively listed in Tables 3.2 and 3.3. These simulations indicate that any fixed values of these integers should give a complete set of eigenstates.

We point out that, due to the Z_2-symmetry (3.2.19) of the R-matrix, the following relations hold:

$$U^i \, t(u) \, U^i = tr_0 \left(U^i \, T_0(u) \, U^i \right) = tr_0 \left(\sigma_0^i \, T_0(u) \, \sigma_0^i \right) = t(u), \qquad (3.2.71)$$

$$U^i = \sigma_1^i \sigma_2^i \cdots \sigma_N^i, \quad i = x, y, z. \qquad (3.2.72)$$

Also note that $\{U^i\}$ form an abelian (nonabelian) group when N is even (odd), i.e.

$$(U^i)^2 = \text{id}, \quad U^i \, U^j = (-1)^N U^j \, U^i, \;\; \text{for } i \neq j, \;\; \text{and } i, j = x, y, z. \qquad (3.2.73)$$

This indicates a double degeneracy of the odd N case, which implies that there exist multiple solutions of the Bethe roots corresponding to one $\Lambda(u)$. Suppose $|\Psi_+\rangle$ and $|\Psi_-\rangle$ are two degenerate eigenstates of $t(u)$ and with $U^x|\Psi_\pm\rangle = \pm|\Psi_\pm\rangle$. The state $|\Psi(\theta)\rangle = \cos\theta|\Psi_+\rangle + \sin\theta|\Psi_-\rangle$ (θ is an arbitrary parameter) must also be an eigenstate of $t(u)$ with the same eigenvalue $\Lambda(u)$. Obviously, $\Lambda(u)$ is independent of θ. However, the Bethe roots describing the eigenstate generally depend on θ. This degeneracy may be lifted to fix the Bethe roots by the common eigenstates of $t(u)$ and U^x.

References

1. H.A. Bethe, On the theory of metals 1. Eigenvalues and eigenfunction of a linear chain of atoms. Z. Phys. **71**, 205 (1931)
2. A. Sommerfeld, H.A. Bethe, Elektronentheorie der Metalle, in *Handbuch der Physik*, ed. Geier and Scheel, vol. 24, Part 2 (Verlag Julius Springer, Berlin, 1933), p. 604
3. L. Hulthén, Über das Austauschproblem eines Kristalles. Ark. Mat. Astron. Fys. B **26A**, 1 (1938)
4. R. Orbach, Linear antiferromagnetic chain with anisotropic coupling. Phys. Rev. **112**, 309 (1958)
5. L.R. Walker, Antiferromagnetic linear chain. Phys. Rev. **116**, 1089 (1959)
6. J. des Cloizeaux, M. Gaudin, Anisotropic linear magnetic chain. J. Math. Phys. **7**, 1384 (1966)
7. C.N. Yang, C.P. Yang, One-dimensional chain of anisotropic spin-spin interactions I. Proof of Bethe's hypothesis for ground state in a finite system. Phys. Rev. **150**, 321 (1966)

8. C.N. Yang, C.P. Yang, One-dimensional chain of anisotropic spin-spin interactions II. Properties of the ground state energy per lattice site for an infinite system. Phys. Rev. **150**, 327 (1966)
9. C.N. Yang, C.P. Yang, One-dimensional chain of anisotropic spin-spin interactions III. Applications. Phys. Rev. **151**, 258 (1966)
10. B.M. McCoy, T.T. Wu, Hydrogen bonded crystals and the anisotropic Heisenberg chain, Il Nuovo Cimento **56**, 311 (1968)
11. E.H. Lieb, Exact solution of the problem of the entropy of two dimensional ice. Phys. Rev. Lett. **18**, 692 (1967)
12. E.H. Lieb, Exact solution of the F model of an antiferroelectric. Phys. Rev. Lett. **18**, 1046 (1967)
13. E.H. Lieb, Exact solution of the two-dimensional Slater KDP model of a ferroelectric. Phys. Rev. Lett. **18**, 1046 (1967)
14. E.H. Lieb, Residual entropy of square ice. Phys. Rev. **162**, 162 (1967)
15. B. Sutherland, Exact solution of a two-dimensional model for hydrogen-bonded crystals. Phys. Rev. Lett. **19**, 103 (1967)
16. B. Sutherland, Two-dimensional hydrogen bonded crystals without the ice rule. J. Math. Phys. **11**, 3183 (1970)
17. R.J. Baxter, 8 vertex model in lattice statistics. Phys. Rev. Lett. **26**, 832 (1971)
18. R.J. Baxter, One-dimensional anisotropic Heisenberg chain. Phys. Rev. Lett. **26**, 834 (1971)
19. R.J. Baxter, One-dimensional anisotropic Heisenberg chain. Ann. Phys. **70**, 323 (1972)
20. R.J. Baxter, Partition function of the eight-vertex model. Ann. Phys. **70**, 193 (1972)
21. R.J. Baxter, Eight-vertex model in lattice statistics and one dimensional anisotropic Heisenberg chain I. Some fundamental eigenvectors. Ann. Phys. **76**, 1 (1973)
22. R.J. Baxter, Eight-vertex model in lattice statistics and one dimensional anisotropic Heisenberg chain II. Equivalence to a generalized ice-type lattice model. Ann. Phys. **76**, 25 (1973)
23. R.J. Baxter, Eight-vertex model in lattice statistics and one dimensional anisotropic Heisenberg chain III. Eigenvectors and eigenvalues of the transfer matrix and hamiltonian. Ann. Phys. **76**, 48 (1973)
24. J.D. Johnson, S. Krinsky, B.M. McCoy, Vertical-arrow correlation length in the eight-vertex model and the low-lying excitations of the XYZ hamiltonian. Phys. Rev. A **8**, 2526 (1973)
25. K. Fabricius, B.M. McCoy, New development in the eight-vertex model. J. Stat. Phys. **111**, 323 (2003)
26. R.J. Baxter, The six and eight-vertex models revisited. J. Stat. Phys. **116**, 43 (2004)
27. V.V. Bazhanov, V.V. Mangazeev, Eight-vertex model and non-stationary Lamé equation. J. Phys. A **38**, L145 (2005)
28. K. Fabricius, B.M. McCoy, New development in the eight-vertex model II. Chains of odd length. J. Stat. Phys. **120**, 37 (2005)
29. K. Fabricius, B.M. McCoy, An elliptic current operator for the eight-vertex model. J. Phys. A **39**, 14869 (2006)
30. V.V. Bazhanov, V.V. Mangazeev, Eight-vertex model and Painlevé VI. J. Phys. A **39**, 12235 (2006)
31. K. Fabricius, A new Q matrix for the eight-vertex model. J. Phys. A **40**, 4075 (2007)
32. S.S. Roan, The Q operator and functional relations for the eight-vertex model at root-of-unity $\eta = 2\text{mK}/\text{N}$ for odd N. J. Phys. A **40**, 11019 (2007)
33. K. Fabricius, B.M. McCoy, The TQ equation of the 8 vertex model for complex elliptic roots of unity. J. Phys. A **40**, 14893 (2007)
34. K. Fabricius, B.M. McCoy, New Q matrices and their functional equations for the eight vertex model at elliptic roots of unity. J. Stat. Phys. **134**, 643 (2009)
35. V.V. Bazhanov, V.V. Mangazeev, Analytic solution of the eight vertex model. Nucl. Phys. B **775**, 225 (2007)
36. O.I. Patu, Free energy of the eight vertex model with an odd number of lattice sites. J. Stat. Mech. P09007, (2007)
37. L.A. Takhtadzhan, L.D. Faddeev, The quantum method of the inverse problem and the Heisenberg XYZ model. Rush. Math. Surveys **34**, 11 (1979)

38. V.E. Korepin, N.M. Bogoliubov, A.G. Izergin, *Quantum Inverse Scattering Method and Correlation Functions* (Cambridge University Press, Cambridge, 1993)
39. N. Kitanine, J.M. Maillet, V. Terras, Form factors of the XXZ Heisenberg spin-1/2 finite chain. Nucl. Phys. B **554**, 647 (1999)
40. J.M. Maillet, V. Terras, On the quantum inverse scattering problem. Nucl. Phys. B **575**, 627 (2000)
41. M. Gaudin, *La Fonction d'Onde de Bethe* (Masson, Paris, 1983)
42. M. Takahashi, *Thermodynamics of One-Dimensional Solvable Models* (Cambridge University Press, Cambridge, 1999)
43. B. Sutherland, *Beautiful Models: 70 Years of Exactly Solved Quantum Many-Body Problems* (World Scientific Publishing, Singapore, 2004)
44. B.M. McCoy, *Advanced Statistical Mechanics* (Oxford Science Publications, Oxford, 2009)
45. J. Cao, S. Cui, W.-L. Yang, K. Shi, Y. Wang, Spin-1/2 XYZ model revisit: general solutions via off-diagonal Bethe Ansatz. Nucl. Phys. B **886**, 185 (2014)
46. J. Cao, W.-L. Yang, K. Shi, Y. Wang, Off-diagonal Bethe Ansatz solution of the aisotropic spin-$\frac{1}{2}$ chain with arbitrary boundary conditions. Nucl. Phys. B **877**, 152 (2013)
47. H. Frahm, J.H. Grelik, A. Seel, T. Wirth, Functional Bethe Ansatz methods for the open XXX chain. J. Phys. A: Math. Theor. **44**, 015001 (2011)
48. S. Niekamp, T. Wirth, H. Frahm, The XXZ model with anti-periodic twisted boundary conditions. J. Phys. A: Math. Theor. **42**, 195008 (2009)
49. A.M. Grabinski, H. Frahm, Non-diagonal boundary conditions for $gl(1|1)$ super spin chains. J. Phys. A **43**, 045207 (2010)
50. G. Niccoli, Non-diagonal open spin-1/2 XXZ quantum chains by separation of variables: complete spectrum and matrix elements of some quasi-local operators, J. Stat. Mech. P10025 (2012)
51. G. Niccoli, Antiperiodic spin-1/2 XXZ quantum chains by separation of variables: complete spectrum and form factors. Nucl. Phys. B **870**, 397 (2013)
52. G. Niccoli, Antiperiodic dynamical 6-vertex model I: complete spectrum by SOV, matrix elements of the identity on separate states and connections to the periodic 8-vertex model. J. Phys. A **46**, 075003 (2013)
53. R.J. Baxter, *Exactly Solved Models in Statistical Mechanics* (Academic Press, London, 1982)
54. R.J. Baxter, Completeness of the Bethe Ansatz for the six and eight vertex models. J. Stat. Phys. **108**, 1 (2002)
55. J. Cao, H.-Q. Lin, K.-J. Shi, Y. Wang, Exact solution of XXZ spin chain with unparallel boundary fields. Nucl. Phys. B **663**, 487 (2003)
56. W.-L. Yang, Y.-Z. Zhang, M. Gould, Exact solution of the XXZ Gaudin model with generic open boundaries. Nucl. Phys. B **698**, 503 (2004)
57. W.-L. Yang, R. Sasaki, Solution of the dual reflection equation for $A_{n-1}^{(1)}$ solid-on-solid model. J. Math. Phys. **45**, 4301 (2004)
58. W.-L. Yang, R. Sasaki, Exact solution of Z_n Belavin model with open boundary condition. Nucl. Phys. B **679**, 495 (2004)
59. W.-L. Yang, Y.-Z. Zhang, T-Q relation and exact solution for the XYZ chain with general nondiagonal boundary terms. Nucl. Phys. B **744**, 312 (2006)
60. R.I. Nepomechie, F. Ravanini, Completeness of the Bethe Ansatz solution of the open XXZ chain with nondiagonal boundary terms. J. Phys. A: Math. Gen. **36**, 11391 (2003)
61. Y.-Y. Li, J. Cao, W.-L. Yang, K. Shi, Y. Wang, Thermodynamic limit and surface energy of the XXZ spin chain with arbitrary boundary fields. Nucl. Phys. B **884**, 17 (2014)

Chapter 4
The Spin-$\frac{1}{2}$ Torus

The spin-$\frac{1}{2}$ torus model describes the anisotropic spin chain with antiperiodic boundary conditions or a Möbius-like topological boundary condition [1–5]. The model itself is physically interesting due to its relevance to the realization of topological states of matter. For example, after a Jordan-Wigner transformation, it describes a p-wave Josephson junction embedded in a spinless Luttinger liquid [6–8]. The integrability of this model is associated with the Z_2-symmetry of the six vertex R matrix [9–12], i.e., the antiperiodic boundary condition mentioned in Chap. 1. Although only one bond is changed from the periodic *XXZ* model, the $U(1)$-symmetry is completely broken, making it difficult to apply the conventional Bethe Ansatz methods for the lack of an obvious reference state.

Historically, the *XXZ* spin torus model has been the first quantum integrable model solved via the ODBA method [13]. This chapter studies the construction of the inhomogeneous $T - Q$ relation, based on the operator product identities of the transfer matrix derived from YBE, and on the intrinsic properties of the R-matrix. It also introduces an alternative derivation of the functional relations and a basis of the Hilbert space used to retrieve the eigenstates and the scalar products [14], as well as some physical properties of the corresponding free fermion model. The last section is attributed to the *XYZ* spin torus model.

4.1 Z_2-symmetry and the Model Hamiltonian

The model Hamiltonian of the *XXZ* spin torus reads

$$H = -\sum_{j=1}^{N}\left[\sigma_j^x\sigma_{j+1}^x + \sigma_j^y\sigma_{j+1}^y + \cosh\eta\,\sigma_j^z\sigma_{j+1}^z\right], \tag{4.1.1}$$

with the anti-periodic boundary conditions $\sigma_{N+1}^{\alpha} = \sigma_1^x\sigma_1^{\alpha}\sigma_1^x$ $(\alpha = x, y, z)$. For such a topological boundary condition, the spin on the N-th site connects with that on the

Y. Wang et al., *Off-Diagonal Bethe Ansatz for Exactly Solvable Models*,
DOI 10.1007/978-3-662-46756-5_4

first site after rotating by an angle π along the x-direction (a kink on the $(N, 1)$ bond) and forms a torus in the spin space. This kink could be shifted to the $(j, j+1)$ bond with the spectrum of the Hamiltonian unchanged

$$\tilde{H}_j = U_j^x H U_j^x, \quad U_j^x = \prod_{l=1}^{j} \sigma_l^x. \tag{4.1.2}$$

Note that the braiding occurs in the quantum space rather than in the real space. The present model therefore describes a quantum Möbius strip. As we showed in Chap. 3 for the *XYZ* chain, in the present case we have $[H, U^x] = 0$, indicating that the present model possesses a global Z_2 invariance.

The integrability of the present model is associated with the R-matrix

$$\begin{aligned} R_{0,j}(u) = & \frac{1}{2}\left[\frac{\sinh(u+\eta)}{\sinh\eta}(1+\sigma_j^z\sigma_0^z) + \frac{\sinh u}{\sinh\eta}(1-\sigma_j^z\sigma_0^z)\right] \\ & + \frac{1}{2}(\sigma_j^x\sigma_0^x + \sigma_j^y\sigma_0^y), \end{aligned} \tag{4.1.3}$$

and the monodromy matrix

$$T_0(u) = \sigma_0^x R_{0,N}(u-\theta_N)\cdots R_{0,1}(u-\theta_1) = \begin{pmatrix} C(u) & D(u) \\ A(u) & B(u) \end{pmatrix}. \tag{4.1.4}$$

Because of the Z_2-symmetry

$$[R_{0,\bar{0}}(u), \sigma_0^x\sigma_{\bar{0}}^x] = 0, \tag{4.1.5}$$

the following relation holds

$$R_{0,\bar{0}}(u-v)T_0(u)T_{\bar{0}}(v) = T_{\bar{0}}(v)T_0(u)R_{0,\bar{0}}(u-v), \tag{4.1.6}$$

which directly gives

$$[t(u), t(v)] = 0, \tag{4.1.7}$$

with the transfer matrix $t(u)$ defined as

$$t(u) = tr_0 T_0(u) = B(u) + C(u). \tag{4.1.8}$$

The first order derivative of the logarithm of the transfer matrix gives the Hamiltonian (4.1.1)

$$H = -2\sinh\eta \left.\frac{\partial \ln t(u)}{\partial u}\right|_{u=0,\{\theta_j=0\}} + N\cosh\eta. \tag{4.1.9}$$

4.2 Operator Product Identities

With the intrinsic properties of the R-matrix given in (3.1.9)–(3.1.14) and the same procedure introduced in Chap. 1, we deduce that

$$\begin{aligned} t(\theta_j) &= tr_0 \left\{ R_{0,N}(\theta_j - \theta_N) \cdots P_{0,j} \cdots R_{0,1}(\theta_j - \theta_1)\sigma_0^x \right\} \\ &= R_{j,j-1}(\theta_j - \theta_{j-1}) \cdots R_{j,1}(\theta_j - \theta_1) \\ &\quad \times \sigma_j^x R_{j,N}(\theta_j - \theta_N) \cdots R_{j,j+1}(\theta_j - \theta_{j+1}), \end{aligned} \tag{4.2.1}$$

and

$$\begin{aligned} t(\theta_j - \eta) &= (-1)^{N-1}\, tr_0 \left\{ R_{0,N}^{t_0}(-\theta_j + \theta_N) \cdots R_{0,1}^{t_0}(-\theta_j + \theta_1)\sigma_0^x \right\} \\ &= (-1)^{N-1}\, tr_0 \left\{ \sigma_0^x R_{0,1}(-\theta_j + \theta_1) \cdots R_{0,N}(-\theta_j + \theta_N) \right\} \\ &= (-1)^{N-1} R_{j,j+1}(-\theta_j + \theta_{j+1}) \cdots R_{j,N}(-\theta_j + \theta_N) \\ &\quad \times \sigma_j^x R_{j,1}(-\theta_j + \theta_1) \cdots R_{j,j-1}(-\theta_j + \theta_{j-1}). \end{aligned} \tag{4.2.2}$$

The unitary property of the R-matrix thus induces the operator product identities

$$t(\theta_j)t(\theta_j - \eta) = -a(\theta_j)d(\theta_j - \eta) \times \text{id}, \quad j = 1, \ldots, N, \tag{4.2.3}$$

with

$$d(u) = a(u - \eta) = \prod_{j=1}^{N} \frac{\sinh(u - \theta_j)}{\sinh \eta}. \tag{4.2.4}$$

By applying the transfer matrix (4.1.8) on an eigenstate $|\Psi\rangle$ we obtain

$$\Lambda(\theta_j)\Lambda(\theta_j - \eta) = -a(\theta_j)d(\theta_j - \eta), \quad j = 1, \ldots, N, \tag{4.2.5}$$

with $\Lambda(u)$ being the corresponding eigenvalue of the transfer matrix.

From the asymptotic expansions of $B(u)$ and $C(u)$ given in (3.1.24), we know that

$$\Lambda(u), \text{ as a function of } u, \\ \text{is a trigonometrical polynomial of degree } N - 1, \tag{4.2.6}$$

with the periodicity property

$$\Lambda(u + i\pi) = (-1)^{N-1}\Lambda(u). \tag{4.2.7}$$

The eigenvalue $\Lambda(u)$ can be parameterized as the following trigonometric polynomial

$$\Lambda(u) = \Lambda_0 \prod_{j=1}^{N-1} \sinh(u - z_j), \tag{4.2.8}$$

and the N equations (4.2.5) determine the N unknowns Λ_0 and $\{z_j | j = 1, \ldots, N-1\}$ completely. The corresponding eigenvalue of the Hamiltonian is thus given by

$$E = 2\sinh\eta \sum_{j=1}^{N-1} \coth z_j + N\cosh\eta. \tag{4.2.9}$$

4.3 The Inhomogeneous $T - Q$ Relation

In this section we show that each solution of (4.2.5)–(4.2.7) can be parameterized in terms of the inhomogeneous $T - Q$ relation [15]

$$\Lambda(u)Q(u) = a(u)e^u Q(u-\eta) - e^{-u-\eta}d(u)Q(u+\eta) - c(u)a(u)d(u), \tag{4.3.1}$$

where $Q(u)$ is a trigonometric polynomial of the type

$$Q(u) = \prod_{j=1}^{N} \frac{\sinh(u-\lambda_j)}{\sinh\eta}, \tag{4.3.2}$$

and $c(u)$ is given by

$$c(u) = e^{u - N\eta + \sum_{l=1}^{N}(\theta_l - \lambda_l)} - e^{-u-\eta-\sum_{l=1}^{N}(\theta_l-\lambda_l)}. \tag{4.3.3}$$

The N parameters $\{\lambda_j\}$ satisfy the associated BAEs

$$e^{\lambda_j}a(\lambda_j)Q(\lambda_j - \eta) - e^{-\lambda_j-\eta}d(\lambda_j)Q(\lambda_j+\eta) - c(\lambda_j)a(\lambda_j)d(\lambda_j) = 0,$$
$$j = 1, \ldots, N. \tag{4.3.4}$$

We follow the same procedure used in Chap. 1 to prove the above statement. Let us introduce a function $f_0(u)$ which is equal to the difference between the left hand side and the right hand side of (4.3.1), namely,

$$f_0(u) = \Lambda(u)Q(u) - a(u)e^u Q(u-\eta) + e^{-u-\eta}d(u)Q(u+\eta) + c(u)a(u)d(u).$$

The relations (4.2.7) and (4.2.6) allow us to derive that $f_0(u)$ satisfies the properties

$$f_0(u+i\pi) = -f_0(u), \tag{4.3.5}$$

$$f_0(u), \text{ as a function of } u, \text{ is a trigonometric polynomial of degree } 2N-1. \tag{4.3.6}$$

This implies that the function $f_0(u)$ is fixed by its values at any $2N$ different points. For each solution of (4.2.5)–(4.2.7), we can always choose a $Q(u)$ of form (4.3.2) such that

$$f_0(\theta_j) = \Lambda(\theta_j)Q(\theta_j) - a(\theta_j)e^{\theta_j}Q(\theta_j-\eta) = 0, \quad j=1,\ldots,N, \tag{4.3.7}$$

$$f_0(\theta_j-\eta) = \Lambda(\theta_j-\eta)Q(\theta_j-\eta) + d(\theta_j-\eta)e^{-\theta_j}Q(\theta_j) = 0, \quad j=1,\ldots,N, \tag{4.3.8}$$

which means $f_0(u)=0$ or (4.3.1) is fulfilled. The relation (4.2.5) implies that only N of the above $2N$ equations are independent, say,

$$\Lambda(\theta_j)Q(\theta_j) = a(\theta_j)e^{\theta_j}Q(\theta_j-\eta), \quad j=1,\ldots,N, \tag{4.3.9}$$

which allow us to determine the $Q(u)$ function of form (4.3.2) by the values of $\Lambda(u)$ at the N points $\{\theta_j|j=1,\ldots,N\}$. Therefore, we are always able to choose $Q(u)$ of form (4.3.2) such that $f_0(u)=0$, provided that $\Lambda(u)$ is an eigenvalue of the transfer matrix (4.1.8). Taking u at the roots of the $Q(u)$ function (i.e., $\{\lambda_j\}$), then $f_0(\lambda_j)=0$ gives rise to the associated BAEs (4.3.4).

In the homogeneous limit $\{\theta_j=0\}$, the eigenvalue of the Hamiltonian in terms of the Bethe roots is given by

Table 4.1 Numerical solutions of the BAEs (4.3.4) for $N=3$, $\eta=1$ and $\{\theta_j=0\}$

λ_1	λ_2	λ_3	E_n	n
$-1.43163-0.00000i\pi$	$-0.50000-0.00000i\pi$	$0.43163-0.00000i\pi$	-3.09636	1
$-1.63742+0.50000i\pi$	$-0.50000+0.50000i\pi$	$0.63742+0.50000i\pi$	-3.09636	1
$-1.18721-0.17145i\pi$	$-0.50000-0.26374i\pi$	$0.18721-0.17145i\pi$	-1.54308	2
$-1.18721+0.17145i\pi$	$-0.50000+0.26374i\pi$	$0.18721+0.17145i\pi$	-1.54308	2
$-1.40287+0.39386i\pi$	$-0.50000+0.28359i\pi$	$0.40287+0.39386i\pi$	-1.54308	2
$-1.40287-0.39386i\pi$	$-0.50000-0.28359i\pi$	$0.40287-0.39386i\pi$	-1.54308	2
$-1.20038-0.50000i\pi$	$-0.50000-0.00000i\pi$	$0.20038-0.50000i\pi$	6.18252	3
$-0.50000-0.50000i\pi$	$-0.50000-0.15964i\pi$	$-0.50000+0.15964i\pi$	6.18252	3

E_n is the nth eigenenergy and n indicates the number of the energy levels. The eigenvalues of the Hamiltonian calculated from (4.3.10) are the same as those calculated with exact diagonalization of the Hamiltonian

Table 4.2 Numerical solutions of the BAEs (4.3.4) for $N = 4$, $\eta = 1$ and $\{\theta_j = 0\}$

λ_1	λ_2	λ_3	λ_4	E_n	n
-2.12754 $+0.50000i\pi$	-1.03389 $+0.50000i\pi$	0.03389 $+0.50000i\pi$	1.12754 $+0.50000i\pi$	-4.27591	1
— — —	— — —	— — —	— — —	-4.27591	1
-1.74449 $+0.15319i\pi$	-0.98935 $+0.15137i\pi$	-0.01065 $+0.15137i\pi$	0.74449 $+0.15319i\pi$	-3.67755	2
-1.74449 $-0.15319i\pi$	-0.98935 $-0.15137i\pi$	-0.01065 $-0.15137i\pi$	0.74449 $-0.15319i\pi$	-3.67755	2
-2.03215 $+0.42071i\pi$	-0.96024 $+0.36347i\pi$	-0.03976 $+0.36347i\pi$	1.03215 $+0.42071i\pi$	-3.67755	2
-2.03215 $-0.42071i\pi$	-0.96024 $-0.36347i\pi$	-0.03976 $-0.36347i\pi$	1.03215 $-0.42071i\pi$	-3.67755	2
-1.86358 $+0.31377i\pi$	-1.02758 $+0.24134i\pi$	0.02758 $+0.24134i\pi$	0.86358 $+0.31377i\pi$	-3.46526	3
-1.86358 $-0.31377i\pi$	-1.02758 $-0.24134i\pi$	0.02758 $-0.24134i\pi$	0.86358 $-0.31377i\pi$	-3.46526	3
-1.60669 $-0.50000i\pi$	-0.50000 $-0.25533i\pi$	-0.50000 $+0.25533i\pi$	0.60669 $+0.50000i\pi$	0.27591	4
— — —	— — —	— — —	— — —	0.27591	4
-1.03760 $-0.12735i\pi$	-0.50000 $-0.41963i\pi$	-0.50000 $+0.15711i\pi$	0.03760 $-0.12735i\pi$	3.67755	5
-1.03760 $+0.12735i\pi$	-0.50000 $+0.41963i\pi$	-0.50000 $-0.15711i\pi$	0.03760 $+0.12735i\pi$	3.67755	5
-1.81891 $-0.48351i\pi$	-0.50000 $-0.47875i\pi$	-0.50000 $-0.06781i\pi$	0.81891 $-0.48351i\pi$	3.67755	5
-1.81891 $+0.48351i\pi$	-0.50000 $+0.47875i\pi$	-0.50000 $+0.06781i\pi$	0.81891 $+0.48351i\pi$	3.67755	5
-1.53412 $-0.41713i\pi$	-0.50000 $-0.21809i\pi$	-0.50000 $+0.02400i\pi$	0.53412 $-0.41713i\pi$	7.46526	6
-1.53412 $+0.41713i\pi$	-0.50000 $+0.21809i\pi$	-0.50000 $-0.02400i\pi$	0.53412 $+0.41713i\pi$	7.46526	6

E_n is the nth eigenenergy and n indicates the number of the energy levels. The eigenvalues of the Hamiltonian calculated from (4.3.10) are the same as those calculated with exact diagonalization of the Hamiltonian. Here "— — —" indicates a degenerate set of Bethe roots

$$E = 2\sinh\eta \sum_{j=1}^{N} [\coth(\lambda_j + \eta) - \coth(\lambda_j)] - N\cosh\eta - 2\sinh\eta. \quad (4.3.10)$$

Numerical results for $N = 3$ and 4 respectively shown in Tables 4.1 and 4.2 indicate that the BAEs (4.3.4) indeed give a complete set of solutions.

4.4 An Alternative Inhomogeneous $T - Q$ Relation

For each eigenvalue $\Lambda(u)$, the functional relations (4.2.5)–(4.2.7) allow us to construct a variety of $T - Q$ relations. Here we introduce another simple $T - Q$ relation for the present model proposed in [13]:

$$\Lambda(u) = e^u a(u)\frac{Q_1(u-\eta)}{Q_2(u)} - e^{-u-\eta}d(u)\frac{Q_2(u+\eta)}{Q_1(u)} - c(u)\frac{a(u)d(u)}{Q_1(u)Q_2(u)}, \qquad (4.4.1)$$

where

$$Q_1(u) = \prod_{j=1}^{M}\sinh(u-\mu_j), \quad Q_2(u) = \prod_{j=1}^{M}\sinh(u-\nu_j), \qquad (4.4.2)$$

$\{\mu_j\}$ and $\{\nu_j\}$ are two sets of Bethe roots, and $c(u)$ is an adjust function. We remark that the above Ansatz satisfies the operator product identities (4.2.5) automatically for arbitrary choices of $Q_{1,2}(u)$ and $c(u)$ because $a(\theta_j - \eta) = 0$ and $d(\theta_j) = 0$. In addition, the exponential factors in the first and second terms are introduced to ensure quasi-periodicity. This implies that the leading terms for $u \rightarrow \pm\infty$ are $e^{\pm(N+1)u}$. The third term in the $T - Q$ relation is included to cancel those unwanted leading terms by proper choices of $c(u)$ and $Q_{1,2}(u)$.

Based on the above considerations, for even N, M may take the value of $\frac{N}{2}$, and

$$c(u) = \sinh^N \eta[e^{i\phi_1+u} - e^{i\phi_2-u-\eta}], \qquad (4.4.3)$$

with

$$i\phi_1 = \sum_{j=1}^{N}\theta_j - M\eta - 2\sum_{j=1}^{M}\mu_j, \qquad (4.4.4)$$

$$-i\phi_2 = \sum_{j=1}^{N}\theta_j - M\eta - 2\sum_{j=1}^{M}\nu_j, \qquad (4.4.5)$$

to cancel the leading terms in Eq. (4.4.1) when $u \rightarrow \pm\infty$. The BAEs determined by the regularity of $\Lambda(u)$ (which ensures $\Lambda(u)$ to be a trigonometric polynomial of degree $N-1$) read

$$d(\nu_j) = \frac{e^{\nu_j}}{c(\nu_j)}Q_1(\nu_j - \eta)Q_1(\nu_j),$$
$$a(\mu_j) = -\frac{e^{-\mu_j-\eta}}{c(\mu_j)}Q_2(\mu_j+\eta)Q_2(\mu_j), \quad j = 1, \ldots, M. \qquad (4.4.6)$$

Now we can safely take the homogeneous limit $\theta_j \to 0$. The BAEs for the homogeneous case thus read

$$\sinh^N(\nu_j) = \frac{e^{\nu_j}\prod_{l=1}^{M}\sinh(\nu_j-\mu_l-\eta)\sinh(\nu_j-\mu_l)}{e^{\nu_j-M\eta-2\sum_{k=1}^{M}\mu_k}-e^{-\nu_j+(M-1)\eta+2\sum_{k=1}^{M}\nu_k}}, \tag{4.4.7}$$

$$\sinh^N(\mu_j+\eta) = -\frac{e^{-\mu_j-\eta}\prod_{l=1}^{M}\sinh(\mu_j-\nu_l+\eta)\sinh(\mu_j-\nu_l)}{e^{\mu_j-M\eta-2\sum_{k=1}^{M}\mu_k}-e^{-\mu_j+(M-1)\eta+2\sum_{k=1}^{M}\nu_k}}, \tag{4.4.8}$$

where $j = 1, \ldots, N/2$.

For odd N, we put $M = (N+1)/2$ and

$$c(u) = \frac{1}{2}\sinh^N\eta\left[e^{i\phi_1+2u}+e^{i\phi_2-2u-2\eta}\right], \tag{4.4.9}$$

where ϕ_1 and ϕ_2 are given by Eqs. (4.4.4) and (4.4.5) with $M = (N+1)/2$. In the homogeneous limit, the BAEs read

$$\sinh^N(\nu_j) = \frac{2e^{\nu_j}\prod_{l=1}^{M}\sinh(\nu_j-\mu_l-\eta)\sinh(\nu_j-\mu_l)}{e^{2\nu_j-M\eta-2\sum_{k=1}^{M}\mu_k}+e^{-2\nu_j+(M-2)\eta+2\sum_{k=1}^{M}\nu_k}}, \tag{4.4.10}$$

$$\sinh^N(\mu_j+\eta) = -\frac{2e^{-\mu_j-\eta}\prod_{l=1}^{M}\sinh(\mu_j-\nu_l+\eta)\sinh(\mu_j-\nu_l)}{e^{2\mu_j-M\eta-2\sum_{k=1}^{M}\mu_k}+e^{-2\mu_j+(M-2)\eta+2\sum_{k=1}^{M}\nu_k}}. \tag{4.4.11}$$

The eigenvalues of the Hamiltonian (4.1.1) in terms of the Bethe roots read

Table 4.3 Numerical solutions of the BAEs (4.4.10)–(4.4.11) for $\eta = 1$, $N = 3$ and $M = 2$

μ_1	μ_2	ν_1	ν_2	E_n	n
−0.75854 − 0.26534$i\pi$	−0.75854 + 0.26534$i\pi$	−0.24146 − 0.26534$i\pi$	−0.24146 + 0.26534$i\pi$	−3.09636	1
− − −	− − −	− − −	− − −	−3.09636	1
−0.68943 + 0.45685$i\pi$	−0.16961 + 0.29120$i\pi$	−0.83039 + 0.29120$i\pi$	−0.31057 + 0.45685$i\pi$	−1.54308	2
−0.68943 − 0.45685$i\pi$	−0.16961 − 0.29120$i\pi$	−0.83039 − 0.29120$i\pi$	−0.31057 − 0.45685$i\pi$	−1.54308	2
−1.40843 − 0.22982$i\pi$	−0.42100 − 0.01823$i\pi$	−0.57900 − 0.01823$i\pi$	0.40843 − 0.22982$i\pi$	−1.54308	2
−1.40843 + 0.22982$i\pi$	−0.42100 + 0.01823$i\pi$	−0.57900 + 0.01823$i\pi$	0.40843 + 0.22982$i\pi$	−1.54308	2
−0.56218 − 0.12659$i\pi$	−0.56218 + 0.12659$i\pi$	−0.43782 − 0.12659$i\pi$	−0.43782 + 0.12659$i\pi$	6.18252	3
− − −	− − −	− − −	− − −	6.18252	3

E_n is the nth eigenenergy and n indicates the number of the energy levels. The eigenvalues of the Hamiltonian calculated from (4.4.12) are the same as those calculated with exact diagonalization of the Hamiltonian. Here "− − −" indicates a degenerate set of Bethe roots

$$E = 2\sinh\eta\sum_{j=1}^{M}\left[\coth(\mu_j+\eta)-\coth(\nu_j)\right]$$
$$-N\cosh\eta - 2\sinh\eta. \tag{4.4.12}$$

Generally, the Bethe roots are distributed across the whole complex plane with the selection rules $\mu_j \neq \mu_l$, $\mu_j \neq \nu_l$ and $\nu_j \neq \nu_l$ to ensure the simplicity of "poles" in the $T-Q$ Ansatz. We list the numerical results for $N=3$ with $\eta=1$ in Table 4.3 and for $N=4$ with $\eta=1$ in Table 4.4.

Table 4.4 Numerical solutions of the BAEs (4.4.7)–(4.4.8) for $\eta=1$, $N=4$ and $M=2$

μ_1	μ_2	ν_1	ν_2	E_n	n
-1.04497 $+0.50000i\pi$	0.81526 $+0.50000i\pi$	-1.81526 $+0.50000i\pi$	0.04497 $+0.50000i\pi$	-4.27591	1
$---$	$---$	$---$	$---$	-4.27591	1
-0.29245 $-0.30672i\pi$	0.46312 $-0.54374i\pi$	-1.49129 $-0.25997i\pi$	0.61334 $-0.38956i\pi$	-3.67755	2
-0.29245 $+0.30672i\pi$	0.46312 $+0.54374i\pi$	-1.49129 $+0.25997i\pi$	0.61334 $+0.38956i\pi$	-3.67755	2
-1.61334 $+0.38956i\pi$	0.49129 $+0.25997i\pi$	-1.46312 $-0.45626i\pi$	-0.70755 $+0.30672i\pi$	-3.67755	2
-1.61334 $-0.38956i\pi$	0.49129 $-0.25997i\pi$	-1.46312 $+0.45626i\pi$	-0.70755 $-0.30672i\pi$	-3.67755	2
-0.25362 $-0.00000i\pi$	0.53960 $+0.50000i\pi$	-1.48317 $-0.00000i\pi$	0.84848 $-0.00000i\pi$	-3.46526	3
-1.84848 $+0.00000i\pi$	0.48317 $+0.00000i\pi$	-1.53960 $-0.50000i\pi$	-0.74638 $-0.00000i\pi$	-3.46526	3
-0.38977 $-0.21972i\pi$	-0.38977 $+0.21972i\pi$	-0.61023 $-0.21972i\pi$	-0.61023 $+0.21972i\pi$	0.27591	4
$---$	$---$	$---$	$---$	0.27591	4
-0.98569 $-0.09329i\pi$	-0.60953 $+0.04232i\pi$	-0.39047 $+0.04232i\pi$	-0.01431 $-0.09329i\pi$	3.67755	5
-0.98569 $+0.09329i\pi$	-0.60953 $-0.04232i\pi$	-0.39047 $-0.04232i\pi$	-0.01431 $+0.09329i\pi$	3.67755	5
-0.50192 $-0.06546i\pi$	-0.10331 $+0.51449i\pi$	-0.89669 $-0.48551i\pi$	-0.49808 $-0.06546i\pi$	3.67755	5
-0.50192 $+0.06546i\pi$	-0.10331 $-0.51449i\pi$	-0.89669 $+0.48551i\pi$	-0.49808 $+0.06546i\pi$	3.67755	5
-0.63365 $-0.16642i\pi$	-0.57469 $+0.03098i\pi$	-0.42531 $+0.03098i\pi$	-0.36635 $-0.16642i\pi$	7.46526	6
-0.63365 $+0.16642i\pi$	-0.57469 $-0.03098i\pi$	-0.42531 $-0.03098i\pi$	-0.36635 $+0.16642i\pi$	7.46526	6

E_n is the nth eigenenergy and n indicates the number of the energy levels. The eigenvalues of the Hamiltonian calculated from (4.4.12) are the same as those calculated with exact diagonalization of the Hamiltonian. Here "$---$" indicates a degenerate set of Bethe roots

4.5 The Scalar Product $F_n(\theta_1, \ldots, \theta_n)$

Historically, an alternating method was first used to derive the functional relations (4.2.5). This method allows us to extract the scalar products from the spectrum even without knowing the exact form of the eigenstate. A scalar product is defined as follows:

$$F_n(\{u_j\}) = \langle \Psi | \prod_{j=1}^{n} B(u_j) | 0 \rangle, \tag{4.5.1}$$

with $\{u_j\}$ indicating the parameter set $\{u_1, \ldots, u_n\}$ for $n = 0, 1, \ldots, N$. Without losing generality, we may put $F_0 = 1$ by properly choosing the normalization of the eigenvector $\langle \Psi |$.

With the same procedure introduced in Chap. 2 we obtain

$$\begin{aligned} &C(u)|0\rangle = 0, \quad A(u)|0\rangle = a(u)|0\rangle, \\ &D(u)|0\rangle = d(u)|0\rangle, \end{aligned} \tag{4.5.2}$$

where $|0\rangle$ is the all-spin-up state as defined in Chap. 2. In addition, the Yang-Baxter relation

$$R_{1,2}(u - v)T_1(u)T_2(v) = T_2(v)T_1(u)R_{1,2}(u - v), \tag{4.5.3}$$

gives rise to the commutation relations

$$\begin{aligned} &[A(u), A(v)] = [D(u), D(v)] = [B(u), B(v)] = [C(u), C(v)] = 0, \\ &A(u)B(v) = \frac{\sinh(u - v - \eta)}{\sinh(u - v)} B(v)A(u) + \frac{\sinh \eta}{\sinh(u - v)} B(u)A(v), \\ &D(u)B(v) = \frac{\sinh(u - v + \eta)}{\sinh(u - v)} B(v)D(u) - \frac{\sinh \eta}{\sinh(u - v)} B(u)D(v), \\ &C(u)A(v) = \frac{\sinh(u - v + \eta)}{\sinh(u - v)} A(v)C(u) - \frac{\sinh \eta}{\sinh(u - v)} A(u)C(v), \\ &C(u)D(v) = \frac{\sinh(u - v - \eta)}{\sinh(u - v)} D(v)C(u) + \frac{\sinh \eta}{\sinh(u - v)} D(u)C(v), \\ &[A(u), D(v)] = \frac{\sinh \eta}{\sinh(u - v)} [C(v)B(u) - C(u)B(v)], \\ &[C(u), B(v)] = \frac{\sinh \eta}{\sinh(u - v)} [D(u)A(v) - D(v)A(u)], \end{aligned} \tag{4.5.4}$$

and the useful formulae [16]

$$
\begin{aligned}
\langle 0|\prod_{l=1}^{n} C(u_l)\, B(u) &= \sum_{l=1}^{n} M_n^l(u, \{u_j\})\langle 0|C_{n-1}^l \\
&\quad + \sum_{k>l} \tilde{M}_n^{kl}(u, \{u_j\})\langle 0|C_{n-1}^{kl}, \\
C(u)\prod_{l=1}^{n} B(u_l)|0\rangle &= \sum_{l=1}^{n} M_n^l(u, \{u_j\}) B_{n-1}^l|0\rangle \\
&\quad + \sum_{k>l} \tilde{M}_n^{kl}(u, \{u_j\}) B_{n-1}^{kl}|0\rangle,
\end{aligned} \tag{4.5.5}
$$

where

$$
\begin{aligned}
C_{n-1}^l &= \prod_{j\neq l}^{n} C(u_j), \quad C_{n-1}^{kl} = C(u)\prod_{j\neq k,l}^{n} C(u_j), \\
B_{n-1}^l &= \prod_{j\neq l}^{n} B(u_j), \quad B_{n-1}^{kl} = B(u)\prod_{j\neq k,l}^{n} B(u_j),
\end{aligned}
$$

and

$$
\begin{aligned}
M_n^l(u, \{u_j\}) &= g(u, u_l)a(u)d(u_l)\prod_{j\neq l} f(u, u_j) f(u_j, u_l) \\
&\quad + g(u_l, u)a(u_l)d(u)\prod_{j\neq l} f(u_j, u) f(u_l, u_j),
\end{aligned} \tag{4.5.6}
$$

$$
\begin{aligned}
\tilde{M}_n^{kl}(u, \{u_j\}) &= g(u, u_k)g(u_l, u) f(u_l, u_k)a(u_l)d(u_k)\prod_{j\neq k,l} f(u_j, u_k) f(u_l, u_j) \\
&\quad + g(u, u_l)g(u_k, u) f(u_k, u_l)a(u_k)d(u_l)\prod_{j\neq k,l} f(u_j, u_l) f(u_k, u_j),
\end{aligned} \tag{4.5.7}
$$

$$
g(u, v) = \frac{\sinh\eta}{\sinh(v-u)}, \quad f(u, v) = \frac{\sinh(u-v-\eta)}{\sinh(u-v)}. \tag{4.5.8}
$$

The above relations lead to the following functional relations based on the calculations of the quantity $\langle\Psi|t(u)\prod_{j=1}^{n} B(u_j)|0\rangle$

$$
\begin{aligned}
&\Lambda(u)F_n = \sum_{l} M_n^l(u)F_{n-1}^l + \sum_{k>l} \tilde{M}_n^{kl}(u)F_{n-1}^{kl} + F_{n+1}, \\
&F_1(u) = \Lambda(u), \quad F_{N+1} \equiv 0.
\end{aligned} \tag{4.5.9}
$$

Note that the notations $F^{l}_{n-1} = F_{n-1}(\{u_j\}_{j\neq l})$, $F^{kl}_{n-1} = F_{n-1}(u, \{u_j\}_{j\neq k,l})$ and $F_n = F_n(\{u_j\})$ are adopted. Since $[B(u_j), B(u_l)] = 0$, function $F_n(\{u_j\})$ is a symmetric function of all the variables $\{u_j\}$. The explicit expression of $F_n(\{u_j\})$ will be given in the next section (see below (4.6.13)).

Since θ_j is a zero of $d(u)$, all functions M^j_n and $\tilde{M}^{jk}_n$ vanish if all their variables take values in the set $\{\theta_1, \ldots, \theta_N\}$ with $\theta_j \neq \theta_k \neq \theta_l \pm \eta$. This implies

$$\Lambda(\theta_1)F_{n-1}(\theta_2, \ldots, \theta_n) = F_n(\theta_1, \ldots, \theta_n), \tag{4.5.10}$$

and the solution

$$F_n(\theta_1, \ldots, \theta_n) = \prod_{j=1}^{n} \Lambda(\theta_j). \tag{4.5.11}$$

Let us consider the case of $n = N$ and $u_j = \theta_j$ in Eq. (4.5.9). Since $\tilde{M}^{jk}_N = 0$, $F_{N+1} = 0$ and

$$M^l_N(u, \{\theta_j\}) = g(\theta_l, u)a(\theta_l)d(u)\prod_{j\neq l}^{N} f(\theta_j, u)f(\theta_l, \theta_j), \tag{4.5.12}$$

with the help of Eq. (4.5.11) we obtain

$$\begin{aligned}\Lambda(u) &= \sum_{j=1}^{N} \frac{a(\theta_j)d(u)}{\Lambda(\theta_j)} g(\theta_j, u)\prod_{l\neq j}^{N} f(\theta_j, \theta_l)f(\theta_l, u)\\ &= -\sum_{j=1}^{N} \frac{a(\theta_j)d(\theta_j-\eta)}{\Lambda(\theta_j)\bar{d}_j(\theta_j)\sinh(u-\theta_j+\eta)} a(u),\end{aligned} \tag{4.5.13}$$

with $\bar{d}_j(u) = (\sinh\eta)^{-N}\prod_{l\neq j}^{N}\sinh(u-\theta_l)$. Taking the limit $u \to \theta_j - \eta$, we readily recover the relation (4.2.5).

An interesting fact is that all the F_n functions can be derived exactly from the recursive relations, for the reason that $\Lambda(u)$ is already completely determined by the $T - Q$ relation and the BAEs.

4.6 Retrieving the Eigenstates

A remarkable ingredient of the SoV approach is the simple basis of the Hilbert space, characterized by N independent variables. Such a basis is useful for constructing SoV eigenstates of the transfer matrix and for computing the correlation functions. In the framework of ODBA, this basis also makes it convenient to retrieve the Bethe states [14].

4.6.1 SoV Basis of the Hilbert Space

Following the procedure introduced in Sect. 2.5, a convenient SoV basis can be constructed for the present model. Let us define the following left and right states

$$\langle \theta_{p_1}, \ldots, \theta_{p_n}| = \langle 0| \prod_{j=1}^{n} C(\theta_{p_j}),$$

$$|\theta_{q_1}, \ldots, \theta_{q_n}\rangle = \prod_{j=1}^{n} B(\theta_{q_j})|0\rangle, \tag{4.6.1}$$

where $q_j, p_j \in \{1, \ldots, N\}$, $p_1 < p_2 < \cdots < p_n$ and $q_1 < q_2 < \cdots < q_n$. The states (4.6.1) are in fact the eigenstates of $D(u)$ [4] and form a basis of the Hilbert space

$$D(u)|\theta_{p_1}, \ldots, \theta_{p_n}\rangle = d(u) \prod_{j=1}^{n} \frac{\sinh(u - \theta_{p_j} + \eta)}{\sinh(u - \theta_{p_j})} |\theta_{p_1}, \ldots, \theta_{p_n}\rangle,$$

$$\langle \theta_{p_1}, \ldots, \theta_{p_n}|D(u) = d(u) \prod_{j=1}^{n} \frac{\sinh(u - \theta_{p_j} + \eta)}{\sinh(u - \theta_{p_j})} \langle \theta_{p_1}, \ldots, \theta_{p_n}|. \tag{4.6.2}$$

Let us introduce the following inner product

$$\langle \theta_1, \ldots, \theta_n| \prod_{k=1}^{m} B(u_k)|0\rangle = \delta_{n,m}\, g_n(\{\theta_j\}|\{u_\alpha\}), \quad g_0 = \langle 0|0\rangle = 1. \tag{4.6.3}$$

The relations (4.5.5) allow us to derive some recursive relations for the function $g_n(\{\theta_j\}|\{u_\alpha\})$:

$$g_n(\{\theta_j\}|\{u_\alpha\}) = \sum_{l=1}^{n} \frac{\sinh\eta\, d(u_1)a(\theta_l)}{\sinh(u_1 - \theta_l)} \prod_{j\neq l}^{n} \frac{\sinh(u_1 - \theta_j + \eta)}{\sinh(u_1 - \theta_j)}$$
$$\times \frac{\sinh(\theta_l - \theta_j - \eta)}{\sinh(\theta_l - \theta_j)} g_{n-1}(\{\theta_j\}_{j\neq l}|\{u_\alpha\}_{\alpha\neq 1}). \tag{4.6.4}$$

The above function can be expressed in terms of certain determinants [16], namely,

$$g_n(\{\theta_j\}|\{u_\alpha\}) = \frac{\prod_{j=1}^{n}\prod_{\alpha=1}^{n} \sinh(u_\alpha - \theta_j + \eta)\, \det \mathcal{N}(\{u_\alpha\}; \{\theta_j\})}{\prod_{j>k} \sinh(\theta_k - \theta_j) \prod_{\alpha>\beta} \sinh(u_\alpha - u_\beta)}, \tag{4.6.5}$$

where the matrix elements of the $n \times n$ matrix $\mathcal{N}(\{u_\alpha\}; \{\theta_j\})$ are given by

$$\mathscr{N}(\{u_\alpha\};\{\theta_j\})_{\alpha,j}=\frac{\sinh\eta\; d(u_\alpha)\,a(\theta_j)}{\sinh(u_\alpha-\theta_j+\eta)\,\sinh(u_\alpha-\theta_j)}. \tag{4.6.6}$$

Similarly, we can derive that

$$\langle 0|\prod_{k=1}^{m}C(u_k)|\theta_1,\ldots,\theta_n\rangle=\delta_{m,n}\,g_n(\{\theta_j\}|\{u_\alpha\}). \tag{4.6.7}$$

The scalar products (4.6.3), (4.6.5) and (4.6.7) lead to

$$\langle\theta_{p_1},\ldots,\theta_{p_n}|\theta_{q_1},\ldots,\theta_{q_m}\rangle=f_n(\theta_{p_1},\ldots,\theta_{p_n})\,\delta_{m,n}\prod_{j=1}^{n}\delta_{p_j,q_j}, \tag{4.6.8}$$

where

$$\begin{aligned}f_n(\theta_{p_1},\ldots,\theta_{p_n})&=g_n(\{\theta_{p_j}\}|\{\theta_{p_j}\})\\&=\prod_{j=1}^{n}a(\theta_{p_j})d_{p_j}(\theta_{p_j})\prod_{k\neq j}^{n}\frac{\sinh(\theta_{p_j}-\theta_{p_k}+\eta)}{\sinh(\theta_{p_j}-\theta_{p_k})}.\end{aligned} \tag{4.6.9}$$

Here function $d_l(u)$ is defined as $d_l(u)=\prod_{j\neq l}^{N}[\sinh(u-\theta_j)/\sinh\eta]$. Note that the total number of the linearly independent right (left) states given in (4.6.1) is

$$\sum_{n=0}^{N}\frac{N!}{(N-n)!n!}=2^N. \tag{4.6.10}$$

Hence these right (left) states form an orthogonal right (left) basis of the Hilbert space. In such sense, the left eigenstate $\langle\Psi|$ can be expressed as

$$\langle\Psi|=\sum_{n=0}^{N}\sum_{\{p_j\}}\chi_n(\theta_{p_1},\ldots,\theta_{p_n})\langle\theta_{p_1},\ldots,\theta_{p_n}|. \tag{4.6.11}$$

The Eqs. (4.5.1), (4.5.11) and (4.6.8) readily give

$$\chi_n(\theta_{p_1},\ldots,\theta_{p_n})=\frac{\prod_{j=1}^{n}\Lambda(\theta_{p_j})}{f_n(\theta_{p_1},\ldots,\theta_{p_n})}. \tag{4.6.12}$$

Note that in the above we have $\chi_0=1$. $\Lambda(u)$ is the eigenvalue of the transfer matrix corresponding to eigenstate $\langle\Psi|$, and function $f_n(\theta_{p_1},\ldots,\theta_{p_n})$ is given by (4.6.9). The Eqs. (4.6.11)–(4.6.12) allow us to compute functions $F_n(\{u_j\})$ defined by (4.5.1) as follows

$$F_n(\{u_j\}) = \sum_{1\le p_1<p_2<\cdots<p_n\le N} \prod_{l=1}^{n} \Lambda(\theta_{p_l}) \frac{g_n(\{\theta_{p_l}\}|\{u_j\})}{g_n(\{\theta_{p_l}\}|\{\theta_{p_l}\})}, \tag{4.6.13}$$

where function $g_n(\{\theta_{p_l}\}|\{u_j\})$ is given by (4.6.5).

Similarly, by considering the quantities $\langle 0|\prod_{j=1}^{n} C(u_j)|\Psi\rangle$, we can derive the right eigenstate

$$|\Psi\rangle = \sum_{n=0}^{N}\sum_{\{p_j\}} \chi_n(\theta_{p_1},\ldots,\theta_{p_n})|\theta_{p_1},\ldots,\theta_{p_n}\rangle, \tag{4.6.14}$$

and the scalar product

$$\langle 0|\prod_{j=1}^{n} C(u_j)|\Psi\rangle = F_n(\{u_j\}). \tag{4.6.15}$$

4.6.2 Bethe States

Since the states $\{\langle\theta_{p_1},\ldots,\theta_{p_n}|\}$ given by (4.6.1) form a left basis of the Hilbert space, an eigenstate $|\Psi\rangle$ of the transfer matrix can be determined (up to an overall factor) by the whole set of scalar products

$$\langle\theta_{p_1},\ldots,\theta_{p_n}|\Psi\rangle = F_n(\theta_{p_1},\ldots,\theta_{p_n}) \overset{(4.5.11)}{=} \prod_{j=1}^{n}\Lambda(\theta_{p_j}),$$
$$n = 0,\ldots,N. \tag{4.6.16}$$

Let us consider the Bethe state

$$|\lambda_1,\ldots,\lambda_N\rangle = \prod_{j=1}^{N} D(\lambda_j)|\Omega;\{\theta_j\}\rangle, \tag{4.6.17}$$

where $\{\lambda_j|j=1,\ldots,N\}$ are the Bethe roots given in the $T-Q$ relation (4.3.1) and the BAEs (4.3.4), and $|\Omega;\{\theta_j\}\rangle$ is a reference state to be determined. We take the reference state so that the following inner products hold:

$$\langle\theta_{q_1},\ldots,\theta_{q_n}|\Omega;\{\theta_j\}\rangle = \prod_{l=1}^{n} a(\theta_{p_l})e^{\theta_{p_l}}, \quad n=0,\ldots,N. \tag{4.6.18}$$

With the help of (4.6.2), a simple calculation induces that

$$\langle \theta_{p_1}, \ldots, \theta_{p_n} | \lambda_1, \ldots, \lambda_N \rangle = \left\{ \prod_{j=1}^{N} d(\lambda_j) \right\} F_n(\theta_{p_1}, \ldots, \theta_{p_n}). \quad (4.6.19)$$

Therefore, the Bethe state (4.6.17) with the reference state satisfying (4.6.18) is an eigenstate (up to an irrelevant normalization constant) of the transfer matrix provided that the parameters $\{\lambda_j\}$ satisfy the associated BAEs (4.3.4).

In order to make (4.6.18) fulfilled, we propose the following Ansatz for the reference state $|\Omega; \{\theta_j\}\rangle$:

$$|\Omega; \{\theta_j\}\rangle = \sum_{l=0}^{\infty} \frac{\left(\tilde{B}^-\right)^l}{[l]_q!} |0\rangle = \sum_{l=0}^{N} \frac{\left(\tilde{B}^-\right)^l}{[l]_q!} |0\rangle, \quad (4.6.20)$$

where the q-integers $\{[l]_q | l = 0, \ldots\}$ and the operator $\tilde{B}^-$ are given by

$$[l]_q = \frac{1-q^{2l}}{1-q^2}, \quad [0]_q = 1,$$

$$[l]_q! = [l]_q \, [l-1]_q \cdots [1]_q, \quad q = e^{\eta}, \quad (4.6.21)$$

$$\tilde{B}^- = \lim_{u \to +\infty} \left\{ \left(2 \sinh \eta \, e^{-u}\right)^{N-1} e^{\sum_{l=1}^{N} \theta_l} \, B(u) \right\}. \quad (4.6.22)$$

The definitions (4.1.3) and (4.1.4) allow us to obtain the explicit expression of the operator $\tilde{B}^-$ as

$$\tilde{B}^- = \sum_{l=1}^{N} e^{\theta_l + \frac{(N-1)\eta}{2}} \, e^{\frac{\eta}{2} \sum_{k=l+1}^{N} \sigma_k^z} \, \sigma_l^- \, e^{-\frac{\eta}{2} \sum_{k=1}^{l-1} \sigma_k^z}. \quad (4.6.23)$$

Now let us compute the following quantity

$$[n]_q! \, \langle \theta_{p_1}, \ldots, \theta_{p_n} | \Omega; \{\theta_j\} \rangle \stackrel{\text{def}}{=} \bar{g}_n(\{\theta_{p_j}\}) = \langle \theta_{p_1}, \ldots, \theta_{p_n} | \left(\tilde{B}^-\right)^n |0\rangle. \quad (4.6.24)$$

The definitions (4.6.3) and (4.6.21) imply that we can calculate the function $\bar{g}_n(\{\theta_j\})$ by the limit

$$\bar{g}_n(\{\theta_j\}) = \lim_{\{u_l \to +\infty\}} \left\{ \left[\prod_{l=1}^{n} \left(2 \sinh \eta \, e^{-u_l}\right)^{N-1} e^{\sum_{k=1}^{N} \theta_k} \right] g_n(\{\theta_j\} | \{u_l\}) \right\}. \quad (4.6.25)$$

Keeping the recursive relations (4.6.4) in mind, we can derive the recursive relations

$$\bar{g}_n(\{\theta_j\}) = \sum_{l=1}^{n} e^{(n-1)\eta}\, a(\theta_l) e^{\theta_l} \prod_{j\neq l}^{n} \frac{\sinh(\theta_l - \theta_j - \eta)}{\sinh(\theta_l - \theta_j)}\, \bar{g}_{n-1}(\{\theta_j\}_{j\neq l}),$$
$$n = 1, \ldots, N,$$

with the initial condition of $\bar{g}_0 = 1$. The above recursive relations uniquely determine the functions $\{\bar{g}_n(\{\theta_j\})|n = 0, \ldots, N\}$:

$$\bar{g}_n(\{\theta_j\}) = \left\{\prod_{l=1}^{n} a(\theta_l) e^{\theta_l}\right\} [n]_q!, \quad n = 0, \ldots, N. \tag{4.6.26}$$

Note that the following identities were used in deriving the above equation

$$\sum_{l=1}^{n} e^{(n-1)\eta} \prod_{j\neq l}^{n} \frac{\sinh(\theta_l - \theta_j - \eta)}{\sinh(\theta_l - \theta_j)} = 1 + e^{2\eta} + \cdots + e^{2(n-1)\eta} = [n]_q. \tag{4.6.27}$$

Substituting (4.6.26) into (4.6.24), we find that the state $|\Omega; \{\theta_j\}\rangle$ given by (4.6.20) indeed satisfies the relations (4.6.18). Then we conclude that the Bethe state (4.6.17) with the corresponding reference state (4.6.20) is an eigenstate of the transfer matrix, provided that the parameters $\{\lambda_j|j = 1, \ldots, N\}$ satisfy the associated BAEs (4.3.4). The corresponding eigenvalue is given by the $T - Q$ relation (4.3.1).

In the homogeneous limit, the reference state (4.6.20) becomes

$$|\Omega\rangle = \lim_{\{\theta_j\to 0\}} |\Omega; \{\theta_j\}\rangle = \sum_{l=0}^{\infty} \frac{(B^-)^l}{[l]_q!}|0\rangle, \tag{4.6.28}$$

where the operator B^- [c.f., (4.6.23)] reads

$$B^- = \lim_{\{\theta_j\to 0\}} \tilde{B}^- = \sum_{l=1}^{N} e^{\frac{(N-1)\eta}{2}}\, e^{\frac{\eta}{2}\sum_{k=l+1}^{N}\sigma_k^z}\, \sigma_l^-\, e^{-\frac{\eta}{2}\sum_{k=1}^{l-1}\sigma_k^z}. \tag{4.6.29}$$

Obviously, the reference state $|\Omega\rangle$ is no longer a pure product state but a highly entangled superposition state (a q-spin coherent state).

Associated with the $T - Q$ relation (4.4.1), we can construct another type of Bethe states

$$|\mu_1, \ldots, \mu_M; \nu_1, \ldots, \nu_M\rangle = \prod_{j=1}^{M} D(\mu_j) D(\nu_j) |\Omega; \{\theta_j\}\rangle, \tag{4.6.30}$$

where the reference state reads

$$|\bar{\Omega}; \{\theta_j\}\rangle = \sum_{n=0}^{N} \sum_{\{p_j\}} f_n^{-1}(\theta_{p_1}, \ldots, \theta_{p_n}) \times \prod_{l=1}^{n} e^{\theta_{p_l}} a(\theta_{p_l}) \frac{Q_1(\theta_{p_l})}{Q_2(\theta_{p_l} - \eta)} |\theta_{p_1}, \ldots, \theta_{p_n}\rangle. \tag{4.6.31}$$

It can be easily checked that

$$\begin{aligned} &\langle \theta_{p_1}, \ldots, \theta_{p_n} | \mu_1, \ldots, \mu_M; \nu_1, \ldots, \nu_M \rangle \\ &= \prod_{j=1}^{M} d(\mu_j) d(\nu_j) \prod_{l=1}^{n} e^{\theta_{p_l}} a(\theta_{p_l}) \frac{Q_1(\theta_{p_l} - \eta)}{Q_2(\theta_{p_l})} \\ &= \prod_{j=1}^{M} d(\mu_j) d(\nu_j) F_n(\theta_{p_1}, \ldots, \theta_{p_n}). \end{aligned} \tag{4.6.32}$$

Therefore, the Bethe state (4.6.30) is an eigenstate of the transfer matrix, provided that the parameters $\{\mu_j\}$ and $\{\nu_j\}$ satisfy the associated BAEs (4.4.6).

In the homogeneous limit, the Bethe state (4.6.30) reads

$$\prod_{j=1}^{M} D(\mu_j) D(\nu_j) |\bar{\Omega}\rangle, \tag{4.6.33}$$

with the reference state defined as

$$|\bar{\Omega}\rangle = \sum_{l=0}^{\infty} \frac{\left(Q_1(0) B^-\right)^l}{[l]_q! \left(Q_2(-\eta)\right)^l} |0\rangle. \tag{4.6.34}$$

Some remarks are in order: The above procedure for deriving the Bethe states is different from that of the algebraic Bethe Ansatz. In the latter scheme, one uses a known reference state and creation operator to derive eigenvalues and eigenstates of the transfer matrix, while in the ODBA scheme, one uses known eigenvalues and creation operator to retrieve the reference state. The key point is that the eigenstates of the creation operator with an arbitrary parameter u form a basis of the Hilbert space. Such a reversed process makes it convenient to approach the eigenstate problem of quantum integrable models without an obvious reference state.

4.6.3 Another Basis

We note that another orthogonal basis spanned by the eigenstates of $A(u)$ also exists. For convenience, let us denote $\bar{\theta}_j = \theta_j - \eta$. The eigenstates of $A(u)$ can be constructed as

$$|\bar{\theta}_{p_1}, \ldots, \bar{\theta}_{p_n}\rangle = \prod_{j=1}^{n} B(\bar{\theta}_{p_j})|0\rangle, \quad \langle\bar{\theta}_{p_1}, \ldots, \bar{\theta}_{p_n}| = \langle 0|\prod_{j=1}^{n} C(\bar{\theta}_{p_j}). \tag{4.6.35}$$

The corresponding eigenvalues are given by

$$A(u)\,|\bar{\theta}_{p_1}, \ldots, \bar{\theta}_{p_n}\rangle = a(u)\prod_{j=1}^{n}\frac{\sinh(u-\bar{\theta}_j-\eta)}{\sinh(u-\bar{\theta}_j)}\,|\bar{\theta}_{p_1}, \ldots, \bar{\theta}_{p_n}\rangle,$$
$$\langle\bar{\theta}_{p_1}, \ldots, \bar{\theta}_{p_n}|\,A(u) = a(u)\prod_{j=1}^{n}\frac{\sinh(u-\bar{\theta}_j-\eta)}{\sinh(u-\bar{\theta}_j)}\,\langle\bar{\theta}_{p_1}, \ldots, \bar{\theta}_{p_n}|.$$

The above states possess the orthogonal properties

$$\langle\bar{\theta}_{p_1}, \ldots, \bar{\theta}_{p_n}|\bar{\theta}_{q_1}, \ldots, \bar{\theta}_{q_m}\rangle = \bar{f}_n(\bar{\theta}_{p_1}, \ldots, \bar{\theta}_{p_n})\delta_{m,n}\prod_{j=1}^{n}\delta_{p_j,q_j}, \tag{4.6.36}$$

with

$$\bar{f}_n(\bar{\theta}_{p_1}, \ldots, \bar{\theta}_{p_n}) = (-1)^n\prod_{j=1}^{n} d(\bar{\theta}_{p_j})d_{p_j}(\bar{\theta}_{p_j}+\eta)$$
$$\times\prod_{k\neq j}^{n}\frac{\sinh(\theta_{p_j}-\theta_{p_k}+\eta)}{\sinh(\theta_{p_j}-\theta_{p_k})}. \tag{4.6.37}$$

The eigenstates can thus be expressed as

$$|\Psi\rangle = \sum_{n=0}^{N}\sum_{\{p_j\}}\bar{\chi}_n(\bar{\theta}_{p_1}, \ldots, \bar{\theta}_{p_n})|\bar{\theta}_{p_1}, \ldots, \bar{\theta}_{p_n}\rangle,$$
$$\langle\Psi| = \sum_{n=0}^{N}\sum_{\{p_j\}}\bar{\chi}_n(\bar{\theta}_{p_1}, \ldots, \bar{\theta}_{p_n})\langle\bar{\theta}_{p_1}, \ldots, \bar{\theta}_{p_n}|, \tag{4.6.38}$$

with

$$\bar{\chi}_n(\bar{\theta}_{p_1}, \ldots, \bar{\theta}_{p_n}) = \frac{\prod_{j=1}^{n}\Lambda(\bar{\theta}_{p_j})}{\bar{f}_n(\bar{\theta}_{p_1}, \ldots, \bar{\theta}_{p_n})}. \tag{4.6.39}$$

4.7 Physical Properties for $\eta = \frac{i\pi}{2}$

At the special point $\eta = \frac{i\pi}{2}$, the model describes the *XX* torus. This model can be transformed to a topological free fermion model via the Jordan-Wigner transformation:

$$\sigma_j^+ = a_j^\dagger e^{i\pi \sum_{l=1}^{j-1} a_l^\dagger a_l}, \quad \sigma_j^- = a_j e^{-i\pi \sum_{l=1}^{j-1} a_l^\dagger a_l}, \quad \sigma_j^z = 2a_j^\dagger a_j - 1. \quad (4.7.1)$$

The resulting fermion Hamiltonian reads

$$H = -2\sum_{j=1}^{N-1} \left[a_j^\dagger a_{j+1} + a_{j+1}^\dagger a_j\right] - 2U^z \left[a_1^\dagger a_N^\dagger + a_N a_1\right], \quad (4.7.2)$$

where $a_j^\dagger$ and a_j are the creation and annihilation operators of fermions, respectively; $n_j = a_j^\dagger a_j$ represents the particle number operator; $U^z = e^{i\pi \sum_{j=1}^{N} n_j}$ is a conserved quantity with the eigenvalues ± 1. The second term of the Hamiltonian is a Cooper-pair term, which obviously breaks the $U(1)$ symmetry.

In this special case, the BAEs satisfied by the roots $\{z_j\}$ can be derived from the recursive equations. Let us check the case of $n = N$ in Eq. (4.5.9)

$$F_N(z, \{u_l\}) = \sum_{j=1,z}^{N-1} M'^j_N F_{N-1}, \quad (4.7.3)$$

with $\{u_l\} = u_1, \ldots, u_{N-1}, z$ being one of the roots of $\Lambda(u)$ and

$$M'^j_N = \lim_{u\to\infty} \Lambda^{-1}(u) M_N^j(u, z, \{u_l\}).$$

In addition, from the recursion relation we also have

$$F_N(z, \{u_l\}) = -\sum_{j=1}^{N-1} M_{N-1}^j(z, \{u_l\}) F_{N-2}^j - \sum_{j<k} \tilde{M}_{N-1}^{jk}(z, \{u_l\}) F_{N-2}^{jk}. \quad (4.7.4)$$

Equations (4.7.3) and (4.7.4) together give rise to

$$F_{N-2}^j = \frac{-1}{M_{N-1}^j(z, \{u_j\})} \left[\sum_l M'^l_N F_{N-1} + \sum_{l\neq j} M_{N-1}^l F_{N-2}^l + \sum_{k>l} \tilde{M}_{N-1}^{kl} F_{N-2}^{kl}\right]. \quad (4.7.5)$$

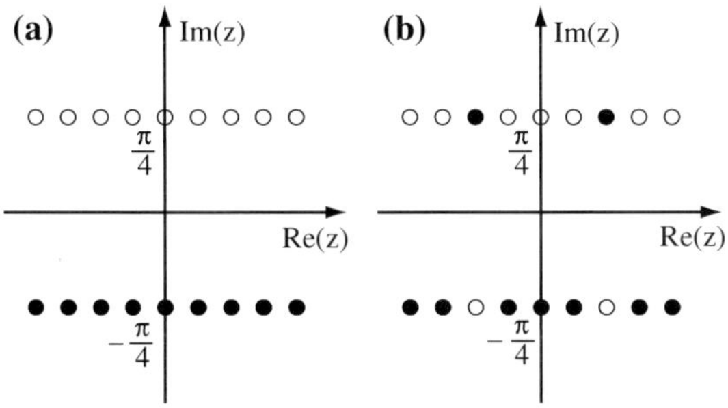

Fig. 4.1 **a** Schematic diagram of the root distribution in the ground state of the *XX* spin torus. The states in the lower solution line are all filled and the upper solution line is unoccupied. **b** The elementary excitation. A "particle" (indicated by the *dot*) in the upper solution line must correspond to a "hole" (indicated by the *circle*) in the lower solution line with the exactly same real part

From the definition of the F-functions we know that all of them are entire functions. This fact requires that the residues of the right hand side of (4.7.5) about $u_j = z \pm i\frac{\pi}{2}$ must be zero. This requirement gives the following BAEs.

$$\coth^{2N}(z_n) = 1, \quad z_j \neq z_k \pm \frac{\pi}{2}i, \tag{4.7.6}$$

or

$$\coth(z_n) = e^{\frac{i\pi n}{N}} \equiv e^{ik_n}, \quad n = \pm 1, \ldots, \pm(N-1). \tag{4.7.7}$$

The $N-1$ pair solutions $\{z_j, z_j + \frac{\pi}{2}i\} \bmod (i\pi)$ of the above BAEs fall in two straight lines parallel to the real axis with the imaginary part $\pm i\pi/4$ as shown in Fig. 4.1. We remark that there is a selection rule to choose $N-1$ z_j from the solution set to form the $N-1$ roots of $\Lambda(u)$, i.e., $z_j \neq z_l \pm i\frac{\pi}{2}$. This selection rule comes from the fact that the poles and the zeros in (4.7.5) satisfy the same Eq. (4.7.6). Therefore, the possible number of $\Lambda(u)$ is 2^{N-1}, which, with the double degeneracy implied by the Z_2-symmetry, constitutes the complete spectrum of the transfer matrix.

The eigenvalue of the Hamiltonian is minimized if we put all the roots z_j in the lower solution line and leave the upper solution line empty as shown in Fig. 4.1a. The ground state energy can be easily calculated as $E_g = -2\cot\frac{\pi}{2N}$, which slightly deviates from that ofthe periodic chain. The elementary excitations of the system are thus formed by the particle-hole pairs as shown in Fig. 4.1b. The selection rule requires that each particle with momentum k must be locked by a hole with momentum $-k$ (Andreev reflection) as shown in Fig. 4.2, revealing the topological nature of the system

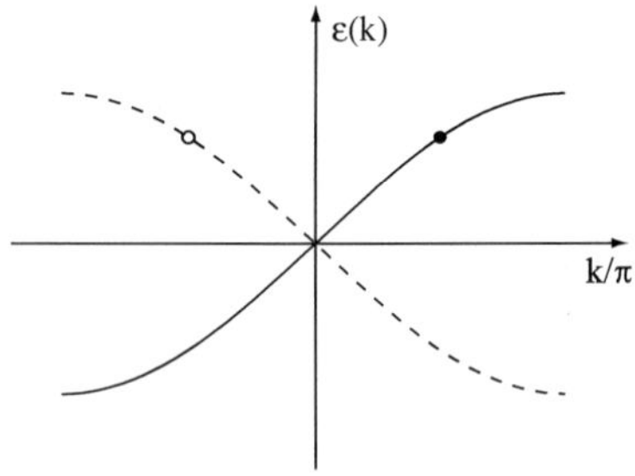

Fig. 4.2 Schematic diagram of the excitation spectrum of the *XX* spin torus. The "particle" and the "hole" carry the exactly same energy and opposite momenta

4.8 The *XYZ* Spin Torus

Following the same procedure, the exact spectrum of the *XYZ* spin torus can also be constructed [17]. The Hamiltonian of the *XYZ* spin torus is given by

$$H = \frac{1}{2}\sum_{n=1}^{N}(J_x\sigma_n^x\sigma_{n+1}^x + J_y\sigma_n^y\sigma_{n+1}^y + J_z\sigma_n^z\sigma_{n+1}^z), \tag{4.8.1}$$

and

$$J_x = e^{i\pi\eta}\frac{\sigma(\eta+\frac{\tau}{2})}{\sigma(\frac{\tau}{2})}, \quad J_y = e^{i\pi\eta}\frac{\sigma(\eta+\frac{1+\tau}{2})}{\sigma(\frac{1+\tau}{2})}, \quad J_z = \frac{\sigma(\eta+\frac{1}{2})}{\sigma(\frac{1}{2})}, \tag{4.8.2}$$

with the antiperiodic boundary condition

$$\sigma_{N+1}^x = \sigma_1^x, \quad \sigma_{N+1}^y = -\sigma_1^y, \quad \sigma_{N+1}^z = -\sigma_1^z. \tag{4.8.3}$$

The corresponding transfer matrix is defined by Eq. (4.1.8) with the R-matrix replaced with the XYZ one (3.2.10). Similarly, we have the following operator identities

$$t(\theta_j)t(\theta_j-\eta) = -a(\theta_j)\,d(\theta_j-\eta), \quad j=1,\ldots,N, \tag{4.8.4}$$

$$\prod_{j=1}^{N} t(\theta_j) = \prod_{j=1}^{N} a(\theta_j)\times U^x, \tag{4.8.5}$$

$$t(u+1) = (-1)^{N-1}\,t(u), \tag{4.8.6}$$

$$t(u+\tau) = (-1)^N e^{-2i\pi\left\{Nu+N\frac{\eta+\tau}{2}-\sum_{l=1}^{N}\theta_l\right\}}\,t(u), \tag{4.8.7}$$

where the functions $a(u)$ and $d(u)$ are given by (3.2.23).

The Z_2-symmetry of the R-matrix indicates that the model Hamiltonian also possesses a Z_2-symmetry [18–20], i.e., $[H, U^x] = 0$. The corresponding functional relations of $\Lambda(u)$ read

$$\Lambda(\theta_j)\Lambda(\theta_j-\eta)=-a(\theta_j)\,d(\theta_j-\eta),\quad j=1,\ldots,N, \tag{4.8.8}$$

$$\prod_{j=1}^{N}\Lambda(\theta_j)=\pm\prod_{j=1}^{N}a(\theta_j), \tag{4.8.9}$$

$$\Lambda(u+1)=(-1)^{N-1}\Lambda(u), \tag{4.8.10}$$

$$\Lambda(u+\tau)=(-1)^{N}e^{-2i\pi\left\{Nu+N\frac{\eta+\tau}{2}-\sum_{l=1}^{N}\theta_l\right\}}\Lambda(u). \tag{4.8.11}$$

Meanwhile,

$$\Lambda(u)\text{ is an entire function of }u. \tag{4.8.12}$$

The relations (4.8.8)–(4.8.12) completely determine the function $\Lambda(u)$. In the homogeneous limit $\theta_j\to 0$, we have the following inhomogeneous $T-Q$ relation for the XYZ spin torus:

$$\begin{aligned}\Lambda(u)=&\,e^{i\pi(2l_1+1)u+i\phi}\frac{\sigma^N(u+\eta)}{\sigma^N(\eta)}\frac{Q_1(u-\eta)}{Q_2(u)}\\&-\frac{e^{-[i\pi(2l_1+1)(u+\eta)+i\phi]}\sigma^N(u)}{\sigma^N(\eta)}\frac{Q_2(u+\eta)}{Q_1(u)}\\&+\frac{c\,e^{i\pi u}\sigma^m(u+\frac{\eta}{2})}{\sigma^m(\eta)Q_1(u)Q_2(u)}\frac{\sigma^N(u+\eta)\sigma^N(u)}{\sigma^N(\eta)\sigma^N(\eta)},\end{aligned} \tag{4.8.13}$$

where l_1 is an integer, and the Q-functions take the forms given by (3.2.37). Note that here the non-negative integer m is even (odd) for even (odd) N and

$$N+m=2M. \tag{4.8.14}$$

The two parameters c, ϕ, and the $2M$ Bethe roots $\{\mu_j\}$ and $\{\nu_j\}$ are determined by the BAEs:

$$\left(\frac{N}{2}-M\right)\eta-\sum_{j=1}^{M}(\mu_j-\nu_j)=\left(l_1+\frac{1}{2}\right)\tau+m_1,\quad l_1,m_1\in Z, \tag{4.8.15}$$

$$M\eta+\sum_{j=1}^{M}(\mu_j+\nu_j)=\frac{1}{2}\tau+m_2,\quad m_2\in Z, \tag{4.8.16}$$

$$\begin{aligned}&\frac{c\,e^{\left[2i\pi(l_1+1)\mu_j+2i\pi(l_1+\frac{1}{2})\eta+i\phi\right]}\sigma^m(\mu_j+\frac{\eta}{2})}{\sigma^m(\eta)}\frac{\sigma^N(\mu_j+\eta)}{\sigma^N(\eta)}\\&=Q_2(\mu_j)Q_2(\mu_j+\eta),\end{aligned} \tag{4.8.17}$$

$$\frac{c\, e^{-2i\pi l_1 \nu_j - i\phi} \sigma^m(\nu_j + \frac{\eta}{2})}{\sigma^m(\eta)} \frac{\sigma^N(\nu_j)}{\sigma^N(\eta)} = -Q_1(\nu_j) Q_1(\nu_j - \eta), \tag{4.8.18}$$

and the selection rule

$$\Lambda(0) = e^{i\phi} \prod_{j=1}^{M} \frac{\sigma(\mu_j + \eta)}{\sigma(\nu_j)} = e^{\frac{i\pi k}{N}}, \quad k = 1, \ldots, 2N. \tag{4.8.19}$$

The eigenvalue of the Hamiltonian reads

$$E = \frac{\sigma(\eta)}{\sigma'(0)} \left\{ \sum_{j=1}^{M} [\zeta(\nu_j) - \zeta(\mu_j + \eta)] + \frac{1}{2} N \zeta(\eta) + 2i\pi \left(l_1 + \frac{1}{2} \right) \right\}, \tag{4.8.20}$$

where function $\zeta(u)$ is defined by (3.2.3). As for the periodic XYZ spin chain, any fixed integers l_1 and m_1 give a complete set of solutions. Numerical results for $N = 3$, $m = 1$ and $N = 4$, $m = 0$ with $l_1 = m_1 = 0$ are shown in Tables 4.5 and 4.6, respectively.

Table 4.5 Numerical solutions of the BAEs (4.8.15)–(4.8.19) for $N = 3$, $\eta = -2/3$ and $\tau = i$

μ_1	μ_2	ν_1		
$0.41667 - 0.24873i$	$0.41667 + 0.24873i$	$-0.47794 - 0.00000i$		
$0.41667 - 0.24873i$	$0.41667 + 0.24873i$	$-0.47794 - 0.00000i$		
$-0.29185 - 0.11151i$	$0.12518 + 0.11151i$	$0.03233 - 0.13153i$		
$-0.29185 - 0.11151i$	$0.12518 + 0.11151i$	$0.03233 - 0.13153i$		
$-0.29185 + 0.11151i$	$0.12518 - 0.11151i$	$0.03233 + 0.13153i$		
$-0.29185 + 0.11151i$	$0.12518 - 0.11151i$	$0.03233 + 0.13153i$		
$-0.08333 - 0.08037i$	$-0.08333 + 0.08037i$	$-0.30406 + 0.50000i$		
$-0.08333 - 0.08037i$	$-0.08333 + 0.08037i$	$-0.30406 + 0.50000i$		
ν_2	c	ϕ	E_n	n
$-0.02206 + 0.50000i$	$0.21550 - 3.10475i$	$1.64010 - 0.57175i$	-1.41032	1
$-0.02206 + 0.50000i$	$-0.21550 + 3.10475i$	$-1.50150 - 0.57175i$	-1.41032	1
$0.46767 + 0.63153i$	$-2.78061 - 2.78148i$	$1.00207 - 1.48298i$	-0.37922	2
$0.46767 + 0.63153i$	$2.78061 + 2.78148i$	$-2.13952 - 1.48298i$	-0.37922	2
$0.46767 + 0.36847i$	$1.43741 - 0.94669i$	$2.34268 - 0.65658i$	-0.37922	2
$0.46767 + 0.36847i$	$-1.43741 + 0.94669i$	$-0.79892 - 0.65658i$	-0.37922	2
$-0.19594 - 0.00000i$	$2.07093 - 1.46458i$	$2.52604 - 0.92817i$	2.16875	3
$-0.19594 - 0.00000i$	$-2.07093 + 1.46458i$	$-0.61555 - 0.92817i$	2.16875	3

E_n is the nth eigenenergy and n indicates the number of the energy levels. The eigenvalues of the Hamiltonian calculated from (4.8.20) are the same as those calculated with exact diagonalization of the Hamiltonian

Table 4.6 Numerical solutions of the BAEs (4.8.15)–(4.8.19) for $N = 4$, $\eta = -2/3$ and $\tau = i$

μ_1	μ_2	ν_1		
$0.33333 - 0.18210i$	$0.33333 + 0.18210i$	$0.33333 - 0.00000i$		
$0.33333 - 0.18210i$	$0.33333 + 0.18210i$	$0.33333 - 0.00000i$		
$-0.16667 - 0.18586i$	$0.33333 + 0.18586i$	$-0.16667 + 0.61266i$		
$-0.16667 + 0.18586i$	$0.33333 - 0.18586i$	$-0.16667 + 0.38734i$		
$-0.16667 - 0.18586i$	$0.33333 + 0.18586i$	$-0.16667 + 0.61266i$		
$-0.16667 + 0.18586i$	$0.33333 - 0.18586i$	$-0.16667 + 0.38734i$		
$0.20335 - 0.00000i$	$0.46332 + 0.00000i$	$-0.16667 + 0.00000i$		
$0.20335 + 0.00000i$	$0.46332 - 0.00000i$	$-0.16667 - 0.00000i$		
$-0.16667 - 0.13615i$	$-0.16667 + 0.13615i$	$0.33333 + 0.00000i$		
$-0.16667 - 0.13615i$	$-0.16667 + 0.13615i$	$0.33333 - 0.00000i$		
$-0.16667 - 0.07684i$	$0.33333 + 0.07684i$	$-0.16667 - 0.04444i$		
$-0.16667 + 0.07684i$	$0.33333 - 0.07684i$	$-0.16667 + 0.04444i$		
$-0.16667 - 0.07684i$	$0.33333 + 0.07684i$	$-0.16667 - 0.04444i$		
$-0.16667 + 0.07684i$	$0.33333 - 0.07684i$	$-0.16667 + 0.04444i$		
$-0.16667 - 0.08893i$	$-0.16667 + 0.08893i$	$-0.16667 + 0.00000i$		
$-0.16667 - 0.08893i$	$-0.16667 + 0.08893i$	$-0.16667 + 0.00000i$		
ν_2	c	ϕ	E_n	n
$0.33333 + 0.50000i$	$-0.69252 + 1.19949i$	$-1.04720 - 0.66489i$	-2.34645	1
$0.33333 + 0.50000i$	$0.69252 - 1.19949i$	$2.09440 - 0.66489i$	-2.34645	1
$0.33333 - 0.11266i$	$1.95137 - 3.37987i$	$-1.04720 - 1.28609i$	-1.01133	2
$0.33333 + 0.11266i$	$-0.96144 + 1.66527i$	$2.09440 - 0.57823i$	-1.01133	2
$0.33333 - 0.11266i$	$-1.95137 + 3.37987i$	$2.09440 - 1.28609i$	-1.01133	2
$0.33333 + 0.11266i$	$0.96144 - 1.66527i$	$-1.04720 - 0.57823i$	-1.01133	2
$-0.16667 + 0.50000i$	$1.21750 - 2.10877i$	$2.09440 - 0.66148i$	-0.64764	3
$-0.16667 + 0.50000i$	$-1.21750 + 2.10877i$	$-1.04720 - 0.66148i$	-0.64764	3
$0.33333 + 0.50000i$	$2.32235 - 4.02243i$	$2.09440 - 1.59613i$	0.06267	4
$0.33333 + 0.50000i$	$-2.32235 + 4.02243i$	$-1.04720 - 1.59613i$	0.06267	4
$0.33333 + 0.54444i$	$1.25919 - 2.18098i$	$-1.04720 - 1.09820i$	1.01133	5
$0.33333 + 0.45556i$	$0.95239 - 1.64960i$	$-1.04720 - 0.81896i$	1.01133	5
$0.33333 + 0.54444i$	$-1.25919 + 2.18098i$	$2.09440 - 1.09820i$	1.01133	5
$0.33333 + 0.45556i$	$-0.95239 + 1.64960i$	$2.09440 - 0.81896i$	1.01133	5
$-0.16667 + 0.50000i$	$-1.86802 + 3.23551i$	$-1.04720 - 1.25303i$	2.93143	6
$-0.16667 + 0.50000i$	$1.86802 - 3.23551i$	$2.09440 - 1.25303i$	2.93143	6

E_n is the nth eigenenergy and n indicates the number of the energy levels. The eigenvalues of the Hamiltonian calculated from (4.8.20) are the same as those calculated with exact diagonalization of the Hamiltonian

In contrast to the periodic XYZ spin chain, the $c = 0$ solution of the BAEs (4.8.15)–(4.8.18) is not allowed for a generic η. However, the $c = 0$ solutions indeed exist for some discrete η values labeled by two integers l_1 and m_1:

$$\left(\frac{N}{2}-M\right)\eta=\left(l_1+\frac{1}{2}\right)\tau+m_1,\quad l_1,m_1\in Z. \tag{4.8.21}$$

In this case, the $T-Q$ relation (4.8.13) is reduced to a conventional one:

$$\begin{aligned}\Lambda(u)&=e^{2i\pi(l_1+\frac{1}{2})u+i\phi}\frac{\sigma^N(u+\eta)}{\sigma^N(\eta)}\frac{Q(u-\eta)}{Q(u)}\\&\quad-e^{-\left[2i\pi(l_1+\frac{1}{2})(u+\eta)+i\phi\right]}\frac{\sigma^N(u)}{\sigma^N(\eta)}\frac{Q(u+\eta)}{Q(u)},\\Q(u)&=\prod_{l=1}^{M}\frac{\sigma(u-\lambda_l)}{\sigma(\eta)}.\end{aligned} \tag{4.8.22}$$

The corresponding BAEs thus read

$$e^{2i\pi\left[(2l_1+1)\lambda_j+(l_1+\frac{1}{2})\eta\right]+2i\phi}\frac{\sigma^N(\lambda_j+\eta)}{\sigma^N(\lambda_j)}=\frac{Q(\lambda_j+\eta)}{Q(\lambda_j-\eta)},\qquad j=1,\ldots,M, \tag{4.8.23}$$

$$e^{i\phi}\prod_{j=1}^{M}\frac{\sigma(\lambda_j+\eta)}{\sigma(\lambda_j)}=e^{\frac{i\pi k}{N}},\quad k=1,\ldots,2N. \tag{4.8.24}$$

As we emphasized in Chap. 3, the degenerate points become dense in the whole complex η-plane in the thermodynamic limit, which allows us to obtain the thermodynamic properties of the model for generic values of η [21] via the conventional thermodynamic Bethe Ansatz methods [22, 23].

References

1. C.M. Yung, M.T. Batchelor, Exact solution for the spin-s XXZ quantum chain with non-diagonal twists. Nucl. Phys. B **446**, 461 (1995)
2. M.T. Batchelor, R.J. Baxter, M.J. O'Rourke, C.M. Yung, Exact solution and interfacial tension of the six-vertex model with anti-periodic boundary conditions. J. Phys. A. Math. Gen. **28**, 2759 (1995)
3. S. Niekamp, T. Wirth, H. Frahm, The XXZ model with anti-periodic twisted boundary conditions. J. Phys. A Math. Theor. **42**, 195008 (2009)
4. G. Niccoli, Antiperiodic spin-1/2 XXZ quantum chains by separation of variables: complete spectrum and form factors. Nucl. Phys. B **870**, 397 (2013)
5. G. Niccoli, Antiperiodic dynamical 6-vertex model I: complete spectrum by SOV, matrix elements of the identity on separate states and connections to the periodic 8-vertex model. J. Phys. A **46**, 075003 (2013)
6. R. Fazio, F.W.J. Hekking, A.A. Odintsov, Josephson current through a Luttinger liquid. Phys. Rev. Lett. **74**, 1843 (1995)
7. C. Winkelholz, R. Fazio, F.W.J. Hekking, G. Schön, Anomalous density of states of a Luttinger liquid in contact with a superconductor. Phys. Rev. Lett. **77**, 3200 (1996)

8. J.-S. Caux, H. Saleur, F. Siano, Josephson current in Luttinger liquid-superconductor junctions. Phys. Rev. Lett. **88**, 106402 (2002)
9. C.N. Yang, Some exact results for the many-body problem in one dimension with repulsive delta-function interaction. Phys. Rev. Lett. **19**, 1312 (1967)
10. C.N. Yang, S matrix for the one-dimensional N-body problem with repulsive or attractive δ-function interaction. Phys. Rev. **168**, 1920 (1968)
11. R.J. Baxter, One-dimensional anisotropic Heisenberg chain, Ann. Phys. **70**, 323 (1972)
12. R.J. Baxter, *Exactly Solved Models in Statistical Mechanics* (Academic Press, London, 1982)
13. J. Cao, W.-L. Yang, K. Shi, Y. Wang, Off-diagonal Bethe Ansatz and exact solution of a topological spin ring. Phys. Rev. Lett. **111**, 137201 (2013)
14. X. Zhang, Y.-Y. Li, J. Cao, W. -L. Yang, K. Shi, Y. Wang, Retrive the Bethe states of quantum integrable models solved via off-diagonal Bethe Ansatz. arXiv:1407.5294
15. J. Cao, W.-L. Yang, K. Shi, Y. Wang, On the complete-spectrum characterization of quantum integrable spin chains via the inhomogeneous $T - Q$ relation. arXiv:1409.5303
16. V.E. Korepin, N.M. Bogoliubov, A.G. Izergin, *Quantum Inverse Scattering Method and Correlation Functions* (Cambridge University Press, Cambridge, 1993)
17. J. Cao, S. Cui, W.-L. Yang, K. Shi, Y. Wang, Spin-$\frac{1}{2}$ XYZ model revisit: general solutions via off-diagonal Bethe Ansatz. Nucl. Phys. B **886**, 185 (2014)
18. R.J. Baxter, 8 vertex model in lattice statistics. Phys. Rev. Lett. **26**, 832 (1971)
19. R.J. Baxter, One-dimensional anisotropic Heisenberg chain. Phys. Rev. Lett. **26**, 834 (1971)
20. R.J. Baxter, Partition function of the eight-vertex lattice model. Ann. Phys. **70**, 193 (1972)
21. Y.-Y. Li, J. Cao, W.-L. Yang, K. Shi, Y. Wang, Thermodynamic limit and surface energy of the XXZ spin chain with arbitrary boundary fields. Nucl. Phys. B **884**, 17 (2014)
22. C.N. Yang, C.P. Yang, Thermodynamics of a one-dimensional system of bosons with repulsive delta-function interaction. J. Math. Phys. **10**, 1115 (1969)
23. M. Takahashi, *Thermodynamics of One-Dimensional Solvable Models* (Cambridge University Press, Cambridge, 1999)

Chapter 5
The Spin-$\frac{1}{2}$ Chains with Arbitrary Boundary Fields

A quantum integrable model with open boundary condition was first solved via the coordinate Bethe Ansatz method by Gaudin [1]. Later, the open Heisenberg chain model with parallel boundary fields was solved by Alcaraz et al. [2]. A significant breakthrough on the open boundary quantum integrable systems was made by Sklyanin [3], who proposed the algebraic Bethe Ansatz method for open boundary quantum integrable models with RE [4]. Interestingly, people found that RE may allow non-diagonal solutions [5–7], and these solutions are generally associated with the existence of unparallel boundary fields, which break the $U(1)$ symmetry of the bulk and make it difficult to use the coordinate Bethe Ansatz method and the algebraic Bethe Ansatz method for the lack of an obvious reference state. After this finding, many efforts were addressed to this unusual problem. Some noteworthy methods are the generalized algebraic Bethe Ansatz [8, 9], the analytic Bethe Ansatz [10–12], the $T-Q$ relation [13, 14], the q-Onsager algebra [15] and the separation of variables method [16–21]. Unfortunately, for a long time, proper BAEs could only be derived for some special cases, until ODBA was used to completely solve the problem [22, 23]. The Bethe states for the *XXX* chain with generic open boundaries were then conjectured in [24] and a systematic method for constructing the Bethe-type eigenstates based on the inhomogeneous $T-Q$ relation was developed in [25, 26].

This chapter elaborates on the systematic applications of ODBA on this model, including construction of the operator product identities, derivation of values of the transfer matrix at some special spectral parameter points, and construction of the inhomogeneous $T-Q$ relations, the BAEs, and the Bethe-type eigenstates. The method for calculating physical quantities in the thermodynamic limit [27] based on the BAEs are also introduced.

Y. Wang et al., *Off-Diagonal Bethe Ansatz for Exactly Solvable Models*,
DOI 10.1007/978-3-662-46756-5_5

5.1 Spectrum of the Open XXX Spin-$\frac{1}{2}$ Chain

5.1.1 The Model Hamiltonian

The spin-$\frac{1}{2}$ chain with arbitrary boundary fields is described by the Hamiltonian

$$H = \sum_{j=1}^{N-1} \left(J_x \sigma_j^x \sigma_{j+1}^x + J_y \sigma_j^y \sigma_{j+1}^y + J_z \sigma_j^z \sigma_{j+1}^z \right) + \mathbf{h}_1 \cdot \sigma_1 + \mathbf{h}_N \cdot \sigma_N, \quad (5.1.1)$$

where J_α ($\alpha = x, y, z$) are the coupling constants, and $\mathbf{h}_1$ and $\mathbf{h}_N$ are the two boundary fields.

Let us first consider the case of $J_x = J_y = J_z = 1$, i.e., the XXX case. Obviously, the bulk possesses a global $SU(2)$-invariance. This allows us to put $\mathbf{h}_1$ along the z direction and $\mathbf{h}_N$ in the $x - z$ plane. Thus we may simplify the Hamiltonian as

$$H = \sum_{j=1}^{N-1} \sigma_j \cdot \sigma_{j+1} + h_1 \sigma_1^z + h_N^x \sigma_N^x + h_N^z \sigma_N^z. \quad (5.1.2)$$

The integrability of the model is associated with the R-matrix (1.5.2) and the reflection rices

$$K^-(u) = \begin{pmatrix} p+u & 0 \\ 0 & p-u \end{pmatrix}, \quad (5.1.3)$$

and

$$K^+(u) = \begin{pmatrix} q+u+\eta & \xi(u+\eta) \\ \xi(u+\eta) & q-u-\eta \end{pmatrix} \equiv \begin{pmatrix} K_{11}^+(u) & K_{12}^+(u) \\ K_{21}^+(u) & K_{22}^+(u) \end{pmatrix}, \quad (5.1.4)$$

which satisfy the following RE and the dual RE respectively

$$\begin{aligned} &R_{1,2}(u-v)K_1^-(u)R_{2,1}(u+v)K_2^-(v) \\ &\quad = K_2^-(v)R_{1,2}(u+v)K_1^-(u)R_{2,1}(u-v), \quad (5.1.5) \\ &R_{1,2}(-u+v)K_1^+(u)R_{2,1}(-u-v-2\eta)K_2^+(v) \\ &\quad = K_2^+(v)R_{1,2}(-u-v-2\eta)K_1^+(u)R_{2,1}(-u+v), \quad (5.1.6) \end{aligned}$$

where p, q, ξ are real numbers to ensure a hermitian Hamiltonian. For the XXX chain, we put $\eta = 1$. In fact, the rational R-matrix possesses the property

$$[R_{1,2}(u), G_1 G_2] = 0, \quad (5.1.7)$$

for arbitrary c-number 2×2 matrix G, which indicates if $K^-(u)$ satisfies RE, $GK^-(u)G^{-1}$ must also satisfy RE.

Let us introduce the one-row monodromy matrices

$$T_0(u) = R_{0,N}(u - \theta_N) \cdots R_{0,1}(u - \theta_1), \tag{5.1.8}$$

$$\hat{T}_0(u) = R_{1,0}(u + \theta_1) \cdots R_{N,0}(u + \theta_N), \tag{5.1.9}$$

and the double-row monodromy matrix $\mathcal{U}_0(u)$

$$\mathcal{U}_0(u) = T_0(u)K_0^-(u)\hat{T}_0(u). \tag{5.1.10}$$

The transfer matrix is given by

$$t(u) = tr_0\{K_0^+(u)\mathcal{U}_0(u)\}. \tag{5.1.11}$$

With the same procedure introduced in Chap. 2, we have $[t(u), t(v)] = 0$. The first order derivative of the logarithm of the transfer matrix $t(u)$ (5.1.11) yields the Hamiltonian (5.1.2)

$$\begin{aligned} H &= \left.\frac{\partial \ln t(u)}{\partial u}\right|_{u=0,\{\theta_j=0\}} - N \\ &= 2\sum_{j=1}^{N-1} P_{j,j+1} + \frac{K_1^{-\prime}(0)}{K_1^-(0)} + 2\frac{K_N^+(0)}{tr_0 K_0^+(0)} - N \\ &= \sum_{j=1}^{N-1} \sigma_j \cdot \sigma_{j+1} + \frac{1}{p}\sigma_1^z + \frac{1}{q}(\sigma_N^z + \xi\sigma_N^x), \end{aligned} \tag{5.1.12}$$

with the parameter correspondences $h_1 = 1/p$, $h_N^x = \xi/q$ and $h_N^z = 1/q$.

5.1.2 Crossing Symmetry of the Transfer Matrix

An important property of the transfer matrix $t(u)$ for open boundary systems is its crossing symmetry which in the present case reads

$$t(u) = t(-u - 1). \tag{5.1.13}$$

After straightforward calculations, we find that the following crossing relations hold:

$$tr_2\left\{P_{1,2}R_{1,2}(-2u-2)[K_2^-(u)]^{t_2}\right\} = -2u\sigma_1^y K_1^-(-u-1)\sigma_1^y, \tag{5.1.14}$$

$$tr_2\left\{P_{1,2}R_{1,2}(2u)[K_2^+(u)]^{t_2}\right\} = 2(u+1)\sigma_1^y K_1^+(-u-1)\sigma_1^y. \tag{5.1.15}$$

These relations together with the duality relation (2.2.34) imply that [14]

$$
\begin{aligned}
t(-u-1) &= tr_0\left\{K_0^+(-u-1)T_0(-u-1)K_0^-(-u-1)\hat{T}_0(-u-1)\right\}\\
&= tr_0\left\{\{K_0^+(-u-1)T_0(-u-1)\}^{t_0}\left\{K_0^-(-u-1)\hat{T}_0(-u-1)\right\}^{t_0}\right\}\\
&= tr_0\left\{T_0^{t_0}(-u-1)[K_0^+(-u-1)]^{t_0}\hat{T}_0^{t_0}(-u-1)[K_0^-(-u-1)]^{t_0}\right\}\\
&= tr_0\left\{\hat{T}_0(u)\left\{\sigma_0^y[K_0^+(-u-1)]^{t_0}\sigma_0^y\right\}T_0(u)\left\{\sigma_0^y[K_0^-(-u-1)]^{t_0}\sigma_0^y\right\}\right\}\\
&= -\frac{1}{4u(u+1)}tr_{0,1,2}\left\{\hat{T}_0(u)P_{0,1}R_{0,1}(2u)K_1^+(u)T_0(u)\right.\\
&\quad\left.\times P_{0,2}R_{0,2}(-2u-2)K_2^-(u)\right\}\\
&= -\frac{1}{4u(u+1)}tr_{0,1,2}\left\{K_1^+(u)T_1(u)P_{0,1}R_{0,1}(2u)\right.\\
&\quad\left.\times P_{0,2}R_{0,2}(-2u-2)K_2^-(u)\hat{T}_1(u)\right\}.
\end{aligned}
\tag{5.1.16}
$$

With the relation

$$
-\frac{1}{4u(u+1)}tr_{0,2}\left\{P_{0,1}R_{0,1}(2u)P_{0,2}R_{0,2}(-2u-2)K_2^-(u)\right\} = K_1^-(u),
\tag{5.1.17}
$$

we arrive at (5.1.13).

5.1.3 Operator Product Identities

Obviously, $t(u)$ is a degree $2N+2$ polynomial of u. In addition, the crossing symmetry (5.1.13) indicates further that $t(u)$ (and therefore its eigenvalue $\Lambda(u)$) is a degree $N+1$ polynomial of $u(u+1)$ with $N+2$ unknown coefficients. Using a procedure similar to that introduced in Chap. 1, we have

$$
\begin{aligned}
t(\theta_j) &= tr_0\left\{K_0^+(\theta_j)R_{0,N}(\theta_j-\theta_N)\cdots R_{0,j+1}(\theta_j-\theta_{j+1})P_{0,j}\right.\\
&\quad\times R_{0,j-1}(\theta_j-\theta_{j-1})\cdots R_{0,1}(\theta_j-\theta_1)K_0^-(\theta_j)R_{1,0}(\theta_1+\theta_j)\cdots\\
&\quad\left.\times R_{j,0}(2\theta_j)R_{j+1,0}(\theta_{j+1}+\theta_j)\cdots R_{N,0}(\theta_N+\theta_j)\right\}\\
&= R_{j,j-1}(\theta_j-\theta_{j-1})\cdots R_{j,1}(\theta_j-\theta_1)K_j^-(\theta_j)R_{1,j}(\theta_1+\theta_j)\cdots\\
&\quad\times R_{j-1,j}(\theta_{j-1}+\theta_j)tr_0\left\{K_0^+(\theta_j)R_{0,N}(\theta_j-\theta_N)\cdots R_{0,j+1}(\theta_j-\theta_{j+1})\right.\\
&\quad\left.\times P_{0,j}R_{j,0}(2\theta_j)R_{j+1,0}(\theta_{j+1}+\theta_j)\cdots R_{N,0}(\theta_N+\theta_j)\right\}.
\end{aligned}
\tag{5.1.18}
$$

From YBE, we obtain

$$\begin{aligned}
&R_{0,j+1}(\theta_j-\theta_{j+1})P_{0,j}R_{j,0}(2\theta_j)R_{j+1,0}(\theta_{j+1}+\theta_j)\\
&\quad=R_{0,j+1}(\theta_j-\theta_{j+1})R_{0,j}(2\theta_j)R_{j+1,j}(\theta_{j+1}+\theta_j)P_{0,j}\\
&\quad=R_{j+1,j}(\theta_{j+1}+\theta_j)R_{0,j}(2\theta_j)R_{0,j+1}(\theta_j-\theta_{j+1})P_{0,j}\\
&\quad=R_{j+1,j}(\theta_{j+1}+\theta_j)P_{0,j}R_{j,0}(2\theta_j)R_{j,j+1}(\theta_j-\theta_{j+1}).
\end{aligned}\tag{5.1.19}$$

This gives rise to

$$\begin{aligned}
t(\theta_j)=\;&R_{j,j-1}(\theta_j-\theta_{j-1})\cdots R_{j,1}(\theta_j-\theta_1)K_j^-(\theta_j)R_{1,j}(\theta_1+\theta_j)\cdots\\
&\times R_{j-1,j}(\theta_{j-1}+\theta_j)R_{j+1,j}(\theta_{j+1}+\theta_j)\cdots R_{N,j}(\theta_N+\theta_j)\\
&\times tr_0\left\{K_0^+(\theta_j)P_{0,j}R_{j,0}(2\theta_j)\right\}R_{j,N}(\theta_j-\theta_N)\cdots R_{j\,j+1}(\theta_j-\theta_{j+1}).
\end{aligned}\tag{5.1.20}$$

The crossing relation (1.5.6) of the R-matrix implies

$$\begin{aligned}
t(\theta_j-1)&=tr_0\{\sigma_0^yK_0^+(\theta_j-1)\sigma_0^yR_{0,N}^{t_0}(-\theta_j+\theta_N)\cdots R_{0,1}^{t_0}(-\theta_j+\theta_1)\\
&\quad\times\sigma_0^yK_0^-(\theta_j-1)\sigma_0^yR_{1,0}^{t_0}(-\theta_1-\theta_j)\cdots R_{0,N}^{t_0}(-\theta_N-\theta_j)\}\\
&=tr_0\{(\sigma_0^yK_0^+(\theta_j-1)\sigma_0^yR_{0,N}^{t_0}(-\theta_j+\theta_N)\cdots R_{0,1}^{t_0}(-\theta_j+\theta_1))^{t_0}\\
&\quad\times(\sigma_0^yK_0^-(\theta_j-1)\sigma_0^yR_{1,0}^{t_0}(-\theta_1-\theta_j)\cdots R_{0,N}^{t_0}(-\theta_N-\theta_j))^{t_0}\}\\
&=tr_0\{(\sigma_0^yK_0^-(\theta_j-1)\sigma_0^y)^{t_0}R_{0,1}(-\theta_j+\theta_1)\cdots R_{0,N}(-\theta_j+\theta_N)\\
&\quad\times(\sigma_0^yK_0^+(\theta_j-1)\sigma_0^y)^{t_0}R_{N,0}(-\theta_N-\theta_j)\cdots R_{1,0}(-\theta_1-\theta_j)\}\\
&=R_{j,j+1}(-\theta_j+\theta_{j+1})\cdots R_{j,N}(-\theta_j+\theta_N)\{\sigma_j^yK_j^+(\theta_j-1)\sigma_j^y\}^{t_j}\\
&\quad\times R_{N,j}(-\theta_N-\theta_j)\cdots R_{j+1,j}(-\theta_{j+1}-\theta_j)R_{j-1,j}(-\theta_{j-1}-\theta_j)\cdots\\
&\quad\times R_{1,j}(-\theta_1-\theta_j)tr_0\left\{(\sigma_0^yK_0^-(\theta_j-1)\sigma_0^y)^{t_0}P_{0,j}R_{j,0}(-2\theta_j)\right\}\\
&\quad\times R_{j,1}(-\theta_j+\theta_1)\cdots R_{j,j-1}(-\theta_j+\theta_{j-1}).
\end{aligned}\tag{5.1.21}$$

With the help of the unitary property (1.5.5) of the R-matrix and the formulae

$$K_j^-(u)tr_0\left\{(\sigma_0^yK_0^-(u-1)\sigma_0^y)^{t_0}P_{0,j}R_{j,0}(-2u)\right\}=-\mathrm{Det}_q\{K^-(u)\},\tag{5.1.22}$$

$$tr_0\left\{K_0^+(u)P_{0,j}R_{j,0}(2u)\right\}\left\{\sigma_j^yK_j^+(u-1)\sigma_j^y\right\}^{t_j}=-\mathrm{Det}_q\left\{K^+(u)\right\},\tag{5.1.23}$$

we find that the transfer matrix satisfies the relations

$$t(\theta_j)t(\theta_j-1)=\frac{\Delta_q(\theta_j)}{(1-2\theta_j)(1+2\theta_j)},\quad j=1,\ldots,N,\tag{5.1.24}$$

with

$$\Delta_q(u) = \mathrm{Det}_q\{T(u)\}\mathrm{Det}_q\{\hat{T}(u)\}\mathrm{Det}_q\{K^-(u)\}\mathrm{Det}_q\{K^+(u)\}, \quad (5.1.25)$$

and

$$\mathrm{Det}_q\{T(u)\} = \prod_{j=1}^{N}(u-\theta_j+1)(u-\theta_j-1), \quad (5.1.26)$$

$$\mathrm{Det}_q\{\hat{T}(u)\} = \prod_{j=1}^{N}(u+\theta_j+1)(u+\theta_j-1), \quad (5.1.27)$$

$$\mathrm{Det}_q\{K^-(u)\} = 2(u-1)(p^2-u^2), \quad (5.1.28)$$

$$\mathrm{Det}_q\{K^+(u)\} = 2(u+1)\left[(1+\xi^2)u^2-q^2\right]. \quad (5.1.29)$$

The above quantum determinants can be easily calculated with the procedure introduced in Sect. 2.4 of Chap. 2.

In addition, by checking the definition of the transfer matrix, it is easily deduced that

$$t(0) = 2pq\prod_{j=1}^{N}(1-\theta_j)(1+\theta_j), \quad (5.1.30)$$

$$t(u) \sim 2u^{2N+2} + \cdots, \quad \text{for } u \to \pm\infty. \quad (5.1.31)$$

The relations (5.1.24)–(5.1.31) together with the crossing symmetry (5.1.13) determine the spectrum of the transfer matrix completely.

5.1.4 The Inhomogeneous T − Q Relation

The properties of the transfer matrix $t(u)$ given by (5.1.13) and (5.1.24)–(5.1.31) imply that the corresponding eigenvalue $\Lambda(u)$, which is a polynomial of u, satisfies the following relations:

$$\text{Crossing symmetry:} \quad \Lambda(-u-1) = \Lambda(u), \quad (5.1.32)$$

$$\text{Initial condition:} \ \Lambda(0) = 2pq\prod_{j=1}^{N}(1-\theta_j)(1+\theta_j) = \Lambda(-1), \quad (5.1.33)$$

$$\text{Asymptotic behavior:} \ \Lambda(u) \sim 2u^{2N+2} + \cdots, \quad u \to \pm\infty, \quad (5.1.34)$$

and

$$\Lambda(\theta_j)\Lambda(\theta_j-1)=\frac{\Delta_q(\theta_j)}{(1-2\theta_j)(1+2\theta_j)} = a(\theta_j)d(\theta_j-1),\quad j=1,\ldots,N, \tag{5.1.35}$$

where

$$a(u)=\frac{2u+2}{2u+1}(u+p)[(1+\xi^2)^{\frac{1}{2}}u+q]\prod_{j=1}^{N}(u+\theta_j+1)(u-\theta_j+1), \tag{5.1.36}$$

$$d(u)=\frac{2u}{2u+1}(u-p+1)[(1+\xi^2)^{\frac{1}{2}}(u+1)-q]\prod_{j=1}^{N}(u+\theta_j)(u-\theta_j) = a(-u-1). \tag{5.1.37}$$

These conditions allow us to construct the following inhomogeneous $T-Q$ relation for each eigenvalue $\Lambda(u)$:

$$\Lambda(u)=a(u)\frac{Q(u-1)}{Q(u)}+d(u)\frac{Q(u+1)}{Q(u)}+2[1-(1+\xi^2)^{\frac{1}{2}}]u(u+1) \times\frac{\prod_{j=1}^{N}(u+\theta_j)(u-\theta_j)(u+\theta_j+1)(u-\theta_j+1)}{Q(u)}. \tag{5.1.38}$$

The function $Q(u)$ is parameterized by N Bethe roots $\{\lambda_j|j=1,\ldots,N\}$ as follows:

$$Q(u)=\prod_{j=1}^{N}(u-\lambda_j)(u+\lambda_j+1). \tag{5.1.39}$$

Such parametrization obviously satisfies the relation (5.1.35) and the asymptotic behavior (5.1.34). To ensure $\Lambda(u)$ to be a polynomial, the residues of $\Lambda(u)$ at the poles λ_j must vanish, i.e., the N Bethe roots must satisfy the BAEs

$$a(\lambda_j)Q(\lambda_j-1)+d(\lambda_j)Q(\lambda_j+1)=-2[1-(1+\xi^2)^{\frac{1}{2}}]\lambda_j(\lambda_j+1) \times\prod_{l=1}^{N}(\lambda_j+\theta_l)(\lambda_j-\theta_l)(\lambda_j+\theta_l+1)(\lambda_j-\theta_l+1),\quad j=1,\ldots,N, \tag{5.1.40}$$

with the selection rules $\lambda_j \neq \lambda_l$ and $\lambda_j \neq -\lambda_l - 1$. Now we can safely take the homogeneous limit $\theta_j \to 0$. In this case, the functional $T - Q$ relation is

$$\begin{aligned}\Lambda(u) &= \frac{2(u+1)^{2N+1}}{2u+1}(u+p)[(1+\xi^2)^{\frac{1}{2}}u+q]\frac{Q(u-1)}{Q(u)} \\ &+\frac{2u^{2N+1}}{2u+1}(u-p+1)[(1+\xi^2)^{\frac{1}{2}}(u+1)-q]\frac{Q(u+1)}{Q(u)} \\ &+2[1-(1+\xi^2)^{\frac{1}{2}}]\frac{[u(u+1)]^{2N+1}}{Q(u)}, \end{aligned} \tag{5.1.41}$$

and the BAEs read

$$\begin{aligned}\left(\frac{\lambda_j+1}{\lambda_j}\right)^{2N+1} &\frac{(\lambda_j+p)[(1+\xi^2)^{\frac{1}{2}}\lambda_j+q]}{(\lambda_j-p+1)[(1+\xi^2)^{\frac{1}{2}}(\lambda_j+1)-q]} = -\frac{Q(\lambda_j+1)}{Q(\lambda_j-1)} \\ &-\frac{[1-(1+\xi^2)^{\frac{1}{2}}](2\lambda_j+1)(\lambda_j+1)^{2N+1}}{(\lambda_j-p+1)[(1+\xi^2)^{\frac{1}{2}}(\lambda_j+1)-q]Q(\lambda_j-1)}, \quad j=1,\ldots,N. \end{aligned} \tag{5.1.42}$$

From the relation (5.1.12) we have the eigenvalue of the Hamiltonian in terms of the Bethe roots given by

$$\begin{aligned}E &= \left.\frac{\partial \ln \Lambda(u)}{\partial u}\right|_{u=0,\{\theta_j=0\}} - N \\ &= \sum_{j=1}^{N}\frac{2}{\lambda_j(\lambda_j+1)} + N - 1 + \frac{1}{p} + \frac{(1+\xi^2)^{\frac{1}{2}}}{q}. \end{aligned} \tag{5.1.43}$$

It has been numerically verified [28] that the BAEs (5.1.42) give a complete set of solutions. Here we list the numerical results for the case of $N = 3$ and 4 in Tables 5.1 and 5.2.

We remark that (5.1.38) is one of the minimal inhomogeneous $T - Q$ relations for the present model given in [22]. For $\xi = 0$, the two boundary fields are parallel. This is the case we studied in Chap. 2. In this case, the inhomogeneous term in the $T - Q$ relation vanishes and part of the N Bethe roots may take the value of infinity. The Q-function is thus reduced to

$$Q(u) = \prod_{j=1}^{M}(u-\lambda_j)(u+\lambda_j+1), \quad M = 0,\ldots,N. \tag{5.1.44}$$

We recover the result obtained via the algebraic Bethe Ansatz method.

Table 5.1 Numerical solutions of (5.1.42) for $N = 3$, $p = -0.6$, $q = -0.3$ and $\xi = 1.2$

λ_1	λ_2	λ_3	E_n	n
$-4.10438 - 0.00000i$	$-3.12345 - 0.00000i$	$-0.50000 - 0.36233i$	-9.66040	1
$-4.46781 - 0.00000i$	$-1.54620 - 1.55846i$	$-1.54620 + 1.55846i$	-5.22656	2
$-4.42536 + 0.00000i$	$-1.73347 + 0.00000i$	$-0.50000 - 1.26655i$	-4.24721	3
$-3.81357 - 0.51692i$	$-3.81357 + 0.51692i$	$-1.61102 + 0.00000i$	-2.49645	4
$-4.42526 - 0.00000i$	$-1.19169 + 0.00000i$	$-0.50000 - 0.87145i$	2.03247	5
$-3.71762 - 0.15797i$	$-3.71762 + 0.15797i$	$-1.19200 + 0.00000i$	4.25829	6
-4.35285	-1.50781	-1.19223	6.60218	7
$-1.43080 - 0.20414i$	$-1.43080 + 0.20414i$	$-1.19144 + 0.00000i$	8.73767	8

E_n is the nth eigenvalue of the Hamiltonian and n denotes the number of the energy levels. The eigenvalues calculated from (5.1.43) are the same as those calculated with exact diagonalization of the Hamiltonian

5.1.5 An Alternative Inhomogeneous $T - Q$ Relation

The operator identities allow us to construct a variety of $T - Q$ relation as we mentioned in Chap. 1. In this section, we introduce another convenient $T - Q$ relation [22]

$$\Lambda(u) = a(u)\frac{Q_1(u-1)}{Q_2(u)} + d(u)\frac{Q_2(u+1)}{Q_1(u)} + 2[(-1)^N - (1+\xi^2)^{\frac{1}{2}}]u(u+1)$$
$$\times \frac{\prod_{j=1}^{N}(u+\theta_j)(u-\theta_j)(u+\theta_j+1)(u-\theta_j+1)}{Q_1(u)Q_2(u)}, \quad (5.1.45)$$

with

$$Q_1(u) = \prod_{j=1}^{N}(u-\lambda_j), \quad Q_2(u) = (-1)^N\prod_{j=1}^{N}(u+\lambda_j+1). \quad (5.1.46)$$

Obviously, this $T - Q$ relation also satisfies equation (5.1.35). The BAEs in the homogeneous limit thus read

$$d(\lambda_j)Q_2(\lambda_j+1)Q_2(\lambda_j) = -2[(-1)^N - (1+\xi^2)^{\frac{1}{2}}][\lambda_j(\lambda_j+1)]^{2N+1}. \quad (5.1.47)$$

The corresponding eigenvalue of the Hamiltonian is

$$E = -\sum_{j=1}^{N}\frac{2}{\lambda_j+1} + N - 1 + \frac{1}{p} + \frac{(1+\xi^2)^{\frac{1}{2}}}{q}. \quad (5.1.48)$$

The numerical results for $N = 3$ and 4 listed in Tables 5.3 and 5.4 indicate that the $T - Q$ relation (5.1.45) and the corresponding BAEs (5.1.47) also give the correct and complete spectrum of the Hamiltonian.

Table 5.2 Numerical solutions of (5.1.42) for $N = 4$, $p = -0.6$, $q = -0.3$ and $\xi = 1.2$

λ_1	λ_2	λ_3	λ_4	E_n	n
$-4.46107 - 0.71343i$	$-4.46107 + 0.71343i$	$-0.50000 - 1.47504i$	$-0.50000 - 0.25996i$	-10.7613	1
$-4.56929 - 1.09496i$	$-4.56929 + 1.09496i$	$-1.88789 - 0.00000i$	$-0.50000 - 0.20948i$	-9.2894	2
$-4.50072 - 0.88253i$	$-4.50072 + 0.88253i$	$-0.04631 - 0.72962i$	$-0.04631 + 0.72962i$	-6.6473	3
$-4.55229 - 1.03934i$	$-4.55229 + 1.03934i$	$-1.66354 - 0.00000i$	$-0.50000 - 0.48843i$	-5.9532	4
$-5.23553 - 0.00000i$	$-2.59403 - 1.82497i$	$-2.59403 + 1.82497i$	$-0.50000 - 2.45317i$	-4.0479	5
$-5.09717 + 0.00000i$	$-1.65621 - 0.00000i$	$-1.17558 - 1.01449i$	$-1.17558 + 1.01449i$	-3.2252	6
$-4.50132 - 0.74750i$	$-4.50132 + 0.74750i$	$-1.60487 + 0.00000i$	$-0.50000 - 1.93378i$	-2.0867	7
$-4.65338 - 1.37662i$	$-4.65338 + 1.37662i$	$-2.83435 - 0.00000i$	$-1.60050 - 0.00000i$	-1.2396	8
$-4.53170 - 0.97733i$	$-4.53170 + 0.97733i$	$-1.19203 - 0.00000i$	$-0.50000 - 0.35150i$	-0.28153	9
$-5.08572 + 0.00000i$	$-1.19200 + 0.00000i$	$-1.04063 - 0.57078i$	$-1.04063 + 0.57078i$	2.5029	10
$-4.47229 - 0.71326i$	$-4.47229 + 0.71326i$	$-1.19205 - 0.00000i$	$-0.50000 + 1.70841i$	4.4645	11
$-5.06505 + 0.00000i$	$-1.51604 - 0.00000i$	$-1.19208 + 0.00000i$	$-0.50000 + 0.83275i$	5.395	12
$-4.61490 - 1.24350i$	$-4.61490 + 1.24350i$	$-2.43548 - 0.00000i$	$-1.19205 + 0.00000i$	5.6159	13
$-4.55981 - 1.07589i$	$-4.55981 + 1.07589i$	$-1.58275 - 0.00000i$	$-1.19206 - 0.00000i$	7.2293	14
$-4.93529 - 0.00000i$	$-1.65630 - 0.08493i$	$-1.65630 + 0.08493i$	$-1.19205 + 0.00000i$	8.5512	15
$-1.57292 - 0.39217i$	$-1.57292 + 0.39217i$	$-1.51678 - 0.00000i$	$-1.19207 + 0.00000i$	9.7732	16

E_n is the nth eigenvalue of the Hamiltonian and n denotes the number of the energy levels. The eigenvalues calculated from (5.1.43) are the same as those calculated with exact diagonalization of the Hamiltonian

Table 5.3 Numerical solutions of (5.1.47) for $N = 3$, $p = -0.6$, $q = -0.3$ and $\xi = 1.2$

λ_1	λ_2	λ_3	E_n	n
$-5.47086 - 0.00000i$	$-0.50127 - 0.36386i$	$-0.50127 + 0.36386i$	-9.66040	1
$-1.10173 - 1.31044i$	$-1.10173 + 1.31044i$	$2.39785 - 0.00000i$	-5.22656	2
$-1.75718 + 0.00000i$	$-0.14959 - 0.98228i$	$-0.14959 + 0.98228i$	-4.24721	3
$-8.50064 + 0.00000i$	$-1.61125 + 0.00000i$	$0.72177 - 0.00000i$	-2.49645	4
$-1.19174 + 0.00000i$	$-0.33280 - 0.55855i$	$-0.33280 + 0.55855i$	2.03247	5
$-6.90555 - 0.00000i$	$-1.19200 + 0.00000i$	$0.23184 + 0.00000i$	4.25829	6
-1.41754	-1.19252	-0.45991	6.60218	7
-1.60035	-1.19205	13.93425	8.73767	8

E_n is the nth eigenvalue of the Hamiltonian and n denotes the number of the energy levels. The eigenvalues calculated from (5.1.48) are the same as those calculated with exact diagonalization of the Hamiltonian

5.2 Bethe States of the Open XXX Spin-$\frac{1}{2}$ Chain

The algebraic Bethe ansatz provides a useful method to construct eigenstates of integrable models. Unfortunately, its application strongly depends on the existence of a proper reference state. For integrable models without such obvious reference state, an elegant method, namely, the separation of variables (SoV) method, was used to derive the eigenstates of these models with inhomogeneity [17] by constructing a simple basis of the Hilbert space. The Bethe states of the open XXX spin chain with generic boundaries were first conjectured in [24] based on the inhomogeneous $T-Q$ relation. A systematic method to retrieve the eigenstates was developed in [25]. This method allows us to construct both the SoV states and the Bethe states, thus solving the homogeneous limit problem of the SoV approach. In this section, we retrieve the eigenstates of the open XXX spin chain with generic boundaries.

5.2.1 Gauge Transformation of the Monodromy Matrices

It is easy to check that $K^+(u)$ can be diagonalized as

$$\begin{aligned}\tilde{K}^+(u) = GK^+(u)G^{-1} &= \begin{pmatrix} q+(1+\xi^2)^{\frac{1}{2}}(u+1) & 0 \\ 0 & q-(1+\xi^2)^{\frac{1}{2}}(u+1) \end{pmatrix} \\ &\equiv \begin{pmatrix} \tilde{K}^+_{11}(u) & 0 \\ 0 & \tilde{K}^+_{22}(u) \end{pmatrix},\end{aligned} \tag{5.2.1}$$

Table 5.4 Numerical solutions of (5.1.47) for $N = 4$, $p = -0.6$, $q = -0.3$ and $\xi = 1.2$

λ_1	λ_2	λ_3	λ_4	E_n	n
$-0.71035 - 1.36767i$	$-0.71035 + 1.36767i$	$-0.50024 - 0.26039i$	$-0.50024 + 0.26039i$	-10.76127	1
$-1.91094 + 0.00000i$	$-0.50021 - 0.20995i$	$-0.50021 + 0.20995i$	1.47382	-9.28939	2
$-0.96346 - 0.70776i$	$-0.96346 + 0.70776i$	$-0.06264 - 0.79468i$	$-0.06264 + 0.79468i$	-6.64726	3
$-1.66631 + 0.00000i$	$-0.49817 - 0.48700i$	$-0.49817 + 0.48700i$	1.04836	-5.95319	4
$-3.89970 - 0.00000i$	$-0.99283 - 1.34713i$	$-0.99283 + 1.34713i$	1.35759	-4.04791	5
$-1.56030 - 0.12746i$	$-1.56030 + 0.12746i$	$-0.39547 - 0.16855i$	$-0.39547 + 0.16855i$	-3.22515	6
$-1.60493 + 0.00000i$	$-0.72696 - 1.77115i$	$-0.72696 + 1.77115i$	0.69601	-2.08666	7
-3.20888	-1.60051	0.69733	3.72062	-1.23956	8
$-1.19203 + 0.00000i$	$-0.49706 - 0.34771i$	$-0.49706 + 0.34771i$	0.38698	-0.28153	9
$-1.71270 - 0.00000i$	$-1.19205 - 0.00000i$	$-0.49072 - 0.19568i$	$-0.49072 + 0.19568i$	2.50287	10
$-1.19205 - 0.00000i$	$-0.72572 - 1.56553i$	$-0.72572 + 1.56553i$	0.21839	4.46454	11
$-1.61832 + 0.00000i$	$-1.19205 - 0.00000i$	$-0.51413 - 0.45569i$	$-0.51413 + 0.45569i$	5.39498	12
-2.59741	-1.19205	0.22830	2.64892	5.61585	13
-1.58150	-1.19206	0.09182	1.17787	7.22925	14
$-1.60145 + 0.00000i$	$-1.19206 - 0.00000i$	$-0.39610 - 1.21382i$	$-0.39610 + 1.21382i$	8.55124	15
-10.43047	-1.60000	-1.19206	5.40224	9.77318	16

E_n is the nth eigenvalue of the Hamiltonian and n denotes the number of the energy levels. The eigenvalues calculated from (5.1.48) are the same as those calculated with exact diagonalization of the Hamiltonian where the matrix G is given by

$$G = \begin{pmatrix} \xi & (1+\xi^2)^{\frac{1}{2}} - 1 \\ \xi & -(1+\xi^2)^{\frac{1}{2}} - 1 \end{pmatrix}. \tag{5.2.2}$$

Accordingly, the gauged $\tilde{K}^-(u)$ matrix reads

$$\begin{aligned}\tilde{K}^-(u) = GK^-(u)G^{-1} &= \begin{pmatrix} p + \frac{1}{(1+\xi^2)^{\frac{1}{2}}}u & \frac{(1+\xi^2)^{\frac{1}{2}}-1}{(1+\xi^2)^{\frac{1}{2}}}u \\ \frac{(1+\xi^2)^{\frac{1}{2}}+1}{(1+\xi^2)^{\frac{1}{2}}}u & p - \frac{1}{(1+\xi^2)^{\frac{1}{2}}}u \end{pmatrix} \\ &\equiv \begin{pmatrix} \tilde{K}^-_{11}(u) & \tilde{K}^-_{12}(u) \\ \tilde{K}^-_{21}(u) & \tilde{K}^-_{22}(u) \end{pmatrix}.\end{aligned} \tag{5.2.3}$$

For convenience, we denote the double-row monodromy matrix as

$$\mathscr{U}(u) = T(u)\,K^-(u)\,\hat{T}(u) = \begin{pmatrix} \mathscr{A}(u) & \mathscr{B}(u) \\ \mathscr{C}(u) & \mathscr{D}(u) \end{pmatrix}, \tag{5.2.4}$$

and its gauged one as

$$\begin{aligned}\tilde{\mathscr{U}}(u) &= G\,T(u)\,K^-(u)\,\hat{T}(u)G^{-1} = GT(u)G^{-1}\,GK^-(u)G^{-1}\,G\hat{T}(u)G^{-1} \\ &\equiv \begin{pmatrix} \tilde{\mathscr{A}}(u) & \tilde{\mathscr{B}}(u) \\ \tilde{\mathscr{C}}(u) & \tilde{\mathscr{D}}(u) \end{pmatrix}.\end{aligned} \tag{5.2.5}$$

The transfer matrix $t(u)$ can be expressed in terms of both double-row monodromy matrices

$$\begin{aligned}t(u) &= K^+_{11}(u)\,\mathscr{A}(u) + K^+_{12}(u)\,\mathscr{C}(u) + K^+_{21}(u)\,\mathscr{B}(u) + K^+_{22}(u)\,\mathscr{D}(u) \\ &= \tilde{K}^+_{11}(u)\,\tilde{\mathscr{A}}(u) + \tilde{K}^+_{22}(u)\,\tilde{\mathscr{D}}(u).\end{aligned} \tag{5.2.6}$$

The explicit expression (5.2.2) of the G-matrix and the relation (5.2.5) between the two double-row monodromy matrices allow us to derive the following relations among their matrix elements:

$$\begin{aligned}\tilde{\mathscr{A}}(u) = \frac{1}{2\xi(1+\xi^2)^{\frac{1}{2}}}\Big\{&\xi(1+(1+\xi^2)^{\frac{1}{2}})\mathscr{A}(u) + \xi^2\mathscr{C}(u) \\ &+ \xi^2\mathscr{B}(u) - \xi(1-(1+\xi^2)^{\frac{1}{2}})\mathscr{D}(u)\Big\},\end{aligned} \tag{5.2.7}$$

$$\tilde{\mathscr{C}}(u) = \frac{1}{2\xi(1+\xi^2)^{\frac{1}{2}}} \left\{ \xi(1+(1+\xi^2)^{\frac{1}{2}})\mathscr{A}(u) - (1+(1+\xi^2)^{\frac{1}{2}})^2\mathscr{C}(u) \right.$$
$$\left. + \xi^2\mathscr{B}(u) - \xi(1+(1+\xi^2)^{\frac{1}{2}})\mathscr{D}(u) \right\}, \tag{5.2.8}$$
$$\tilde{\mathscr{D}}(u) = \frac{1}{2\xi(1+\xi^2)^{\frac{1}{2}}} \left\{ \xi((1+\xi^2)^{\frac{1}{2}} - 1)\mathscr{A}(u) - \xi^2\mathscr{C}(u) \right.$$
$$\left. - \xi^2\mathscr{B}(u) + \xi(1+(1+\xi^2)^{\frac{1}{2}})\mathscr{D}(u) \right\}. \tag{5.2.9}$$

Both the double-row monodromy matrix and its gauged form satisfy the reflection algebra (2.2.5). Due to the invariance (5.1.7) of $R(u)$, the commutation relations among $\tilde{\mathscr{A}}(u)$, $\tilde{\mathscr{B}}(u)$, $\tilde{\mathscr{C}}(u)$ and $\tilde{\mathscr{D}}(u)$ take the same forms as those among $\mathscr{A}(u)$, $\mathscr{B}(u)$, $\mathscr{C}(u)$ and $\mathscr{D}(u)$ (see Sect. 2.2 of Chap. 2):

$$\tilde{\mathscr{C}}(u)\tilde{\mathscr{A}}(v) = \frac{(u+v)(u-v+1)}{(u-v)(u+v+1)}\tilde{\mathscr{A}}(v)\tilde{\mathscr{C}}(u) - \frac{1}{u+v+1}\tilde{\mathscr{D}}(u)\tilde{\mathscr{C}}(v)$$
$$- \frac{u+v}{(u-v)(u+v+1)}\tilde{\mathscr{A}}(u)\tilde{\mathscr{C}}(v), \tag{5.2.10}$$
$$\tilde{\mathscr{D}}(v)\tilde{\mathscr{C}}(u) = \frac{(u+v)(u-v+1)}{(u-v)(u+v+1)}\tilde{\mathscr{C}}(u)\tilde{\mathscr{D}}(v) - \frac{1}{u+v+1}\tilde{\mathscr{C}}(v)\tilde{\mathscr{A}}(u)$$
$$- \frac{u+v}{(u-v)(u+v+1)}\tilde{\mathscr{C}}(v)\tilde{\mathscr{D}}(u), \tag{5.2.11}$$
$$\tilde{\mathscr{A}}(u)\tilde{\mathscr{A}}(v) = \tilde{\mathscr{A}}(v)\tilde{\mathscr{A}}(u) + \frac{1}{u+v+1}\tilde{\mathscr{B}}(v)\tilde{\mathscr{C}}(u)$$
$$- \frac{1}{u+v+1}\tilde{\mathscr{B}}(u)\tilde{\mathscr{C}}(v), \tag{5.2.12}$$
$$\tilde{\mathscr{D}}(u)\tilde{\mathscr{D}}(v) = \tilde{\mathscr{D}}(v)\tilde{\mathscr{D}}(u) + \frac{1}{u+v+1}\tilde{\mathscr{C}}(v)\tilde{\mathscr{B}}(u)$$
$$- \frac{1}{u+v+1}\tilde{\mathscr{C}}(u)\tilde{\mathscr{B}}(v), \tag{5.2.13}$$
$$\tilde{\mathscr{D}}(u)\tilde{\mathscr{A}}(v) = \tilde{\mathscr{A}}(v)\tilde{\mathscr{D}}(u) - \frac{u+v+2}{(u-v)(u+v+1)}\tilde{\mathscr{B}}(u)\tilde{\mathscr{C}}(v)$$
$$+ \frac{u+v+2}{(u-v)(u+v+1)}\tilde{\mathscr{B}}(v)\tilde{\mathscr{C}}(u), \tag{5.2.14}$$
$$[\tilde{\mathscr{C}}(u), \tilde{\mathscr{C}}(v)] = [\tilde{\mathscr{B}}(u), \tilde{\mathscr{B}}(v)] = 0. \tag{5.2.15}$$

5.2.2 SoV Basis

Let us introduce two local states of site n

$$|1\rangle_n = \frac{(1+\xi^2)^{\frac{1}{2}}+1}{2\xi(1+\xi^2)^{\frac{1}{2}}}|\uparrow\rangle_n + \frac{1}{2(1+\xi^2)^{\frac{1}{2}}}|\downarrow\rangle_n, \quad n=1,\ldots,N, \tag{5.2.16}$$

$$|2\rangle_n = \frac{(1+\xi^2)^{\frac{1}{2}}-1}{2\xi(1+\xi^2)^{\frac{1}{2}}}|\uparrow\rangle_n - \frac{1}{2(1+\xi^2)^{\frac{1}{2}}}|\downarrow\rangle_n, \quad n=1,\ldots,N, \tag{5.2.17}$$

and their dual states

$$\begin{aligned} {}_n\langle 1| &= \xi\ {}_n\langle\uparrow| + ((1+\xi^2)^{\frac{1}{2}}-1)\ {}_n\langle\downarrow|, \quad n=1,\ldots,N, \\ {}_n\langle 2| &= \xi\ {}_n\langle\uparrow| - ((1+\xi^2)^{\frac{1}{2}}+1)\ {}_n\langle\downarrow|, \quad n=1,\ldots,N. \end{aligned} \tag{5.2.18}$$

These states satisfy the orthogonal relations

$${}_j\langle a|b\rangle_k = \delta_{a,b}\,\delta_{j,k}, \quad a,b=1,2, \quad j,k=1,\ldots,N.$$

Based on the above local states, we define two global product states

$$|\Omega\rangle = \otimes_{j=1}^{N}|1\rangle_j, \quad \langle\bar{\Omega}| = \otimes_{j=1}^{N}\ {}_j\langle 2|. \tag{5.2.19}$$

For convenience, let us introduce the notation for the gauged one-row monodromy matrix

$$G\,T(u)\,G^{-1} = \begin{pmatrix} \tilde{A}(u) & \tilde{B}(u) \\ \tilde{C}(u) & \tilde{D}(u) \end{pmatrix}. \tag{5.2.20}$$

It is easy to check that

$$\tilde{A}(u)|\Omega\rangle = \tilde{a}(u)|\Omega\rangle, \quad \tilde{D}(u)|\Omega\rangle = \tilde{d}(u)|\Omega\rangle, \tag{5.2.21}$$

$$\langle\bar{\Omega}|\tilde{A}(u) = \tilde{a}(u)\langle\bar{\Omega}|, \quad \langle\bar{\Omega}|\tilde{D}(u) = \tilde{d}(u)\langle\bar{\Omega}|, \tag{5.2.22}$$

$$\tilde{C}(u)|\Omega\rangle = \langle\bar{\Omega}|\tilde{B}(u) = 0, \tag{5.2.23}$$

with

$$\tilde{d}(u) = \tilde{a}(u-1) = \prod_{j=1}^{N}(u-\theta_j). \tag{5.2.24}$$

Adopting a similar procedure to derive (2.2.37), by expanding $\tilde{\mathscr{C}}(u)$ in terms of elements of the gauged one-row monodromy matrices, we can easily deduce

$$\tilde{\mathscr{C}}(u)|\Omega\rangle = (-1)^N \tilde{K}^-_{21}(u)\tilde{d}(-u-1)\tilde{d}(u)|\Omega\rangle, \tag{5.2.25}$$

$$\langle\bar{\Omega}|\tilde{\mathscr{C}}(u) = (-1)^N \tilde{K}^-_{21}(u)\tilde{a}(u)\tilde{a}(-u-1)\langle\bar{\Omega}|. \tag{5.2.26}$$

Noting that $\tilde{\mathscr{C}}(u)$ forms a commuting family, i.e., $[\tilde{\mathscr{C}}(u), \tilde{\mathscr{C}}(v)] = 0$, therefore, its eigenstates do not depend on the spectral parameter and form a basis of the Hilbert space for generic $\{\theta_j\}$. Particularly

$$\tilde{\mathscr{C}}(\theta_j)|\Omega\rangle = \tilde{\mathscr{C}}(-\theta_j-1)|\Omega\rangle = 0, \tag{5.2.27}$$

$$\langle\bar{\Omega}|\tilde{\mathscr{C}}(-\theta_j) = \langle\bar{\Omega}|\tilde{\mathscr{C}}(\theta_j-1) = 0, \tag{5.2.28}$$

allow us to construct the SoV states as

$$|\theta_{p_1}, \ldots, \theta_{p_n}\rangle = \tilde{\mathscr{A}}(\theta_{p_1})\cdots\tilde{\mathscr{A}}(\theta_{p_n})|\Omega\rangle, \tag{5.2.29}$$

$$\langle\theta_{q_1}, \ldots, \theta_{q_n}| = \langle\bar{\Omega}|\tilde{\mathscr{D}}(-\theta_{q_1})\cdots\tilde{\mathscr{D}}(-\theta_{q_n}), \tag{5.2.30}$$

where $q_j, p_j \in \{1, \ldots, N\}$, $p_1 < p_2 < \cdots < p_n$ and $q_1 < q_2 < \cdots < q_n$. Using the relations (5.2.10)–(5.2.14) and the relations (5.2.27) and (5.2.28), we conclude that the above states are exactly the eigenstates of $\tilde{\mathscr{C}}(u)$

$$\tilde{\mathscr{C}}(u)|\theta_{p_1}, \ldots, \theta_{p_n}\rangle = h(u, \{\theta_{p_1}, \ldots, \theta_{p_n}\})|\theta_{p_1}, \ldots, \theta_{p_n}\rangle, \tag{5.2.31}$$

$$\langle\theta_{q_1}, \ldots, \theta_{q_n}|\tilde{\mathscr{C}}(u) = \bar{h}(u, \{\theta_{q_1}, \ldots, \theta_{q_n}\})\langle\theta_{q_1}, \ldots, \theta_{q_n}|, \tag{5.2.32}$$

with the corresponding eigenvalues being

$$\begin{aligned} h(u, \{\theta_{p_1}, \ldots, \theta_{p_n}\}) &= (-1)^N \tilde{K}^-_{21}(u)\tilde{d}(u)\tilde{d}(-u-1) \\ &\quad \times \prod_{j=1}^{n} \frac{(u+\theta_{p_j})(u-\theta_{p_j}+1)}{(u-\theta_{p_j})(u+\theta_{p_j}+1)}, \end{aligned} \tag{5.2.33}$$

$$\begin{aligned} \bar{h}(u, \{\theta_{q_1}, \ldots, \theta_{q_n}\}) &= (-1)^N \tilde{K}^-_{21}(u)\tilde{a}(u)\tilde{a}(-u-1) \\ &\quad \times \prod_{j=1}^{n} \frac{(u-\theta_{q_j})(u+\theta_{q_j}+1)}{(u+\theta_{q_j})(u-\theta_{q_j}+1)}. \end{aligned} \tag{5.2.34}$$

It follows from (5.2.12) that the operators $\tilde{\mathscr{A}}(u)$ with different generic spectral parameters are not mutually commuting. However, the commutation relation (5.2.12) and the relations (5.2.31) and (5.2.33) imply that the state $|\theta_{p_1}, \ldots, \theta_{p_n}\rangle$ does not depend on the order of $\tilde{\mathscr{A}}(\theta_{p_j})$ in the right hand side of (5.2.29). Similarly, the state $\langle\theta_{q_1}, \ldots, \theta_{q_n}|$ is independent of the order of $\tilde{\mathscr{D}}(-\theta_{q_j})$ in the right hand side of (5.2.30).

For generic inhomogeneous parameters $\{\theta_j\}$, (5.2.31)–(5.2.32) imply that the left states and right states satisfy the relations

$$\langle\theta_{q_1},\ldots,\theta_{q_m}|\theta_{p_1},\ldots,\theta_{p_n}\rangle = f_n(\theta_{p_1},\ldots,\theta_{p_n})\delta_{m+n,N}\delta_{\{q\},\{p\}}, \tag{5.2.35}$$

where $\delta_{\{q\},\{p\}}$ is defined as

$$\delta_{\{q\},\{p\}} = \begin{cases} 1 & \text{if } \{q_1,\ldots,q_m,p_1,\ldots,p_n\} = \{1,\ldots,N\}, \\ 0 & \text{otherwise,} \end{cases} \tag{5.2.36}$$

and $f_n(\theta_{p_1},\ldots,\theta_{p_n})$ is given by

$$\begin{aligned} f_n(\theta_{p_1},\ldots,\theta_{p_n}) &= \langle\theta_{p_{n+1}},\ldots,\theta_{p_N}|\theta_{p_1},\ldots,\theta_{p_n}\rangle \\ &= \prod_{j=1}^{n}(-1)^N\tilde{K}_{21}^{-}(\theta_{p_j})\tilde{d}(-\theta_{p_j}-1)\tilde{a}(\theta_{p_j})\prod_{k=n+1}^{N}(-1)^N\tilde{K}_{21}^{-}(-\theta_{p_k}) \\ &\quad\times\tilde{a}(-\theta_{p_k})\tilde{d}(\theta_{p_k}-1)\prod_{j=1}^{n}\prod_{l>j}^{n}\frac{\theta_{p_j}+\theta_{p_l}}{\theta_{p_j}+\theta_{p_l}+1}\prod_{j=n+1}^{N}\prod_{l>j}^{N}\frac{\theta_{p_j}+\theta_{p_l}}{\theta_{p_j}+\theta_{p_l}-1} \\ &\quad\times\prod_{j=1}^{n}\prod_{l=n+1}^{N}\frac{\theta_{p_l}-\theta_{p_j}}{\theta_{p_l}-\theta_{p_j}-1}, \end{aligned} \tag{5.2.37}$$

with the convention $p_1 < \cdots < p_n$ and $p_{n+1} < \cdots < p_N$.

The right states $\{|\theta_{p_1},\ldots,\theta_{p_n}\rangle\}$ given by (5.2.29) (or the left states $\{\langle\theta_{p_1},\ldots,\theta_{p_n}|\}$ given by (5.2.30)) form a right (or left) basis of the Hilbert space. Therefore, any right (or left) state can be decomposed as a unique linear combination of the basis.

5.2.3 The Scalar Product $F_n(\theta_{p_1},\ldots,\theta_{p_n})$

Assume $\langle\Psi|$ to be a common eigenstate of the transfer matrix $t(u)$, namely,

$$\langle\Psi|\,t(u) = \langle\Psi|\,\Lambda(u),$$

where the eigenvalue $\Lambda(u)$ is given by (5.1.38). Following the method used in Sect. 4.5 of Chap. 4, we introduce

$$F_n(\theta_{p_1},\ldots,\theta_{p_n}) = \langle\Psi|\theta_{p_1},\ldots,\theta_{p_n}\rangle \tag{5.2.38}$$

Let us consider the quantity $\langle\Psi|t(\theta_{p_{n+1}})|\theta_{p_1},\ldots,\theta_{p_n}\rangle$. Acting $t(\theta_{p_{n+1}})$ to the left and to the right alternately, we obtain the relation

$$\begin{aligned}
&\Lambda(\theta_{p_{n+1}})F_n(\theta_{p_1},\ldots,\theta_{p_n}) = \langle\Psi|t(\theta_{p_{n+1}})|\theta_{p_1},\ldots,\theta_{p_n}\rangle \\
&= \tilde{K}^+_{11}(\theta_{p_{n+1}})\, F_{n+1}(\theta_{p_1},\ldots,\theta_{p_{n+1}}) \\
&\quad + \tilde{K}^+_{22}(\theta_{p_{n+1}})\,\langle\Psi|\tilde{\mathscr{D}}(\theta_{p_{n+1}})\prod_{j=1}^{n}\tilde{\mathscr{A}}(\theta_{p_j})|\Omega\rangle.
\end{aligned}$$

Using (5.2.14), (5.2.31) and (5.2.33), we have

$$\begin{aligned}
\Lambda(\theta_{p_{n+1}})F_n(\theta_{p_1},\ldots,\theta_{p_n}) &= \tilde{K}^+_{11}(\theta_{p_{n+1}})\, F_{n+1}(\theta_{p_1},\ldots,\theta_{p_{n+1}}) \\
&\quad + \tilde{K}^+_{22}(\theta_{p_{n+1}})\,\langle\Psi|\prod_{j=1}^{n}\tilde{\mathscr{A}}(\theta_{p_j})\tilde{\mathscr{D}}(\theta_{p_{n+1}})|\Omega\rangle. \qquad (5.2.39)
\end{aligned}$$

With a similar procedure to derive (2.2.37) and (5.2.25), we can easily derive the equation

$$\tilde{\mathscr{D}}(\theta_{p_j})|\Omega\rangle = \frac{1}{2\theta_{p_j}+1}\tilde{\mathscr{A}}(\theta_{p_j})|\Omega\rangle, \quad j=1,\ldots,N. \qquad (5.2.40)$$

Substituting (5.2.40) into the relation (5.2.39), we obtain

$$\begin{aligned}
\Lambda(\theta_{p_{n+1}})F_n(\theta_{p_1},\ldots,\theta_{p_n}) &= \frac{(2\theta_{p_{n+1}}+1)\tilde{K}^+_{11}(\theta_{p_{n+1}}) + \tilde{K}^+_{22}(\theta_{p_{n+1}})}{2\theta_{p_{n+1}}+1} \\
&\quad \times F_{n+1}(\theta_{p_1},\ldots,\theta_{p_{n+1}}). \qquad (5.2.41)
\end{aligned}$$

The above recursive relation allows us to determine $\{F_n(\theta_{p_1},\ldots,\theta_{p_n})\}$ as

$$F_n(\theta_{p_1},\ldots,\theta_{p_n}) = \prod_{j=1}^{n}\frac{(2\theta_{p_j}+1)\Lambda(\theta_{p_j})}{(2\theta_{p_j}+1)\tilde{K}^+_{11}(\theta_{p_j}) + \tilde{K}^+_{22}(\theta_{p_j})}F_0,$$

where $F_0 = \langle\Psi|\Omega\rangle$ is an overall scalar factor.

Keeping the explicit expression of the eigenvalue $\Lambda(u)$ given by (5.1.38) in mind, we further rewrite the above expression of $\{F_n(\theta_{p_1},\ldots,\theta_{p_n})\}$ as follows

$$\begin{aligned}
&F_n(\theta_{p_1},\ldots,\theta_{p_n}) = \langle\Psi|\theta_{p_1},\ldots,\theta_{p_n}\rangle \\
&\quad = \prod_{j=1}^{n}(-1)^N(\theta_{p_j}+p)\,\tilde{a}(\theta_{p_j})\tilde{d}(-\theta_{p_j}-1)\frac{Q(\theta_{p_j}-1)}{Q(\theta_{p_j})}F_0, \\
&\quad n=0,\ldots,N. \qquad (5.2.42)
\end{aligned}$$

The eigenstates are thus expressed as

$$\langle\Psi| = \sum_{n=0}^{N}\sum_{\{p_j\}} \frac{F_n(\theta_{p_1},\ldots,\theta_{p_n})}{f_n(\theta_{p_1},\ldots,\theta_{p_n})}\langle\theta_{p_{n+1}},\ldots,\theta_{p_N}|, \tag{5.2.43}$$

with $p_1 < \cdots < p_n$ and $p_{n+1} < \cdots < p_N$.

5.2.4 The Inner Product $\langle 0|\theta_{p_1},\ldots,\theta_{p_n}\rangle$

An important quantity we shall use is $\langle 0|\theta_{p_1},\ldots,\theta_{p_n}\rangle$. From the definition (5.2.29) we have

$$\langle 0|\theta_{p_1},\ldots,\theta_{p_{n+1}}\rangle = \langle 0|\,\tilde{\mathscr{A}}(\theta_{p_{n+1}})\,|\theta_{p_1},\ldots,\theta_{p_n}\rangle.$$

With a similar procedure to derive (2.2.37), we find that the following relations hold:

$$\langle 0|\,\mathscr{A}(u) = (-1)^N K_{11}^-(u)\,\tilde{a}(u)\,\tilde{d}(-u-1)\,\langle 0|, \tag{5.2.44}$$

$$\begin{aligned}\langle 0|\,\mathscr{D}(u) &= (-1)^N \frac{1}{2u+1} K_{11}^-(u)\,\tilde{a}(u)\,\tilde{d}(-u-1)\,\langle 0| \\ &\quad + (-1)^N \frac{(2u+1)K_{22}^-(u) - K_{11}^-(u)}{2u+1}\,\tilde{d}(u)\,\tilde{a}(-u-1)\,\langle 0|,\end{aligned} \tag{5.2.45}$$

$$\langle 0|\,\mathscr{B}(u) = 0. \tag{5.2.46}$$

From (5.2.7)–(5.2.8) we deduce that $\langle 0|\tilde{\mathscr{A}}(\theta_{p_{n+1}})$ can be expressed as a linear combination of $\langle 0|$ and $\langle 0|\tilde{\mathscr{C}}(\theta_{p_{n+1}})$, which induces the recursive relation

$$\begin{aligned}\langle 0|\theta_{p_1},\ldots,\theta_{p_{n+1}}\rangle &= (-1)^N K_{11}^-(\theta_{p_{n+1}})\tilde{a}(\theta_{p_{n+1}})\tilde{d}(-\theta_{p_{n+1}}-1) \\ &\quad \times\langle 0|\theta_{p_1},\ldots,\theta_{p_n}\rangle, \quad n = 0,\ldots,N-1,\end{aligned} \tag{5.2.47}$$

and gives rise to

$$\begin{aligned}\langle 0|\theta_{p_1},\ldots,\theta_{p_n}\rangle &= \prod_{j=1}^{n}(-1)^N(\theta_{p_j}+p)\,\tilde{a}(\theta_{p_j})\tilde{d}(-\theta_{p_j}-1)\langle 0|\Omega\rangle, \\ n &= 0,\ldots,N.\end{aligned} \tag{5.2.48}$$

With the help of (5.2.16) and (5.2.19), it is easy to check that for a generic nonzero ξ the overall constant $\langle 0|\Omega\rangle$ does not vanish.

5.2.5 Bethe States

For each solution of the BAEs (5.1.40), let us introduce the left Bethe state

$$\langle\lambda_1,\ldots,\lambda_N| = \langle 0|\prod_{j=1}^{N}\tilde{\mathscr{C}}(\lambda_j). \tag{5.2.49}$$

The relations (5.2.31), (5.2.33) and (5.2.48) imply that

$$\begin{aligned}\langle\lambda_1,\ldots,\lambda_N|\theta_{p_1},\ldots,\theta_{p_n}\rangle &= \prod_{j=1}^{n}(-1)^N(\theta_{p_j}+p)\tilde{a}(\theta_{p_j})\tilde{d}(-\theta_{p_j}-1)\\ &\times\frac{Q(\theta_{p_j}-1)}{Q(\theta_{p_j})}\prod_{k=1}^{N}(-1)^N\tilde{K}^{-}_{21}(\lambda_k)\tilde{d}(\lambda_k)\tilde{d}(-\lambda_k-1)\langle 0|\Omega\rangle,\\ n&=0,\ldots,N.\end{aligned} \tag{5.2.50}$$

Comparing the above expression with (5.2.42), we conclude that the Bethe state $\langle\lambda_1,\ldots,\lambda_N|$ given by (5.2.49) is an eigenstate (up to an irrelevant normalization factor) of the transfer matrix $t(u)$ with the corresponding eigenvalue (5.1.38), provided that the parameters $\{\lambda_j\}$ satisfy the BAEs (5.1.40).

It follows from their definition that the two "reference" states $|0\rangle$ and $\langle 0|$ are independent of the inhomogeneous parameters $\{\theta_j\}$. It is easy to check that all of the operator $\tilde{\mathscr{C}}(u)$ (also $\tilde{\mathscr{A}}(u)$, $\tilde{\mathscr{B}}(u)$ and $\tilde{\mathscr{D}}(u)$), the $T-Q$ relation (5.1.38) and the associated BAEs (5.1.40) have well-defined homogeneous limits of $\{\theta_j\to 0\}$. This implies that the homogeneous limit of the Bethe state $\langle\lambda_1,\ldots,\lambda_N|$ gives rise to a left eigenstate of the transfer matrix for the homogeneous open XXX spin-$\frac{1}{2}$ chain.

It should be remarked that the common eigenstates of $\tilde{\mathscr{B}}(u)$ also span the Hilbert space and the Bethe states associated with the $T-Q$ relation (5.1.45) and the BAEs (5.1.48) can be constructed with the same procedure as

$$|\lambda_1,\ldots,\lambda_N\rangle = \prod_{j=1}^{N}\tilde{\mathscr{B}}(\lambda_j)|0\rangle, \tag{5.2.51}$$

which is an eigenstate of the transfer matrix $t(u)$ with the corresponding eigenvalue (5.1.38), provided that the parameters $\{\lambda_j\}$ satisfy the BAEs (5.1.40). The corresponding eigenvalue and BAEs are given by the homogeneous limits of (5.1.38) and (5.1.40).

We note that for $\xi=0$, the present method does not work, since the eigenstates of $\mathscr{C}(u)$ and $\mathscr{B}(u)$ can not form SoV basis. In that case, the eigenstates can be obtained via the algebraic Bethe Ansatz method introduced in Chap. 2.

5.3 Spectrum of the Open *XXZ* Spin-$\frac{1}{2}$ Chain

5.3.1 The Model Hamiltonian

For the *XXZ* spin-$\frac{1}{2}$ chain, the corresponding monodromy matrices and transfer matrix are still defined by (5.1.8)–(5.1.11) but with the R-matrix (4.1.3) and the K-matrices

$$\begin{aligned}
K^-(u) &= \begin{pmatrix} K_{11}^-(u) & K_{12}^-(u) \\ K_{21}^-(u) & K_{22}^-(u) \end{pmatrix}, \\
K_{11}^-(u) &= 2\left[\sinh(\alpha_-)\cosh(\beta_-)\cosh(u) + \cosh(\alpha_-)\sinh(\beta_-)\sinh(u)\right], \\
K_{22}^-(u) &= 2\left[\sinh(\alpha_-)\cosh(\beta_-)\cosh(u) - \cosh(\alpha_-)\sinh(\beta_-)\sinh(u)\right], \\
K_{12}^-(u) &= e^{\theta_-}\sinh(2u), \quad K_{21}^-(u) = e^{-\theta_-}\sinh(2u),
\end{aligned} \tag{5.3.1}$$

and

$$K^+(u) = K^-(-u-\eta)\big|_{(\alpha_-,\beta_-,\theta_-)\to(-\alpha_+,-\beta_+,\theta_+)}, \tag{5.3.2}$$

which respectively satisfy RE (5.1.5) and the dual RE (5.1.6) associated with the R-matrix (4.1.3). Here $\alpha_\mp$, $\beta_\mp$, $\theta_\mp$ are the boundary parameters associated with the boundary field terms. The Hamiltonian reads

$$\begin{aligned}
H &= \sinh\eta\,\frac{\partial \ln t(u)}{\partial u}\bigg|_{u=0,\{\theta_j=0\}} - N\cosh\eta - \tanh\eta\sinh\eta \\
&= 2\sinh\eta\sum_{j=1}^{N-1} P_{j,j+1}R'_{j,j+1}(0) + \sinh\eta\frac{tr_0 K_0^{+\prime}(0)}{tr_0 K_0^+(0)} \\
&\quad +2\sinh\eta\frac{tr_0 K_0^+(0)P_{N,0}R'_{N,0}(0)}{tr_0 K_0^+(0)} + \sinh\eta\frac{K_1^{-\prime}(0)}{K_1^-(0)} \\
&\quad -N\cosh\eta - \tanh\eta\sinh\eta \\
&= \sum_{j=1}^{N-1}\left[\sigma_j^x\sigma_{j+1}^x + \sigma_j^y\sigma_{j+1}^y + \cosh\eta\,\sigma_j^z\sigma_{j+1}^z\right] \\
&\quad +\frac{\sinh\eta}{\sinh\alpha_-\cosh\beta_-}(\cosh\alpha_-\sinh\beta_-\sigma_1^z + \cosh\theta_-\sigma_1^x + i\sinh\theta_-\sigma_1^y) \\
&\quad -\frac{\sinh\eta}{\sinh\alpha_+\cosh\beta_+}(\cosh\alpha_+\sinh\beta_+\sigma_N^z - \cosh\theta_+\sigma_N^x - i\sinh\theta_+\sigma_N^y).
\end{aligned} \tag{5.3.3}$$

5.3.2 Functional Relations

With the same procedure introduced in Sect. 5.1, we can deduce the quantum determinant $\Delta_q(u)$ for the present case as

$$\begin{aligned}\Delta_q(u) = &-2^4 \frac{\sinh(2u-2\eta)\sinh(2u+2\eta)}{\sinh^2\eta}\sinh(u+\alpha_-)\sinh(u-\alpha_-)\\ &\times\cosh(u+\beta_-)\cosh(u-\beta_-)\sinh(u+\alpha_+)\sinh(u-\alpha_+)\\ &\times\cosh(u+\beta_+)\cosh(u-\beta_+)\sinh^{-4N}\eta\prod_{l=1}^{N}\sinh(u+\theta_l+\eta)\\ &\times\sinh(u-\theta_l+\eta)\sinh(u+\theta_l-\eta)\sinh(u-\theta_l-\eta). \end{aligned} \tag{5.3.4}$$

The operator product identities are

$$t(\theta_j)t(\theta_j-\eta) = \frac{\Delta_q(\theta_j)\sinh\eta\sinh\eta}{\sinh(\eta-2\theta_j)\sinh(\eta+2\theta_j)}\times \mathrm{id},\quad j=1,\ldots,N. \tag{5.3.5}$$

In addition, the following crossing relation holds:

$$t(-u-\eta) = t(u). \tag{5.3.6}$$

The quasi-periodicity of the R-matrix and K-matrices

$$R_{1,2}(u+i\pi) = -\sigma_1^z R_{1,2}(u)\sigma_1^z = -\sigma_2^z R_{1,2}(u)\sigma_2^z, \tag{5.3.7}$$
$$K^{\pm}(u+i\pi) = -\sigma^z K^{\pm}(u)\sigma^z, \tag{5.3.8}$$

and the following properties of $K^-(u)$

$$K^-(0) = \frac{1}{2}tr[K^-(0)]\times \mathrm{id}, \tag{5.3.9}$$
$$K^-\left(\frac{i\pi}{2}\right) = \frac{1}{2}tr\left[K^-\left(\frac{i\pi}{2}\right)\sigma^z\right]\times\sigma^z, \tag{5.3.10}$$

allow us to derive the equations of $t(u)$

$$t(u+i\pi) = t(u), \tag{5.3.11}$$
$$\begin{aligned} t(0) = &-2^3\sinh\alpha_-\cosh\beta_-\sinh\alpha_+\cosh\beta_+\cosh\eta\\ &\times\prod_{l=1}^{N}\frac{\sinh(\eta-\theta_l)\sinh(\eta+\theta_l)}{\sinh^2\eta}\times \mathrm{id}, \end{aligned} \tag{5.3.12}$$
$$t\left(\frac{i\pi}{2}\right) = -2^3\cosh\alpha_-\sinh\beta_-\cosh\alpha_+\sinh\beta_+\cosh\eta$$

$$\times \prod_{l=1}^{N} \frac{\sinh\left(\frac{i\pi}{2}+\theta_l+\eta\right)\sinh\left(\frac{i\pi}{2}+\theta_l-\eta\right)}{\sinh^2\eta} \times \mathrm{id}, \quad (5.3.13)$$

$$\lim_{u\to\pm\infty} t(u) = -\frac{\cosh(\theta_- - \theta_+)e^{\pm[(2N+4)u+(N+2)\eta]}}{2^{2N+1}\sinh^{2N}\eta} \times \mathrm{id} + \cdots. \quad (5.3.14)$$

The relations (5.3.5)–(5.3.6) and (5.3.11)–(5.3.14) can completely determine the eigenvalue function $\Lambda(u)$. These equations imply the following functional relations of $\Lambda(u)$:

$$\Lambda(\theta_j)\Lambda(\theta_j-\eta) = \frac{\Delta_q(\theta_j)\sinh\eta\sinh\eta}{\sinh(\eta-2\theta_j)\sinh(\eta+2\theta_j)}, \quad j=1,\ldots,N, \quad (5.3.15)$$

$$\Lambda(-u-\eta) = \Lambda(u), \quad \Lambda(u+i\pi) = \Lambda(u), \quad (5.3.16)$$

$$\Lambda(0) = -2^3 \sinh\alpha_- \cosh\beta_- \sinh\alpha_+ \cosh\beta_+ \cosh\eta \times \prod_{l=1}^{N} \frac{\sinh(\eta-\theta_l)\sinh(\eta+\theta_l)}{\sinh^2\eta}, \quad (5.3.17)$$

$$\Lambda\left(\frac{i\pi}{2}\right) = -2^3 \cosh\alpha_- \sinh\beta_- \cosh\alpha_+ \sinh\beta_+ \cosh\eta \times \prod_{l=1}^{N} \frac{\sinh\left(\frac{i\pi}{2}+\theta_l+\eta\right)\sinh\left(\frac{i\pi}{2}+\theta_l-\eta\right)}{\sinh^2\eta}, \quad (5.3.18)$$

$$\lim_{u\to\pm\infty} \Lambda(u) = -\frac{\cosh(\theta_- - \theta_+)e^{\pm[(2N+4)u+(N+2)\eta]}}{2^{2N+1}\sinh^{2N}\eta} + \cdots. \quad (5.3.19)$$

A simple analysis, as we used in the previous sections, leads to the fact that

$\Lambda(u)$, as an entire function of u, is a trigonometric polynomial of degree $2N+4$. (5.3.20)

As we proved for the XXX spin-$\frac{1}{2}$ chain in Chap. 1, we can show that the functional relations (5.3.15)–(5.3.20) completely determine the function $\Lambda(u)$ and imply the inhomogeneous $T-Q$ relation formalism.

5.3.3 The Inhomogeneous $T-Q$ Relation

For convenience, let us introduce the notation

$$\bar{A}(u) = \prod_{l=1}^{N} \frac{\sinh(u-\theta_l+\eta)\sinh(u+\theta_l+\eta)}{\sinh^2\eta}, \quad (5.3.21)$$

$$a(u) = -2^2 \frac{\sinh(2u+2\eta)}{\sinh(2u+\eta)} \sinh(u-\alpha_-)\cosh(u-\beta_-) \times \sinh(u-\alpha_+)\cosh(u-\beta_+)\bar{A}(u), \tag{5.3.22}$$

$$d(u) = a(-u-\eta), \tag{5.3.23}$$

and the Q-functions

$$Q_1(u) = \prod_{j=1}^{N+m} \frac{\sinh(u-\mu_j)}{\sinh\eta}, \tag{5.3.24}$$

$$Q_2(u) = \prod_{j=1}^{N+m} \frac{\sinh(u+\mu_j+\eta)}{\sinh\eta} = Q_1(-u-\eta), \tag{5.3.25}$$

where $m = 0$ for an even N and $m = 1$ for an odd N. The inhomogeneous $T-Q$ relation for the present case reads

$$\Lambda(u) = a(u)\frac{Q_1(u-\eta)}{Q_2(u)} + d(u)\frac{Q_2(u+\eta)}{Q_1(u)} + \frac{\sinh^m u \sinh^m(u+\eta)}{\sinh^{2m}\eta} \times \frac{2\bar{c}\sinh(2u)\sinh(2u+2\eta)}{Q_1(u)Q_2(u)}\bar{A}(u)\bar{A}(-u-\eta). \tag{5.3.26}$$

To match the asymptotic behavior (5.3.19), the parameter $\bar{c}$ must take the value

$$\bar{c} = \cosh\left[(N+1+2m)\eta + \alpha_- + \beta_- + \alpha_+ + \beta_+ + 2\sum_{j=1}^{N+m}\mu_j\right] - \cosh(\theta_- - \theta_+). \tag{5.3.27}$$

If the $N+m$ parameters $\{\mu_j | j = 1, \ldots, N+m\}$ satisfy the BAEs

$$\frac{2\bar{c}\sinh(2\mu_j)\sinh(2\mu_j+2\eta)\,\bar{A}(\mu_j)\bar{A}(-\mu_j-\eta)}{d(\mu_j)Q_2(\mu_j)Q_2(\mu_j+\eta)} = -\frac{\sinh^{2m}\eta}{\sinh^m\mu_j \sinh^m(\mu_j+\eta)}, \tag{5.3.28}$$

the function $\Lambda(u)$ parameterized by the $T-Q$ relation (5.3.26) is a solution of (5.3.15)–(5.3.20). Note that the selection rules

$$\mu_j \neq \mu_l, \quad \text{and} \quad \mu_j \neq -\mu_l - \eta, \tag{5.3.29}$$

must hold to ensure the simplicity of the poles in (5.3.26).

In the homogeneous limit $\theta_j = 0$, the BAEs (5.3.28) are reduced to

$$\frac{\bar{c}\sinh(2\mu_j+\eta)\sinh(2\mu_j+2\eta)\sinh^m\mu_j\sinh^m(\mu_j+\eta)\sinh^{2N}(\mu_j+\eta)}{2\sinh(\mu_j+\alpha_-+\eta)\cosh(\mu_j+\beta_-+\eta)\sinh(\mu_j+\alpha_++\eta)\cosh(\mu_j+\beta_++\eta)}$$
$$=\prod_{l=1}^{N+m}\sinh(\mu_j+\mu_l+\eta)\sinh(\mu_j+\mu_l+2\eta),\quad j=1,\ldots,N+m. \tag{5.3.30}$$

The eigenvalue of the Hamiltonian (5.3.3) in terms of the Bethe roots is given by

$$E = -\sinh\eta\left[\coth\alpha_- + \tanh\beta_- + \coth\alpha_+ + \tanh\beta_+\right]$$
$$-2\sinh\eta\sum_{j=1}^{N+m}\coth(\mu_j+\eta) + (N-1)\cosh\eta. \tag{5.3.31}$$

To check the completeness of the solutions, numerical solution of the BAEs (5.3.30) and exact diagonalization of the Hamiltonian were performed in [23] for small N and a randomly chosen η and the boundary parameters $\alpha_\pm$, $\beta_\pm$ and $\theta_\pm$. Here we list the numerical results for $N = 3$ and 4 in Tables 5.5 and 5.6, respectively. The numerical results indicate that the BAEs (5.3.30) indeed give the complete spectrum of the Hamiltonian.

Now let us seek the $\bar{c} = 0$ solutions of the BAEs. In this case, a constraint among the boundary parameters is needed [9, 10, 29, 30], and a proper reference state may exist, which allows us to apply the conventional Bethe Ansatz methods. For $\bar{c} = 0$, from the BAEs (5.3.28) we find that the Bethe roots $\{\mu_j\}$ might form two types of pairs:

$$(\mu_l, -\mu_l-\eta),\quad (\mu_l, -\mu_l-2\eta), \tag{5.3.32}$$

and the $T - Q$ relation (5.3.26) is reduced to a conventional one

$$\Lambda(u) = a(u)\frac{\bar{Q}(u-\eta)}{\bar{Q}(u)} + d(u)\frac{\bar{Q}(u+\eta)}{\bar{Q}(u)}, \tag{5.3.33}$$

with

$$\bar{Q}(u) = \prod_{j=1}^{M}\frac{\sinh(u-\mu_l)\sinh(u+\mu_l+\eta)}{\sinh^2\eta},$$

and M is the number of the first type of pairs in (5.3.32). The constraint (5.3.27) allows us to fix the integer M by the condition

Table 5.5 Numerical solutions of the BAEs (5.3.30) for $N = 3$, $\eta = 0.5$, $\alpha_+ = 1$, $\alpha_- = 0.8$, $\beta_+ = 0.4$, $\beta_- = 0.3$, $\theta_+ = 0.7i$ and $\theta_- = 0.9i$

μ_1	μ_2	μ_3	μ_4	E_n	n
$-0.5276 - 0.3652i$	$-0.5276 + 0.3652i$	$-0.2481 - 0.1756i$	$-0.2481 + 0.1756i$	-4.8590	1
$-2.9056 + 0.0000i$	$-1.1969 - 0.0000i$	$-0.2500 - 0.1261i$	$-0.2500 + 0.1261i$	-3.5939	2
$-0.6974 - 0.5166i$	$-0.6974 + 0.5166i$	$-0.2826 - 0.4381i$	$-0.2826 + 0.4381i$	-0.1251	3
$-0.9296 - 0.0000i$	$-0.2637 - 0.4333i$	$-0.2637 + 0.4333i$	$0.6919 + 1.5708i$	-0.0479	4
$-0.9741 + 4.7124i$	$-0.7424 - 4.7124i$	$-0.5001 - 0.5664i$	$-0.5001 + 0.5664i$	1.1449	5
$-1.1498 + 0.0000i$	$-0.5212 - 2.5079i$	$-0.5212 - 0.6337i$	$0.0230 + 1.5708i$	1.8855	6
$-1.1060 - 0.1659i$	$-1.1060 + 0.1659i$	$-0.5216 - 1.5708i$	$0.3535 + 0.0000i$	2.5676	7
$-1.5205 + 0.0000i$	$-1.2965 - 0.0000i$	$-1.1030 + 1.5708i$	$0.8003 - 1.5708i$	3.0278	8

E_n is the nth eigenvalue of the Hamiltonian, and n denotes the number of the energy levels. The eigenvalues calculated from (5.3.31) are the same as those calculated with exact diagonalization of the Hamiltonian (reproduced from [23])

Table 5.6 Numerical solutions of the BAEs (5.3.30) for the $N = 4$ case with parameters: $\eta = 0.5$, $\alpha_+ = 1$, $\alpha_- = 0.8$, $\beta_+ = 0.4$, $\beta_- = 0.3$, $\theta_+ = 0.7i$ and $\theta_- = 0.9i$

μ_1	μ_2	μ_3	μ_4	E_n	n
$-0.3330 - 0.4622i$	$-0.3330 + 0.4622i$	$-0.2506 - 0.1242i$	$-0.2506 + 0.1242i$	-6.8670	1
$-1.0988 - 1.5708i$	$-0.6931 + 1.5708i$	$-0.2500 - 0.1095i$	$-0.2500 + 0.1095i$	-4.8468	2
$-1.0171 + 0.0000i$	$-0.2501 - 0.0969i$	$-0.2501 + 0.0969i$	$-0.2457 + 1.5708i$	-3.9266	3
$-2.0689 - 1.5708i$	$-1.1954 - 0.0000i$	$-0.2500 - 0.0945i$	$-0.2500 + 0.0945i$	-3.2170	4
$-1.1075 - 1.5708i$	$-0.6883 + 1.5708i$	$-0.2500 - 0.2762i$	$-0.2500 + 0.2762i$	-1.6077	5
$-1.0558 - 0.0000i$	$-0.2497 - 0.2356i$	$-0.2497 + 0.2356i$	$-0.2458 - 1.5708i$	-1.2212	6
$-2.1748 + 1.5708i$	$-1.2064 + 0.0000i$	$-0.2499 - 0.2179i$	$-0.2499 + 0.2179i$	-0.6645	7
$-0.6387 - 0.5472i$	$-0.6387 + 0.5472i$	$-0.1043 + 0.5711i$	$-0.1043 + 2.5705i$	0.5747	8
$-1.2297 - 1.5708i$	$-0.5885 + 1.5708i$	$-0.2591 + 0.9163i$	$-0.2591 + 2.2253i$	1.5474	9
$-1.1140 + 0.0000i$	$-0.2950 - 0.6757i$	$-0.2950 + 0.6757i$	$0.1594 - 1.5708i$	1.8547	10
$-3.2361 - 1.5708i$	$-1.2319 - 0.0000i$	$-0.2509 - 0.4386i$	$-0.2509 + 0.4386i$	2.0475	11
$-2.9114 - 3.1416i$	$-0.9120 - 1.5708i$	$-0.7921 + 1.5708i$	$1.9542 + 1.5708i$	2.2990	12
$-1.1848 + 0.0000i$	$-1.0107 + 1.5708i$	$-0.7379 - 1.5708i$	$0.7058 - 0.0000i$	2.8030	13
$-1.1547 + 0.0000i$	$-0.9587 - 0.0000i$	$-0.4986 - 1.5708i$	$-0.0459 + 0.0000i$	3.3542	14
$-1.5932 + 0.0000i$	$-1.2867 + 0.0000i$	$-0.4656 + 1.5708i$	$0.1759 - 1.5708i$	3.8081	15
$-1.5191 - 0.0000i$	$-1.2970 - 0.0000i$	$-1.0092 - 1.5708i$	$0.8832 + 1.5708i$	4.0622	16

E_n is the nth eigenvalue of the Hamiltonian and n denotes the number of the energy levels. The eigenvalues calculated from (5.3.31) are the same as those calculated with exact diagonalization of the Hamiltonian (reproduced from [23])

$$(N-1-2M)\eta = \alpha_- + \beta_- + \alpha_+ + \beta_+ \pm (\theta_- - \theta_+) \ \mathrm{mod}(2i\pi). \quad (5.3.34)$$

This constraint condition was originally found in [9, 10].

In fact, another set of $\bar{c}=0$ solutions of the BAEs (5.3.30) also exists, namely,

$$(\mu_l, -\mu_l-\eta), \ \left(\alpha_- - \eta, \alpha_+ - \eta, \beta_- - \eta + \frac{i\pi}{2}, \beta_+ - \eta + \frac{i\pi}{2}\right), \ (\mu_l, -\mu_l - 2\eta).$$

In this case, the constraint condition (5.3.34) can be rewritten with another integer $\bar{M} = N-1-M$ as

$$(2\bar{M}+1-N)\eta = \alpha_- + \beta_- + \alpha_+ + \beta_+ \pm (\theta_- - \theta_+) \ \mathrm{mod}(2i\pi). \quad (5.3.35)$$

The resulting $T-Q$ relation thus reads

$$\Lambda(u) = \bar{a}(u)\frac{\bar{Q}(u-\eta)}{\bar{Q}(u)} + \bar{d}(u)\frac{\bar{Q}(u+\eta)}{\bar{Q}(u)}, \quad (5.3.36)$$

with

$$\bar{a}(u) = -2^2\frac{\sinh(2u+2\eta)}{\sinh(2u+\eta)}\sinh(u+\alpha_-)\cosh(u+\beta_-)$$
$$\times \sinh(u+\alpha_+)\cosh(u+\beta_+)\bar{A}(u), \quad (5.3.37)$$
$$\bar{d}(u) = \bar{a}(-u-\eta), \quad (5.3.38)$$
$$\bar{Q}(u) = \prod_{j=1}^{\bar{M}} \frac{\sinh(u-\mu_l)\sinh(u+\mu_l+\eta)}{\sinh^2\eta}. \quad (5.3.39)$$

The two resulting conventional $T-Q$ relations (5.3.33) and (5.3.36) with the constraints (5.3.34) recover the Bethe Ansatz solutions [9–11, 14]. We remark that in the diagonal boundary limit $\alpha_\pm$ or $\beta_\pm \to \infty$, the $U(1)$-symmetry is recovered and the resulting $T-Q$ relation is reduced to the conventional form with $M=0,\ldots,N$. It should also be remarked that there are in fact several choices for the functions $a(u)$ and $d(u)$, depending on the decomposition of the quantum determinant. Naturally, we can respectively replace the boundary parameters $\alpha_+, \beta_+, \alpha_-$ and β_- by $\varepsilon_1\alpha_+, \varepsilon_2\beta_+, \varepsilon_3\alpha_-$ and $\varepsilon_4\beta_-$ in $a(u)$, $d(u)$ and $\bar{c}$ with $\{\varepsilon_i = \pm 1 | i=1,2,3,4\}$ and $\prod_{i=1}^4 \varepsilon_i = 1$. Accordingly, the degenerate condition (5.3.34) becomes

$$(N-1-2M)\eta = \varepsilon_1\alpha_- + \varepsilon_2\beta_- + \varepsilon_3\alpha_+ + \varepsilon_4\beta_+ \pm (\theta_- - \theta_+) \ \mathrm{mod}(2i\pi), \quad (5.3.40)$$

under which a homogeneous $T-Q$ relation exists and the method developed in [9] is applicable.

5.3.4 An Alternative Inhomogeneous $T - Q$ Relation

As for the XXX case, an alternative $T - Q$ relation similar to equation (5.1.38) also exists:

$$\Lambda(u) = a(u)\frac{Q(u-\eta)}{Q(u)} + d(u)\frac{Q(u+\eta)}{Q(u)} + \frac{2c\sinh(2u)\sinh(2u+2\eta)}{Q(u)}\bar{A}(u)\bar{A}(-u-\eta), \qquad (5.3.41)$$

where functions $\bar{A}(u)$, $a(u)$ and $d(u)$ are given by (5.3.21)–(5.3.23), and the Q-function reads

$$Q(u) = \prod_{j=1}^{N} \frac{\sinh(u-\mu_j)\sinh(u+\mu_j+\eta)}{\sinh^2\eta}. \qquad (5.3.42)$$

To match the asymptotic behavior of (5.3.19), the parameter c must take the value

$$c = \cosh\left[(N+1)\eta + \alpha_- + \beta_- + \alpha_+ + \beta_+\right] - \cosh(\theta_- - \theta_+). \qquad (5.3.43)$$

Obviously, the above $T - Q$ relation satisfies the functional relations (5.3.15)–(5.3.19). The regular property (5.3.20) induces the following BAEs

$$a(\mu_j)Q(\mu_j-\eta) + d(\mu_j)Q(\mu_j+\eta) + 2c\sinh(2\mu_j)\sinh(2\mu_j+2\eta)\bar{A}(\mu_j)\bar{A}(-\mu_j-\eta) = 0. \qquad (5.3.44)$$

The corresponding eigenvalue of the Hamiltonian in terms of the Bethe roots $\{\mu_j|j = 1,\ldots,N\}$ with $\{\theta_j = 0|j = 1,\ldots,N\}$ reads

$$E = -\sinh\eta[\coth\alpha_- + \tanh\beta_- + \coth\alpha_+ + \tanh\beta_+] + 2\sum_{j=1}^{N}\frac{\sinh^2\eta}{\sinh\mu_j\sinh(\mu_j+\eta)} + (N-1)\cosh\eta. \qquad (5.3.45)$$

The numerical results for $N = 3$ and 4 shown in Tables 5.7 and 5.8 imply that the $T - Q$ relation (5.3.41) and the BAEs (5.3.44) indeed give the correct and complete solutions.

Table 5.7 Numerical solutions of the BAEs (5.3.44) for $N = 3$, $\{\theta_j = 0\}$, $\eta = 0.5$, $\alpha_+ = 1$, $\alpha_- = 0.8$, $\beta_+ = 0.4$, $\beta_- = 0.3$, $\theta_+ = 0.7i$ and $\theta_- = 0.9i$

μ_1	μ_2	μ_3	E_n	n
$-1.06915 - 0.24629i$	$-1.06915 + 0.24629i$	$-0.25000 + 0.15380i$	-4.85901	1
$-1.14894 - 0.00000i$	$-0.56599 + 0.00000i$	$-0.06600 + 0.00000i$	-3.59385	2
$-1.16389 - 0.27755i$	$-1.16389 + 0.27755i$	$-0.25000 + 0.65753i$	-0.12505	3
$-1.00194 - 0.02084i$	$-1.00194 + 0.02084i$	$-0.25000 - 0.43639i$	-0.04790	4
$-1.14715 + 0.00000i$	$-1.02769 - 0.48866i$	$-1.02769 + 0.48866i$	1.14491	5
-0.98811	$-0.97869 - 0.37467i$	$-0.97869 + 0.37467i$	1.88549	6
$-0.92025 - 0.00000i$	$-0.88426 - 0.29853i$	$-0.88426 + 0.29853i$	2.56765	7
$-0.87087 + 0.00000i$	$-0.82191 - 0.27455i$	$-0.82191 + 0.27455i$	3.02776	8

E_n is the nth eigenvalue of the Hamiltonian, and n denotes the number of the energy levels. The eigenvalues calculated from (5.3.45) are the same as those calculated with exact diagonalization of the Hamiltonian

5.4 Thermodynamic Limit and Surface Energy

The Bethe root distribution of the ODBA equations is somewhat complicated compared to that of the conventional BAEs. This makes it difficult to use these equations to approach the thermodynamic limit of such models. Nevertheless, a method to address this problem was developed in [27] based on the reduced BAEs. The central point of this approach is that in the thermodynamic limit the two boundaries are decoupled and a fixed M in the reduced BAEs may give the complete set of solutions. We introduce this method by using the open *XXZ* spin chain with generic boundary fields as an example.

5.4.1 Reduced BAEs for Imaginary η

We consider the imaginary η and $\theta_\pm$ case. To ensure the Hamiltonian to be hermitian, we assume $\alpha_\pm$ to be imaginary and $\beta_\pm$ to be real. Let us examine the degenerate points of η in (5.3.34) with $M = N$ and $\beta_\pm = \pm\beta$,

$$\eta_m = -\frac{\alpha_- + \alpha_+ \pm (\theta_- - \theta_+) + 2\pi i m}{N+1}, \tag{5.4.1}$$

where m is an arbitrary integer. For convenience, we introduce the notations $\lambda_j = \mu_j + \frac{\eta}{2}$, $ia_\pm = \alpha_\pm + \frac{\eta}{2}$, $\eta = i\gamma$, with $a_\pm, \gamma \in (0, \pi)$. The reduced BAEs for $\eta = \eta_m$ read

Table 5.8 Numerical solutions of the BAEs (5.3.44) for $N = 4$, $\{\theta_j = 0\}$, $\eta = 0.5$, $\alpha_+ = 1$, $\alpha_- = 0.8$, $\beta_+ = 0.4$, $\beta_- = 0.3$, $\theta_+ = 0.7i$ and $\theta_- = 0.9i$

μ_1	μ_2	μ_3	μ_4	E_n	n
$-1.16661 - 0.25604i$	$-1.16661 + 0.25604i$	$-0.25000 - 0.43682i$	$-0.25000 - 0.12153i$	-6.86698	1
$-1.18142 - 0.00000i$	$-1.03374 - 0.45689i$	$-1.03374 + 0.45689i$	$-0.25000 - 0.10980i$	-4.84682	2
$-1.00736 - 0.00000i$	$-0.96655 - 0.26968i$	$-0.96655 + 0.26968i$	$-0.25000 - 0.09869i$	-3.92660	3
$-1.14233 - 0.00000i$	$-0.78174 - 0.22239i$	$-0.78174 + 0.22239i$	$-0.25000 - 0.11265i$	-3.21698	4
$-1.18285 + 0.00000i$	$-1.04297 - 0.44910i$	$-1.04297 + 0.44910i$	$-0.25000 - 0.27492i$	-1.60767	5
$-1.00070 + 0.00000i$	$-0.98604 - 0.27419i$	$-0.98604 + 0.27419i$	$-0.25000 - 0.23247i$	-1.22121	6
$-1.10287 - 0.00000i$	$-0.79160 - 0.13526i$	$-0.79160 + 0.13526i$	$-0.25000 + 0.17083i$	-0.66450	7
$-1.21105 - 0.25885i$	$-1.21105 + 0.25885i$	$-0.59130 - 0.54567i$	$-0.59130 + 0.54567i$	0.57472	8
$-1.29514 - 0.00000i$	$-1.19594 - 0.48073i$	$-1.19594 + 0.48073i$	$-0.25000 - 0.43007i$	1.54744	9
$-1.09301 - 0.32244i$	$-1.09301 + 0.32244i$	$-1.04507 - 0.00000i$	$-0.25000 - 0.70456i$	1.85466	10
$-1.01712 - 0.00000i$	$-0.96751 - 0.21434i$	$-0.96751 + 0.21434i$	$-0.25000 + 0.53712i$	2.04748	11
$-1.14900 - 0.18919i$	$-1.14900 + 0.18919i$	$-0.98954 - 0.59828i$	$-0.98954 + 0.59828i$	2.29902	12
$-1.06648 - 0.13337i$	$-1.06648 + 0.13337i$	$-0.98391 - 0.53394i$	$-0.98391 + 0.53394i$	2.80302	13
$-0.99018 - 0.12406i$	$-0.99018 + 0.12406i$	$-0.93657 - 0.45093i$	$-0.93657 + 0.45093i$	3.35418	14
$-0.94064 - 0.11534i$	$-0.94064 + 0.11534i$	$-0.87595 - 0.39752i$	$-0.87595 + 0.39752i$	3.80806	15
$-0.91292 - 0.11111i$	$-0.91292 + 0.11111i$	$-0.84077 - 0.37645i$	$-0.84077 + 0.37645i$	4.06217	16

E_n is the nth eigenvalue of the Hamiltonian, and n denotes the number of the energy levels. The eigenvalues calculated from (5.3.45) are the same as those calculated with exact diagonalization of the Hamiltonian

$$\left[\frac{\sinh(\lambda_j - i\frac{\gamma}{2})}{\sinh(\lambda_j + i\frac{\gamma}{2})}\right]^{2N} \frac{\sinh(2\lambda_j - i\gamma)\sinh(\lambda_j + ia_+)}{\sinh(2\lambda_j + i\gamma)\sinh(\lambda_j - ia_+)}$$
$$\times \frac{\sinh(\lambda_j + ia_-)\cosh(\lambda_j + \beta + i\frac{\gamma}{2})\cosh(\lambda_j - \beta + i\frac{\gamma}{2})}{\sinh(\lambda_j - ia_-)\cosh(\lambda_j + \beta - i\frac{\gamma}{2})\cosh(\lambda_j - \beta - i\frac{\gamma}{2})}$$
$$= -\prod_{l=1}^{N} \frac{\sinh(\lambda_j - \lambda_l - i\gamma)\sinh(\lambda_j + \lambda_l - i\gamma)}{\sinh(\lambda_j - \lambda_l + i\gamma)\sinh(\lambda_j + \lambda_l + i\gamma)}, \quad j = 1, \ldots, N. \quad (5.4.2)$$

The corresponding eigenvalue of the Hamiltonian in terms of the Bethe roots is thus

$$E = -\sum_{j=1}^{N} \frac{4\sin^2\gamma}{\cosh(2\lambda_j) - \cos\gamma} - \sin\gamma\left[\cot\left(a_+ - \frac{\gamma}{2}\right)\right.$$
$$\left. + \cot\left(a_- - \frac{\gamma}{2}\right)\right] + (N-1)\cos\gamma. \quad (5.4.3)$$

It is important to note that the reduced BAEs (5.4.2) give a complete set of solutions as verified numerically in [27, 31]. Here we list the numerical solutions of (5.4.2) for $N = 3$ and 4 with randomly chosen boundary parameters in Tables 5.9 and 5.10.

5.4.2 Surface Energy in the Critical Phase

Based on the reduced BAEs (5.4.2), the thermodynamic limit of the present model can be constructed with the conventional methods introduced in Chap. 2. To show the procedure clearly, let us introduce the derivation of the surface energy at the degenerate crossing parameter points given in (5.4.1). From (5.4.3) we know that the

Table 5.9 The numerical solutions of (5.4.2) for $N = 3$ with the parameters $\eta = -i$, $\alpha_+ = 2i$, $\alpha_- = 3i$, $\beta_+ = 1$, $\beta_- = -1$, $\theta_+ = 2i$, $\theta_- = i$ and $m = 0$

λ_1	λ_2	λ_3	E_n	n
$-1.48510 + 0.67075i$	$-1.48510 + 2.47085i$	$0.36994 - 0.00000i$	-9.10664	1
$-0.63430 - 1.57080i$	$-0.38556 + 0.50089i$	$-0.38556 - 0.50089i$	-5.80407	2
$-1.11069 + 1.00247i$	$-1.11069 + 2.13912i$	$-1.10123 - 0.00000i$	-5.30177	3
$-1.68396 - 0.65365i$	$0.59457 + 1.57080i$	$1.68396 - 0.65365i$	-4.08354	4
$-0.51260 - 1.57080i$	$-0.38055 + 0.00000i$	$-0.00000 + 0.64158i$	3.46000	5
$-1.56515 - 0.66501i$	$-0.00000 + 0.64158i$	$1.56515 - 0.66501i$	5.73191	6
$-0.00000 + 0.64159i$	$0.25391 - 1.57080i$	$1.09544 + 0.00000i$	6.81205	7
$-0.94157 + 1.57080i$	$-0.00000 + 0.64159i$	$0.20977 + 1.57080i$	8.29206	8

The symbol n indicates the level number. The eigenvalues E_n calculated from (5.4.3) are the same as those from the exact diagonalization of the Hamiltonian (reproduced from [27])

Table 5.10 The numerical solutions of (5.4.2) for $N = 4$ with the parameters $\gamma = 1, a_+ = 2.5, a_- = 1.5, \beta = 1, \theta_- = 3i, \theta_+ = -5i$ and $m = 0$

λ_1	λ_2	λ_3	λ_4	E_n	n
$-1.66762 - 0.00000i$	$-1.21632 + 1.57080i$	$-0.55018 + 0.00000i$	$-0.21316 + 0.00000i$	−5.93342	1
$-0.92025 - 0.00000i$	$-0.91893 + 0.99518i$	$0.20501 - 3.14159i$	$0.91893 + 0.99518i$	−3.85243	2
$-0.94690 - 2.65582i$	$-0.68831 - 1.57080i$	$-0.19733 - 0.00000i$	$0.94690 + 3.62737i$	−3.58123	3
$-0.46931 - 1.57080i$	$-0.17931 - 0.00000i$	$1.43232 + 1.57080i$	$1.90426 - 0.00000i$	−2.35148	4
$-0.91410 + 0.00000i$	$-0.91160 - 0.99287i$	$-0.91160 + 0.99287i$	$-0.50119 + 0.00000i$	−1.47531	5
$-0.87562 - 3.62850i$	$-0.87562 + 0.48690i$	$-0.57755 - 1.57080i$	$-0.45893 - 0.00000i$	−1.36888	6
$0.39497 - 3.14159i$	$0.39554 + 1.57080i$	$1.40894 + 1.57080i$	$1.87940 + 0.00000i$	−0.43122	7
$-1.77682 + 0.00000i$	$-0.48316 + 0.50058i$	$0.48316 + 0.50058i$	$1.32868 - 1.57080i$	0.08195	8
$-1.40329 + 0.00000i$	$-0.89579 - 2.13792i$	$0.89236 + 0.00000i$	$0.89579 + 1.00367i$	0.98414	9
$-0.55149 + 0.49964i$	$-0.38646 - 1.57080i$	$0.55149 + 0.49964i$	$1.39986 + 0.00000i$	1.15594	10
$-1.80115 - 0.00000i$	$-1.32916 + 1.57080i$	$-0.72884 - 0.00000i$	$-0.28785 + 1.57080i$	1.68692	11
$-0.92875 - 1.57080i$	$-0.90786 + 0.99956i$	$-0.90786 + 2.14204i$	$-0.90693 + 0.00000i$	2.08877	12
$-0.86225 + 1.57080i$	$-0.46738 - 2.64162i$	$0.33674 - 1.57080i$	$0.46738 + 0.49997i$	2.26194	13
$-0.93579 + 0.99864i$	$-0.93536 - 0.00000i$	$0.23087 + 1.57080i$	$0.93579 - 2.14295i$	3.06940	14
$0.22280 + 1.57080i$	$0.92385 - 1.57080i$	$1.13160 - 0.48933i$	$1.13160 + 0.48933i$	3.33186	15
$-1.56849 - 1.57080i$	$-0.75387 + 1.57080i$	$0.19162 + 1.57080i$	$2.05390 + 0.00000i$	4.33306	16

The eigenvalues E_n calculated from (5.4.3) are the same as those from the exact diagonalization of the Hamiltonian (reproduced from [27])

contribution of a real Bethe root λ_j to the energy is negative. Therefore, in the ground state the Bethe roots should occupy the real axis as much as possible to minimize the energy. However, due to the selection rule, the real axis cannot accommodate all the N Bethe roots. So some of the Bethe roots must be located in the complex plane and form some strings as we discussed in Chap. 2.

Let us consider the case of a k-string [32] in the ground state configuration with

$$\lambda_l^s = \lambda^r + i(k+1-2l)\frac{\gamma}{2} + O(e^{-\delta N}), \quad l = 1, \ldots, k, \tag{5.4.4}$$

where λ^r is the real part of the string and δ is a positive number to account for the small finite size corrections. By omitting the exponentially small corrections we can rewrite the BAEs (5.4.2) as

$$\begin{aligned}
&\left[\frac{\sinh\left(\lambda_j - i\frac{\gamma}{2}\right)}{\sinh\left(\lambda_j + i\frac{\gamma}{2}\right)}\right]^{2N} \frac{\sinh(2\lambda_j - i\gamma)}{\sinh(2\lambda_j + i\gamma)} \frac{\sinh(\lambda_j + ia_+)}{\sinh(\lambda_j - ia_+)} \\
&\quad \times \frac{\sinh(\lambda_j + ia_-)}{\sinh(\lambda_j - ia_-)} \frac{\cosh\left(\lambda_j + \beta + i\frac{\gamma}{2}\right)}{\cosh\left(\lambda_j + \beta - i\frac{\gamma}{2}\right)} \frac{\cosh\left(\lambda_j - \beta + i\frac{\gamma}{2}\right)}{\cosh\left(\lambda_j - \beta - i\frac{\gamma}{2}\right)} \\
&= -\prod_{l=1}^{N-k} \frac{\sinh(\lambda_j - \lambda_l - i\gamma)\sinh(\lambda_j + \lambda_l - i\gamma)}{\sinh(\lambda_j - \lambda_l + i\gamma)\sinh(\lambda_j + \lambda_l + i\gamma)} \\
&\quad \times \frac{\sinh\left(\lambda_j + \lambda^r - i(k+1)\frac{\gamma}{2}\right)\sinh\left(\lambda_j + \lambda^r - i(k-1)\frac{\gamma}{2}\right)}{\sinh\left(\lambda_j + \lambda^r + i(k+1)\frac{\gamma}{2}\right)\sinh\left(\lambda_j + \lambda^r + i(k-1)\frac{\gamma}{2}\right)} \\
&\quad \times \frac{\sinh\left(\lambda_j - \lambda^r - i(k+1)\frac{\gamma}{2}\right)\sinh\left(\lambda_j - \lambda^r - i(k-1)\frac{\gamma}{2}\right)}{\sinh\left(\lambda_j - \lambda^r + i(k+1)\frac{\gamma}{2}\right)\sinh\left(\lambda_j - \lambda^r + i(k-1)\frac{\gamma}{2}\right)},
\end{aligned} \tag{5.4.5}$$

with $j = 1, \ldots, N-k$. For $a_\pm \in (\frac{\pi}{2}, \pi)$, the logarithmic version of (5.4.5) reads

$$\begin{aligned}
&\phi_1(\lambda_j) + \frac{1}{2N}\Big[\phi_2(2\lambda_j) - \phi_{2a_+/\gamma}(\lambda_j) - \phi_{2a_-/\gamma}(\lambda_j) + w(\lambda_j + \beta) + w(\lambda_j - \beta) \\
&\quad -\pi - \phi_{k+1}(\lambda_j - \lambda^r) - \phi_{k-1}(\lambda_j - \lambda^r) - \phi_{k+1}(\lambda_j + \lambda^r) - \phi_{k-1}(\lambda_j + \lambda^r)\Big] \\
&= \frac{\pi I_j}{N} + \frac{1}{2N}\sum_{l=1}^{N-k}\left[\phi_2(\lambda_j - \lambda_l) + \phi_2(\lambda_j + \lambda_l)\right],
\end{aligned} \tag{5.4.6}$$

with I_j being an integer and

$$\phi_m(\lambda_j) = -i \ln \frac{\sinh(\lambda_j - i\frac{m\gamma}{2})}{\sinh(\lambda_j + i\frac{m\gamma}{2})}, \tag{5.4.7}$$

$$w(\lambda_j) = -i \ln \frac{\cosh(\lambda_j + i\frac{\gamma}{2})}{\cosh(\lambda_j - i\frac{\gamma}{2})}. \tag{5.4.8}$$

Following the same procedure introduced in Sect. 2.2.2, we define the counting function $Z(\lambda)$ as

$$
\begin{aligned}
Z(\lambda) = \frac{1}{2\pi}\Bigg\{ \phi_1(\lambda) + \frac{1}{2N}\Bigg[\phi_2(2\lambda) - \phi_{2a_+/\gamma}(\lambda) - \phi_{2a_-/\gamma}(\lambda) + w(\lambda+\beta) \\
+ w(\lambda-\beta) - \phi_{k+1}(\lambda-\lambda^r) - \phi_{k-1}(\lambda-\lambda^r) - \phi_{k+1}(\lambda+\lambda^r) \\
- \phi_{k-1}(\lambda+\lambda^r) - \pi - \sum_{l=1}^{N-k} [\phi_2(\lambda-\lambda_l) + \phi_2(\lambda+\lambda_l)] \Bigg]\Bigg\}.
\end{aligned}
\tag{5.4.9}
$$

In the thermodynamic limit $N \to \infty$, the density of the real Bethe roots $\rho(\lambda)$ is

$$
\begin{aligned}
\rho(\lambda) &= \frac{dZ(\lambda)}{d\lambda} - \frac{1}{2N}\delta(\lambda) \\
&= a_1(\lambda) + \frac{1}{2N}[2a_2(2\lambda) - a_{2a_+/\gamma}(\lambda) - a_{2a_-/\gamma}(\lambda) + b(\lambda+\beta) \\
&\quad + b(\lambda-\beta) - a_{k+1}(\lambda-\lambda^r) - a_{k-1}(\lambda-\lambda^r) - a_{k+1}(\lambda+\lambda^r) \\
&\quad - a_{k-1}(\lambda+\lambda^r) - \delta(\lambda)] - \int_{-\infty}^{\infty} a_2(\lambda-\nu)\rho(\nu)d\nu,
\end{aligned}
\tag{5.4.10}
$$

with

$$
a_m(\lambda) = \frac{1}{2\pi}\frac{d\phi_m(\lambda)}{d\lambda} = \frac{1}{\pi}\frac{\sin(m\gamma)}{\cosh(2\lambda) - \cos(m\gamma)}, \tag{5.4.11}
$$

$$
b(\lambda) = \frac{1}{2\pi}\frac{dw(\lambda)}{d\lambda} = \frac{1}{\pi}\frac{\sin\gamma}{\cosh(2\lambda) + \cos\gamma}. \tag{5.4.12}
$$

By using Fourier transformations, we obtain

$$
\tilde{\rho}(\omega) = \tilde{\rho}_0(\omega) + \tilde{\rho}_b(\omega), \tag{5.4.13}
$$

where

$$
\tilde{\rho}_0(\omega) = \frac{\tilde{a}_1(\omega)}{1 + \tilde{a}_2(\omega)}, \tag{5.4.14}
$$

$$
\begin{aligned}
\tilde{\rho}_b(\omega) = \frac{1}{2N[1+\tilde{a}_2(\omega)]}\Big\{ \tilde{a}_2(\frac{\omega}{2}) - \tilde{a}_{2a_+/\gamma}(\omega) - \tilde{a}_{2a_-/\gamma}(\omega) \\
+ 2\cos(\beta\omega)\tilde{b}(\omega) - 2\cos(\lambda^r\omega)\big[\tilde{a}_{k+1}(\omega) + \tilde{a}_{k-1}(\omega)\big] - 1 \Big\},
\end{aligned}
\tag{5.4.15}
$$

$$
\tilde{a}_m(\omega) = \frac{\sinh(\pi\omega/2 - \delta_m\pi\omega)}{\sinh(\pi\omega/2)}, \quad \tilde{b}(\omega) = \frac{\sinh(\gamma\omega/2)}{\sinh(\pi\omega/2)}, \tag{5.4.16}
$$

where $\delta_m \equiv \frac{m\gamma}{2\pi} - \lfloor \frac{m\gamma}{2\pi} \rfloor$ denotes the fractional part of $\frac{m\gamma}{2\pi}$. The total number of the real Bethe roots is thus given by

$$N \int_{-\infty}^{\infty} \rho(\lambda) d\lambda = N - k, \tag{5.4.17}$$

with which we can determine the length of the string

$$k = \frac{N}{2} - \frac{a_+ + a_- + 2\pi(\delta_{k+1} + \delta_{k-1}) - 3\pi}{2(\pi - \gamma)}. \tag{5.4.18}$$

Obviously, both k and the number of real Bethe roots are in the order of $N/2$.

We remark that $\lambda^r \to \infty$ is indeed a solution of the BAEs in the thermodynamic limit $N \to \infty$ and minimizes the energy. In such a sense, the ground state energy in the thermodynamic limit takes the following form

$$\begin{aligned} E &= -4\pi N \sin\gamma \int_{-\infty}^{\infty} a_1(\lambda)\rho(\lambda)d\lambda - \sin\gamma[4\pi a_k(\lambda^r) \\ &\quad + \cot(a_+ - \gamma/2) + \cot(a_- - \gamma/2) - (N-1)\cot\gamma] \\ &= Ne_g + e_b, \end{aligned} \tag{5.4.19}$$

with

$$\begin{aligned} e_g &= -\int_{-\infty}^{\infty} \frac{\sin\gamma \sinh(\pi\omega/2 - \gamma\omega/2)}{\sinh(\pi\omega/2)\cosh(\gamma\omega/2)} d\omega + \cos\gamma, \\ e_b &= e_b^0 + I_1(a_+) + I_1(a_-) + 2I_2(\beta), \end{aligned} \tag{5.4.20}$$

where e_g is the ground state energy density of the periodic chain and e_b is the surface energy induced by the open boundary and the boundary fields with

$$\begin{aligned} e_b^0 &= -\sin\gamma \int_{-\infty}^{\infty} \frac{\tilde{a}_1(\omega)}{1+\tilde{a}_2(\omega)} [\tilde{a}_2(\omega/2) - 1] d\omega - \cos\gamma, \\ I_1(\alpha) &= \sin\gamma \int_{-\infty}^{\infty} \frac{\tilde{a}_1(\omega)}{1+\tilde{a}_2(\omega)} \tilde{a}_{2\alpha/\gamma}(\omega) d\omega - \sin\gamma \cot(\alpha - \gamma/2), \\ I_2(\beta) &= -\sin\gamma \int_{-\infty}^{\infty} \frac{\tilde{a}_1(\omega)}{1+\tilde{a}_2(\omega)} \cos(\beta\omega)\tilde{b}(\omega) d\omega. \end{aligned} \tag{5.4.21}$$

Note that in the above derivations we restrict ourselves to the $\beta_+ = -\beta_-$ case. For arbitrary $\beta_\pm$, the degenerate points of η given by the constraint condition (5.3.34) take complex values and the above derivations are invalid. Obviously, the two boundaries are decoupled completely. This allows us to make the following arguments. The surface energy in the thermodynamic limit $N \to \infty$ must take the form

$$\varepsilon_b = \varepsilon_b^0 + \bar{\varepsilon}_b(\alpha_+, \beta_+, \theta_+) + \bar{\varepsilon}_b(\alpha_-, \beta_-, \theta_-), \tag{5.4.22}$$

where $\bar{\varepsilon}_b(\alpha_+, \beta_+, \theta_+)$ and $\bar{\varepsilon}_b(\alpha_-, \beta_-, \theta_-)$ are the contributions of the left and right boundary fields, respectively. Let us denote further

$$\bar{\varepsilon}_b(\alpha_\pm, \beta_\pm, \theta_\pm) = I_1(a_\pm) + \bar{I}(a_\pm, \beta_\pm, \theta_\pm). \tag{5.4.23}$$

When $\beta_\pm = \pm\beta$, Eq. (5.4.20) gives

$$\bar{I}(a_+, \beta, \theta_+) + \bar{I}(a_-, -\beta, \theta_-) = 2I_2(\beta). \tag{5.4.24}$$

Therefore, $\bar{I}(\alpha_\pm, \beta_\pm, \theta_\pm)$ are independent of $\alpha_\pm$ and $\theta_\pm$. Based on the fact that for $\alpha_- = i\pi/2$ the boundary field is an even function of β_-, we conclude that $\bar{I}(\alpha_-, \beta_-, \theta_-) = I_2(\beta_-)$ must be an even function of β_-. Similarly, we also have $\bar{I}(\alpha_+, \beta_+, \theta_+) = I_2(\beta_+)$. Therefore the surface energy for arbitrary $\beta_\pm$ and $\eta = \eta_m$ should be

$$\varepsilon_b = \varepsilon_b^0 + I_1(a_+) + I_1(a_-) + I_2(\beta_+) + I_2(\beta_-). \tag{5.4.25}$$

In fact, η_m form a dense spectrum in the thermodynamic limit and the above equation is valid for arbitrary imaginary η.

It should be remarked that in the thermodynamic limit the k string does not con tribute to the energy. However, for a finite system, the string should contribute exponentially small corrections. Though we only considered the $a_\pm \in (\pi/2, \pi)$ case, the surface energy is in fact a smooth function as was studied in [33–37] for the diagonal boundary case. The formulae (5.4.20)–(5.4.21) can be naturally extrapolated to the region of $a_\pm \in (0, \pi/2)$. In the thermodynamic limit, the two boundary fields are decoupled completely and the surface energy does not depend on $\theta_\pm$ at all. In such a sense, we can always adjust $\theta_\pm$ to match the constraint condition (5.3.34) for arbitrary η and M. We can easily check that with a proper $M \sim N/2$, the reduced BAEs also give the same surface energy without using the k-string. We note that the surface energy does rely on the directions of the boundary fields. The surface energy of the *XXX* spin chain with arbitrary boundary fields can be obtained by taking the limit $\beta_\pm \to 0$ and $\eta \to 0$, which obviously does not depend on the angles $\theta_\pm$ and coincides exactly with the result of the parallel boundary fields introduced in Chap. 2.

5.4.3 Finite-Size Corrections

Now let us examine the corrections implied by the deviation of η from the degenerate points η_m. Given an arbitrary physical quantity $F(\eta)$

$$F(\eta_m) = Nf_0(\eta_m) + f_1(\eta_m) + \frac{1}{N}f_2(\eta_m) + O(N^{-2}), \tag{5.4.26}$$

let us treat $f_n(\eta)$ $(n = 0, 1, 2)$ as known functions because they can be derived from the reduced BAEs with conventional methods. For a generic $i\eta_{m+1} \geq i\eta \geq i\eta_m$, $F(\eta)$ reads

$$F(\eta) = N\bar{f}_0(\eta) + \bar{f}_1(\eta) + \frac{1}{N}\bar{f}_2(\eta) + O(N^{-2}), \tag{5.4.27}$$

where $\bar{f}_n(\eta)$ are some unknown functions. If both $f_n(\eta)$ and $\bar{f}_n(\eta)$ are smooth functions about η, we have

$$\bar{f}_n(\eta_m) = f_n(\eta_m). \tag{5.4.28}$$

In addition, since $\bar{f}_0(\eta)$ and $f_0(\eta)$ are both boundary-field independent and can be calculated from the corresponding periodic system, they must be identical, i.e., $\bar{f}_0(\eta) \equiv f_0(\eta)$. With the Taylor expansions respectively around η_m and η_{m+1} $(n = 1, 2)$ we readily have

$$\bar{f}_n(\eta) = f_n(\eta) + O(N^{-2}). \tag{5.4.29}$$

The above relation implies that the quantity $F(\eta)$ can be derived from the reduced BAEs up to the order of $O(N^{-2})$.

We remark that in deriving the surface energy, we put the integral limits to infinity. To calculate the finite size corrections (e.g., the Casimir effect or central charge term on the order of $1/N$), a finite cutoff of the integrals should be adopted and the standard Wiener-Hopf methods [38–40] are applicable.

5.4.4 Surface Energy in the Gapped Phase

In this subsection, we calculate the surface energy in the case of real η (antiferromagnetic Ising regime). The surface energy for the constrained boundary parameters was calculated in [41]. Let us first make the following physically reasonable arguments. To ensure a hermitian Hamiltonian (5.3.3), $\alpha_\pm$, $\beta_\pm$ must take values either on the real axis or on the $\frac{i\pi}{2}$ line. Assuming that only one boundary field exists, e.g., $\alpha_+ \to \infty$ and $\beta_+ \to 0$, the contribution of the boundary field to the surface energy should not change when its direction is reversed and rotated along the z-axis. From the definition of the boundary field we conclude that the surface energy is an even function of both α_- and β_- and is irrelevant to the value of θ_-. On the other hand, the two boundary fields decouple completely in the thermodynamic limit $N \to \infty$ and their contributions to the energy are additive, since the correlations are exponentially decayed in this gapped phase. It is indeed the case as verified for both boundary fields along the z axis [35]. Based on the above arguments, we can calculate the surface energy with the constraint condition (5.3.34) due to the fact that for any given left boundary field we can always choose a proper right boundary field to match that constraint condition. For example, to calculate the contributions of α_+ and β_+, a natural choice is to take $\alpha_- \to \infty$ and $\beta_- \to -\infty$ with (5.3.34) fulfilled. In such a limit, the boundary field associated with α_- and β_- is fixed.

Without losing generality, we consider the odd N case. Taking $2M = N-1$, and $\alpha_+ + \alpha_- + \beta_+ + \beta_- + \theta_+ - \theta_- = 0$, the constraint condition (5.3.34) is fulfilled. For convenience, we put $\mu_j = i\frac{\lambda_j}{2} - \frac{\eta}{2}$, $a_\pm = \alpha_\pm + \frac{\eta}{2}$ and $b_\pm = \beta_\pm + \frac{\eta}{2}$ with $\eta > 0$, $\lambda_j \in (-\pi, \pi]$, $a_\pm, b_+ > 0$ and $b_- < 0$ with $a_- + a_+ + b_- + b_+ = 2\eta$. The reduced BAEs are

$$\left[\frac{\sin(\frac{\lambda_j}{2} - i\frac{\eta}{2})}{\sin(\frac{\lambda_j}{2} + i\frac{\eta}{2})}\right]^{2N} \frac{\sin(\frac{\lambda_j}{2} + ia_-)}{\sin(\frac{\lambda_j}{2} - ia_-)} \frac{\sin(\frac{\lambda_j}{2} + ia_+)}{\sin(\frac{\lambda_j}{2} - ia_+)} \frac{\cos(\frac{\lambda_j}{2} + ib_-)}{\cos(\frac{\lambda_j}{2} - ib_-)} \frac{\cos(\frac{\lambda_j}{2} + ib_+)}{\cos(\frac{\lambda_j}{2} - ib_+)}$$
$$= -\frac{\sin(\lambda_j + i\eta)}{\sin(\lambda_j - i\eta)} \prod_{l=1}^{M} \frac{\sin(\frac{\lambda_j}{2} - \frac{\lambda_l}{2} - i\eta)\sin(\frac{\lambda_j}{2} + \frac{\lambda_l}{2} - i\eta)}{\sin(\frac{\lambda_j}{2} - \frac{\lambda_l}{2} + i\eta)\sin(\frac{\lambda_j}{2} + \frac{\lambda_l}{2} + i\eta)}, \; j = 1, \ldots, M. \tag{5.4.30}$$

The corresponding eigenvalue reads

$$E = \sum_{j=1}^{M} \frac{4\sinh^2\eta}{\cos\lambda_j - \cosh\eta} + N\cosh\eta + E_0, \tag{5.4.31}$$

with

$$E_0 = -\sinh\eta(\coth\alpha_- + \tanh\beta_- + \coth\alpha_+ + \tanh\beta_+) - \cosh\eta. \tag{5.4.32}$$

Obviously, each real λ_j contributes negative energy. Therefore, in the ground state, the Bethe roots should fill the real axis in the range of $(-\pi, \pi]$ as long as it is possible. Let us consider the case of all λ_j, $a_\pm$ and $b_\pm$ being real. Taking the logarithm of (5.4.30) gives

$$2N\phi_1(\lambda_j) + \phi_2(2\lambda_j) - \phi_{2a_-/\eta}(\lambda_j) - \phi_{2a_+/\eta}(\lambda_j) + w_+(\lambda_j) + w_-(\lambda_j) + \pi$$
$$= 2\pi I_j + \sum_{l=1}^{M}\Big[\phi_2(\lambda_j - \lambda_l) + \phi_2(\lambda_j + \lambda_l)\Big], \quad j = 1, \ldots, M, \tag{5.4.33}$$

where I_j is an integer and

$$\phi_m(\lambda_j) = -i\ln\frac{\sin(\frac{\lambda_j}{2} - i\frac{m\eta}{2})}{\sin(\frac{\lambda_j}{2} + i\frac{m\eta}{2})}, \quad w_\pm(\lambda_j) = -i\ln\frac{\cos(\frac{\lambda_j}{2} + ib_\pm)}{\cos(\frac{\lambda_j}{2} - ib_\pm)}. \tag{5.4.34}$$

Define the counting function $Z(\lambda)$ as

$$Z(\lambda) = \frac{1}{2\pi}\left\{\phi_1(\lambda) + \frac{1}{2N}\Big[\phi_2(2\lambda) - \phi_{2a_-/\eta}(\lambda) - \phi_{2a_+/\eta}(\lambda) + w_+(\lambda) + w_-(\lambda) + \pi - \sum_{l=1}^{M}(\phi_2(\lambda-\lambda_l) + \phi_2(\lambda+\lambda_l))\Big]\right\}. \quad (5.4.35)$$

In the thermodynamic limit $N \to \infty$, the density of the roots $\rho(\lambda)$ satisfies

$$\begin{aligned}\rho(\lambda) &= \frac{dZ(\lambda)}{d\lambda} - \frac{1}{2N}\delta(\lambda) - \frac{1}{2N}\delta(\lambda-\pi)\\ &= g_1(\lambda) + \frac{1}{2N}\Big[2q(\lambda) - g_{2a_-/\eta}(\lambda) - g_{2a_+/\eta}(\lambda) + h_+(\lambda) + h_-(\lambda)\\ &\quad -\delta(\lambda) - \delta(\lambda-\pi)\Big] - \int_{-\pi}^{\pi} g_2(\lambda-\nu)\rho(\nu)d\nu, \end{aligned} \quad (5.4.36)$$

with

$$g_m(\lambda) = \frac{1}{2\pi}\frac{d\phi_m(\lambda)}{d\lambda} = \frac{1}{2\pi}\frac{\sinh(m\eta)}{\cosh(m\eta) - \cos\lambda}, \quad (5.4.37)$$

$$h_\pm(\lambda) = \frac{1}{2\pi}\frac{dw_\pm(\lambda)}{d\lambda} = -\frac{1}{2\pi}\frac{\sinh(2b_\pm)}{\cos\lambda + \cosh(2b_\pm)}, \quad (5.4.38)$$

$$q(\lambda) \equiv g_2(2\lambda). \quad (5.4.39)$$

With Fourier transformation defined in Sect. 3.1 of Chap. 3, we have

$$\tilde{g}_m(\omega) = e^{-m\eta|\omega|}, \quad \tilde{q}(\omega) = \frac{e^{-\eta|\omega|}}{2}(1 + (-1)^\omega), \quad (5.4.40)$$

$$\tilde{h}_\pm(\omega) = \mp(-1)^\omega e^{-2|b_\pm\omega|}. \quad (5.4.41)$$

Accordingly, we get

$$\tilde{\rho}(\omega) = \tilde{\rho}_0(\omega) + \tilde{\rho}_b^0(\omega) + \tilde{\rho}_{a_+}(\omega) + \tilde{\rho}_{a_-}(\omega) + \tilde{\rho}_{b_+}(\omega) + \tilde{\rho}_{b_-}(\omega), \quad (5.4.42)$$

where

$$\tilde{\rho}_0(\omega) = \frac{\tilde{g}_1(\omega)}{1 + \tilde{g}_2(\omega)} = \frac{1}{2\cosh(\eta\omega)}, \quad (5.4.43)$$

$$\tilde{\rho}_b^0(\omega) = \frac{1}{2N(1 + \tilde{g}_2(\omega))}\Big[2\tilde{q}(\omega) - (-1)^\omega - 1\Big], \quad (5.4.44)$$

$$\tilde{\rho}_{a_\pm}(\omega) = -\frac{\tilde{g}_{2a_\pm/\eta}(\omega)}{2N(1 + \tilde{g}_2(\omega))}, \quad \tilde{\rho}_{b_\pm}(\omega) = \frac{\tilde{h}_\pm(\omega)}{2N(1 + \tilde{g}_2(\omega))}. \quad (5.4.45)$$

Note that the following relation holds

$$N \int_{-\pi}^{\pi} \rho(\lambda) d\lambda = \frac{N-1}{2} = M. \tag{5.4.46}$$

The energy associated with this root distribution can be expressed as

$$E = Ne_0 + e_b^0 + e_{a_+} + e_{a_-} + e_{b_+} + e_{b_-}, \tag{5.4.47}$$

where

$$e_0 = -2\sinh\eta \sum_{\omega=-\infty}^{\infty} \frac{e^{-\eta|\omega|}}{\cosh(\eta\omega)} + \cosh\eta, \tag{5.4.48}$$

$$e_b^0 = -\cosh\eta - \sum_{\omega=-\infty}^{\infty} \frac{\sinh\eta}{\cosh(\eta\omega)} \Big[2\tilde{q}(\omega) - (-1)^{\omega} - 1 \Big], \tag{5.4.49}$$

represent the energy density of the bulk and the energy of the free boundary respectively, and

$$e_{a_\pm} = -\sinh\eta \coth\alpha_\pm + \sinh\eta \sum_{\omega=-\infty}^{\infty} \frac{\tilde{g}_{2a_\pm/\eta}(\omega)}{\cosh(\eta\omega)}, \tag{5.4.50}$$

$$e_{b_\pm} = -\sinh\eta \tanh\beta_\pm - \sinh\eta \sum_{\omega=-\infty}^{\infty} \frac{\tilde{h}_\pm(\omega)}{\cosh(\eta\omega)}. \tag{5.4.51}$$

We remark that the Bethe root distribution considered above may not correspond to the true ground state but the ground state energy can be extracted from it with possible boundary excitations considered. It is obvious that $\lambda_j = 2ia_+$ $(\alpha_+ > 0)$ is a solution of the reduced BAEs (5.4.30) for $N \to \infty$. For $0 < \alpha_+ < \frac{\eta}{2}$, the boundary mode must be accompanied by a bulk hole (say, at λ^h) to ensure the condition $2M = N - 1$. This boundary mode induces a deviation of the density by

$$\begin{aligned} \delta\tilde{\rho}(\omega) = &-\frac{1}{2N(1+\tilde{g}_2(\omega))}[\tilde{g}_{2-2a_+/\eta}(\omega) + \tilde{g}_{2a_+/\eta+2}(\omega) \\ &+2\cos(\lambda^h\omega)], \end{aligned} \tag{5.4.52}$$

and the excitation energy

$$\begin{aligned} \delta e_{a_+} - &\frac{4\sinh^2\eta}{\cosh(2\alpha_+ + \eta) - \cosh\eta} + 2\sinh\eta \sum_{\omega=-\infty}^{\infty} \frac{e^{-2\eta|\omega|}\cosh(2a_+\omega)}{\cosh(\eta\omega)} \\ &+2\sinh\eta \sum_{\omega=-\infty}^{\infty} \frac{e^{-i\omega\lambda^h}}{\cosh(\omega\eta)} > 0. \end{aligned} \tag{5.4.53}$$

For $\alpha_+ > \frac{\eta}{2}$, in addition to the imaginary mode, two holes (say, at λ_1^h and λ_2^h respectively) must exist in the bulk to ensure the condition $2M = N - 1$. The corresponding excitation energy can be calculated as

$$\delta e_{a_+} = 2\sinh\eta \sum_{\omega=-\infty}^{\infty} \frac{e^{-i\omega\lambda_1^h} + e^{-i\omega\lambda_2^h}}{\cosh(\omega\eta)} > 0. \tag{5.4.54}$$

Therefore, (5.4.50) gives the lowest energy of the boundary parameters $\alpha_\pm > 0$.

Similarly, $\lambda_j = \pi + 2ib_+$ is also a solution of (5.4.30) for $N \to \infty$. This boundary mode (also accompanied by a hole at λ^h) contributes excitation energy for $0 < \beta_+ < \frac{\eta}{2}$

$$\delta e_{b_+} = -\frac{4\sinh^2\eta}{\cosh(2\beta_+ + \eta) + \cosh\eta} + 2\sinh\eta \sum_{\omega=-\infty}^{\infty} \frac{(-1)^\omega e^{-2\eta|\omega|}\cosh(2b_+\omega)}{\cosh(\eta\omega)}$$
$$+2\sinh\eta \sum_{\omega=-\infty}^{\infty} \frac{e^{-i\omega\lambda^h}}{\cosh(\omega\eta)} > 0, \tag{5.4.55}$$

and for $\beta_+ > \frac{\eta}{2}$, the excitation energy of the imaginary mode (associated with two holes at λ_1^h and λ_2^h respectively) reads

$$\delta e_{b_+} = 2\sinh\eta \sum_{\omega=-\infty}^{\infty} \frac{e^{-i\omega\lambda_1^h} + e^{-i\omega\lambda_2^h}}{\cosh(\omega\eta)} > 0, \tag{5.4.56}$$

indicating that the lowest energy of the boundary parameter $\beta_+ > 0$ is given by e_{b_+}.

For convenience, let us define the following functions

$$H_1(x) = -\sinh\eta\coth x + \sinh\eta \sum_{\omega=-\infty}^{\infty} \frac{e^{-|(2x+\eta)\omega|}}{\cosh(\eta\omega)}, \tag{5.4.57}$$

$$H_2(x) = -\sinh\eta\tanh x + \sinh\eta \sum_{\omega=-\infty}^{\infty} \frac{(-1)^\omega e^{-|(2x+\eta)\omega|}}{\cosh(\eta\omega)}. \tag{5.4.58}$$

Based on the above discussions and the parameter duality $\beta_\pm + \frac{i\pi}{2} \to \alpha_\pm$, we conclude that the surface energy reads

$$e_b = e_b^0 + e_{\alpha_+} + e_{\beta_+} + e_{\alpha_-} + e_{\beta_-}, \tag{5.4.59}$$

with $e_{\alpha_\pm} = H_1(|\alpha_\pm|)$ for real $\alpha_\pm$ and $e_{\alpha_\pm} = H_2(|Re(\alpha_\pm)|)$ for $Im(\alpha_\pm) = \frac{i\pi}{2}$, while $e_{\beta_\pm} = H_2(|\beta_\pm|)$ for real $\beta_\pm$, and $e_{\beta_\pm} = H_1(|Re(\beta_\pm)|)$ for $Im(\beta_\pm) = \frac{i\pi}{2}$.

5.5 Bethe States of the Open *XXZ* Spin Chain

The Bethe statesof the open *XXZ* spin chain are more complicated than those of the *XXX* spin chain. The first nontrivial result about the Bethe states of the open *XXZ* chain with non-diagonal boundary fields (under the constraint condition (5.3.34)) was obtained in [9], where a local gauge transformation for the *XXZ* chain, an analogue of Baxter's vertex-face correspondence of the eight-vertex model, was used. In this section, we show that such local gauge transformation combining with the inhomogeneous $T-Q$ relation introduced in Sect. 5.3, allow us to construct the Bethe-type eigenstates of the open *XXZ* spin chain with generic boundary fields in a rather compact form [26]. The key points are: (1) with one set of gauge transformation, $K^+(u)$ can be diagonalized and the double-row transfer matrix can be expressed as a linear combination of the diagonal elements of the gauged monodromy matrix, while the off-diagonal elements of the gauged monodromy matrix can be treated as the generators of the Bethe states; (2) with another set of local gauge transformation $K^-(u)$ can be trigonalized, which allows us to construct a proper reference state; (3) applying N generators to the reference state, we readily obtain the Bethe states provided that the entries of the generators are the Bethe roots satisfying the BAEs; (4) such states can be proven to be eigenstates of the transfer matrix by employing Sklyanin's SoV basis.

5.5.1 Local Gauge Matrices

Let us introduce the column matrices

$$X_m(u)=\begin{pmatrix} e^{-[u+(\alpha+m)\eta]} \\ 1 \end{pmatrix},\qquad Y_m(u)=\begin{pmatrix} e^{-[u+(\alpha-m)\eta]} \\ 1 \end{pmatrix},\tag{5.5.1}$$

where α and m are two arbitrary complex parameters. Accordingly, we define the following gauge matrices:

$$\bar{G}_m(u)=\big(X_m(u),\, Y_m(u)\big),\quad \bar{G}_m^{-1}(u)=\begin{pmatrix} \bar{Y}_m(u) \\ \bar{X}_m(u) \end{pmatrix},\tag{5.5.2}$$

$$\tilde{G}_m(u)=\big(X_{m+1}(u),\, Y_{m-1}(u)\big),\quad G_m^{-1}(u)=\begin{pmatrix} \tilde{Y}_{m-1}(u) \\ \tilde{X}_{m+1}(u) \end{pmatrix},\tag{5.5.3}$$

$$\hat{G}_m(u)=\big(\hat{X}_{m-1}(u),\, \hat{Y}_{m+1}(u)\big),\quad \hat{G}_m^{-1}(u)=\begin{pmatrix} \bar{Y}_{m+1}(u) \\ \bar{X}_{m-1}(u) \end{pmatrix},\tag{5.5.4}$$

where

$$\bar{X}_m(u) = \frac{e^{u+\alpha\eta}}{2\sinh(m\eta)}\left(1,\ -e^{-[u+(\alpha+m)\eta]}\right), \tag{5.5.5}$$

$$\bar{Y}_m(u) = \frac{e^{u+\alpha\eta}}{2\sinh(m\eta)}\left(-1,\ e^{-[u+(\alpha-m)\eta]}\right), \tag{5.5.6}$$

$$\tilde{X}_m(u) = \frac{e^{\eta}\sinh(m\eta)}{\sinh(m-1)\eta}\bar{X}_m(u), \quad \tilde{Y}_m(u) = \frac{e^{\eta}\sinh(m\eta)}{\sinh(m+1)\eta}\bar{Y}_m(u), \tag{5.5.7}$$

$$\hat{X}_m(u) = \frac{e^{-\eta}\sinh(m+2)\eta}{\sinh(m+1)\eta}X_m(u), \tag{5.5.8}$$

$$\hat{Y}_m(u) = \frac{e^{-\eta}\sinh(m-2)\eta}{\sinh(m-1)\eta}Y_m(u). \tag{5.5.9}$$

By simple computations we can obtain the useful orthonormal relations:

$$\begin{aligned}
&\bar{Y}_m(u)X_m(u) = 1, \qquad \bar{Y}_m(u)Y_m(u) = 0,\\
&\bar{X}_m(u)X_m(u) = 0, \qquad \bar{X}_m(u)Y_m(u) = 1,\\
&X_m(u)\bar{Y}_m(u) + Y_m(u)\bar{X}_m(u) = \begin{pmatrix}1&0\\0&1\end{pmatrix},
\end{aligned} \tag{5.5.10}$$

$$\begin{aligned}
&\tilde{Y}_{m-1}(u)X_{m+1}(u) = 1, \qquad \tilde{Y}_{m-1}(u)Y_{m-1}(u) = 0,\\
&\tilde{X}_{m+1}(u)X_{m+1}(u) = 0, \qquad \tilde{X}_{m+1}(u)Y_{m-1}(u) = 1,\\
&X_{m+1}(u)\tilde{Y}_{m-1}(u) + Y_{m-1}(u)\tilde{X}_{m+1}(u) = \begin{pmatrix}1&0\\0&1\end{pmatrix},
\end{aligned} \tag{5.5.11}$$

$$\begin{aligned}
&\bar{Y}_{m+1}(u)\hat{X}_{m-1}(u) = 1, \qquad \bar{Y}_{m+1}(u)\hat{Y}_{m+1}(u) = 0,\\
&\bar{X}_{m-1}(u)\hat{X}_{m-1}(u) = 0, \qquad \bar{X}_{m-1}(u)\hat{Y}_{m+1}(u) = 1,\\
&\hat{X}_{m-1}(u)\bar{Y}_{m+1}(u) + \hat{Y}_{m+1}(u)\bar{X}_{m-1}(u) = \begin{pmatrix}1&0\\0&1\end{pmatrix}.
\end{aligned} \tag{5.5.12}$$

Moreover, we can easily check that the following commutation relations hold:

$$R_{1,2}(u_1-u_2)X^1_{m+2}(u_1)X^2_{m+1}(u_2) = \frac{\sinh(u_1-u_2+\eta)}{\sinh\eta}X^2_{m+2}(u_2)X^1_{m+1}(u_1), \tag{5.5.13}$$

$$\begin{aligned}
R_{1,2}(u_1-u_2)X^1_m(u_1)Y^2_{m-1}(u_2) &= \frac{\sinh(u_1-u_2)\sinh(m-1)\eta}{\sinh\eta\sinh(m\eta)}Y^2_m(u_2)X^1_{m+1}(u_1)\\
&+\frac{\sinh(m\eta+u_1-u_2)}{\sinh(m\eta)}X^2_m(u_2)Y^1_{m-1}(u_1),
\end{aligned} \tag{5.5.14}$$

$$\begin{aligned}
R_{1,2}(u_1-u_2)Y^1_m(u_1)X^2_{m+1}(u_2) &= \frac{\sinh(u_1-u_2)\sinh(m+1)\eta}{\sinh\eta\sinh(m\eta)}X^2_m(u_2)Y^1_{m-1}(u_1)\\
&+\frac{\sinh(m\eta-u_1+u_2)}{\sinh(m\eta)}Y^2_m(u_2)X^1_{m+1}(u_1),
\end{aligned} \tag{5.5.15}$$

$$R_{1,2}(u_1-u_2)Y^1_{m-2}(u_1)Y^2_{m-1}(u_2) = \frac{\sinh(u_1-u_2+\eta)}{\sinh\eta}Y^2_{m-2}(u_2)Y^1_{m-1}(u_1), \tag{5.5.16}$$

$$R_{1,2}(u_1-u_2)\hat{X}^2_{m-1}(u_2)\hat{X}^1_m(u_1) = \frac{\sinh(u_1-u_2+\eta)}{\sinh\eta}\hat{X}^2_m(u_2)\hat{X}^1_{m-1}(u_1), \tag{5.5.17}$$

$$\begin{aligned}R_{1,2}(u_1-u_2)\hat{X}^2_{m-1}(u_2)\hat{Y}^1_{m+2}(u_1) &= \frac{\sinh(u_1-u_2)\sinh(m+1)\eta}{\sinh\eta\sinh(m\eta)}\hat{X}^2_{m-2}(u_2)\hat{Y}^1_{m+1}(u_1)\\ &+\frac{\sinh(m\eta-u_1+u_2)}{\sinh(m\eta)}\hat{Y}^2_{m+2}(u_2)\hat{X}^1_{m-1}(u_1),\end{aligned} \tag{5.5.18}$$

$$\begin{aligned}R_{1,2}(u_1-u_2)\hat{Y}^2_{m+1}(u_2)\hat{X}^1_{m-2}(u_1) &= \frac{\sinh(u_1-u_2)\sinh(m-1)\eta}{\sinh\eta\sinh(m\eta)}\hat{Y}^2_{m+2}(u_2)\hat{X}^1_{m-1}(u_1)\\ &+\frac{\sinh(m\eta+u_1-u_2)}{\sinh(m\eta)}\hat{X}^2_{m-2}(u_2)\hat{Y}^1_{m+1}(u_1),\end{aligned} \tag{5.5.19}$$

$$R_{1,2}(u_1-u_2)\hat{Y}^2_{m+1}(u_2)\hat{Y}^1_m(u_1) = \frac{\sinh(u_1-u_2+\eta)}{\sinh\eta}\hat{Y}^2_m(u_2)\hat{Y}^1_{m+1}(u_1), \tag{5.5.20}$$

$$\bar{X}^1_{m-1}(u_1)\bar{X}^2_{m-2}(u_2)R_{1,2}(u_1-u_2) = \frac{\sinh(u_1-u_2+\eta)}{\sinh\eta}\bar{X}^2_{m-1}(u_2)\bar{X}^1_{m-2}(u_1), \tag{5.5.21}$$

$$\begin{aligned}\bar{X}^1_{m-1}(u_1)\bar{Y}^2_m(u_2)R_{1,2}(u_1-u_2) &= \frac{\sinh(u_1-u_2)\sinh(m+1)\eta}{\sinh\eta\sinh(m\eta)}\bar{Y}^2_{m+1}(u_2)\bar{X}^1_m(u_1)\\ &+\frac{\sinh(m\eta+u_1-u_2)}{\sinh(m\eta)}\bar{X}^2_{m-1}(u_2)\bar{Y}^1_m(u_1),\end{aligned} \tag{5.5.22}$$

$$\begin{aligned}\bar{Y}^1_{m+1}(u_1)\bar{X}^2_m(u_2)R_{1,2}(u_1-u_2) &= \frac{\sinh(u_1-u_2)\sinh(m-1)\eta}{\sinh\eta\sinh(m\eta)}\bar{X}^2_{m-1}(u_2)\bar{Y}^1_m(u_1)\\ &+\frac{\sinh(m\eta-u_1+u_2)}{\sinh(m\eta)}\bar{Y}^2_{m+1}(u_2)\bar{X}^1_m(u_1),\end{aligned} \tag{5.5.23}$$

$$\bar{Y}^1_{m+1}(u_1)\bar{Y}^2_{m+2}(u_2)R_{1,2}(u_1-u_2) = \frac{\sinh(u_1-u_2+\eta)}{\sinh\eta}\bar{Y}^2_{m+1}(u_2)\bar{Y}^1_{m+2}(u_1), \tag{5.5.24}$$

$$\tilde{X}^1_{m+1}(u_1)\tilde{X}^2_m(u_2)R_{1,2}(u_1-u_2) = \frac{\sinh(u_1-u_2+\eta)}{\sinh\eta}\tilde{X}^2_{m+1}(u_2)\tilde{X}^1_m(u_1), \tag{5.5.25}$$

$$\begin{aligned}\tilde{X}^1_{m+1}(u_1)\tilde{Y}^2_{m-2}(u_2)R_{1,2}(u_1-u_2) &= \frac{\sinh(u_1-u_2)\sinh(m+1)\eta}{\sinh\eta\sinh(m\eta)}\tilde{Y}^2_{m-1}(u_2)\tilde{X}^1_{m+2}(u_1)\\ &+\frac{\sinh(m\eta+u_1-u_2)}{\sinh(m\eta)}\tilde{X}^2_{m+1}(u_2)\tilde{Y}^1_{m-2}(u_1),\end{aligned} \tag{5.5.26}$$

$$\begin{aligned}\tilde{Y}^1_{m-1}(u_1)\tilde{X}^2_{m+2}(u_2)R_{1,2}(u_1-u_2) &= \frac{\sinh(u_1-u_2)\sinh(m-1)\eta}{\sinh\eta\sinh(m\eta)}\tilde{X}^2_{m+1}(u_2)\tilde{Y}^1_{m-2}(u_1)\\ &+\frac{\sinh(m\eta-u_1+u_2)}{\sinh(m\eta)}\tilde{Y}^2_{m-1}(u_2)\tilde{X}^1_{m+2}(u_1),\end{aligned} \tag{5.5.27}$$

$$\tilde{Y}^1_{m-1}(u_1)\tilde{Y}^2_m(u_2)R_{1,2}(u_1-u_2) = \frac{\sinh(u_1-u_2+\eta)}{\sinh\eta}\tilde{Y}^2_{m-1}(u_2)\tilde{Y}^1_m(u_1), \tag{5.5.28}$$

$$\bar{X}^2_m(u_2)R_{1,2}(u_1-u_2)X^1_m(u_1) = \frac{\sinh(u_1-u_2)\sinh(m-1)\eta}{\sinh\eta\sinh(m\eta)}\bar{X}^2_{m-1}(u_2)X^1_{m+1}(u_1), \tag{5.5.29}$$

$$\begin{aligned}\bar{X}^2_m(u_2)R_{1,2}(u_1-u_2)Y^1_m(u_1) &= \frac{\sinh(u_1-u_2+\eta)}{\sinh\eta}\bar{X}^2_{m+1}(u_2)Y^1_{m+1}(u_1)\\ &+\frac{\sinh(m\eta-u_1+u_2)}{\sinh(m\eta)}\bar{Y}^2_{m+1}(u_2)X^1_{m+1}(u_1),\end{aligned} \tag{5.5.30}$$

$$\begin{aligned}\bar{Y}^2_m(u_2)R_{1,2}(u_1-u_2)X^1_m(u_1) &= \frac{\sinh(u_1-u_2+\eta)}{\sinh\eta}\bar{Y}^2_{m-1}(u_2)X^1_{m-1}(u_1)\\ &+\frac{\sinh(m\eta+u_1-u_2)}{\sinh(m\eta)}\bar{X}^2_{m-1}(u_2)Y^1_{m-1}(u_1),\end{aligned} \tag{5.5.31}$$

$$\bar{Y}^2_m(u_2)R_{1,2}(u_1-u_2)Y^1_m(u_1)=\frac{\sinh(u_1-u_2)\sinh(m+1)\eta}{\sinh\eta\sinh(m\eta)}\bar{Y}^2_{m+1}(u_2)Y^1_{m-1}(u_1), \tag{5.5.32}$$

$$\tilde{X}^1_{m+1}(u_1)R_{1,2}(u_1-u_2)X^2_{m+1}(u_2)=\frac{\sinh(u_1-u_2)\sinh(m+1)\eta}{\sinh\eta\sinh(m\eta)}X^2_m(u_2)\tilde{X}^1_{m+2}(u_1), \tag{5.5.33}$$

$$\begin{aligned}\tilde{X}^1_{m+1}(u_1)R_{1,2}(u_1-u_2)Y^2_{m-1}(u_2)&=\frac{\sinh(u_1-u_2+\eta)}{\sinh\eta}Y^2_{m-2}(u_2)\tilde{X}^1_m(u_1)\\&+\frac{\sinh(m\eta+u_1-u_2)}{\sinh(m\eta)}X^2_m(u_2)\tilde{Y}^1_{m-2}(u_1),\end{aligned} \tag{5.5.34}$$

$$\begin{aligned}\tilde{Y}^1_{m-1}(u_1)R_{1,2}(u_1-u_2)X^2_{m+1}(u_2)&=\frac{\sinh(u_1-u_2+\eta)}{\sinh\eta}X^2_{m+2}(u_2)\tilde{Y}^1_m(u_1)\\&+\frac{\sinh(m\eta-u_1+u_2)}{\sinh(m\eta)}Y^2_m(u_2)\tilde{X}^1_{m+2}(u_1),\end{aligned} \tag{5.5.35}$$

$$\tilde{Y}^1_{m-1}(u_1)R_{1,2}(u_1-u_2)Y^2_{m-1}(u_2)=\frac{\sinh(u_1-u_2)\sinh(m-1)\eta}{\sinh\eta\sinh(m\eta)}Y^2_m(u_2)\tilde{Y}^1_{m-2}(u_1), \tag{5.5.36}$$

$$\bar{X}^1_{m-1}(u_1)R_{1,2}(u_1-u_2)\hat{X}^2_{m-1}(u_2)=\frac{\sinh(u_1-u_2)\sinh(m+1)\eta}{\sinh\eta\sinh(m\eta)}\hat{X}^2_{m-2}(u_2)\bar{X}^1_m(u_1), \tag{5.5.37}$$

$$\begin{aligned}\bar{X}^1_{m-1}(u_1)R_{1,2}(u_1-u_2)\hat{Y}^2_{m+1}(u_2)&=\frac{\sinh(u_1-u_2+\eta)}{\sinh\eta}\hat{Y}^2_m(u_2)\bar{X}^1_{m-2}(u_1)\\&+\frac{\sinh(m\eta+u_1-u_2)}{\sinh(m\eta)}\hat{X}^2_{m-2}(u_2)\bar{Y}^1_m(u_1),\end{aligned} \tag{5.5.38}$$

$$\begin{aligned}\bar{Y}^1_{m+1}(u_1)R_{1,2}(u_1-u_2)\hat{X}^2_{m-1}(u_2)&=\frac{\sinh(u_1-u_2+\eta)}{\sinh\eta}\hat{X}^2_m(u_2)\bar{Y}^1_{m+2}(u_1)\\&+\frac{\sinh(m\eta-u_1+u_2)}{\sinh(m\eta)}\hat{Y}^2_{m+2}(u_2)\bar{X}^1_m(u_1),\end{aligned} \tag{5.5.39}$$

$$\bar{Y}^1_{m+1}(u_1)R_{1,2}(u_1-u_2)\hat{Y}^2_{m+1}(u_2)=\frac{\sinh(u_1-u_2)\sinh(m-1)\eta}{\sinh\eta\sinh(m\eta)}\hat{Y}^2_{m+2}(u_2)\bar{Y}^1_m(u_1). \tag{5.5.40}$$

Here the R-matrix is given by (4.1.3) and the superscripts $1, 2$ indicate the space indices.

5.5.2 Commutation Relations

Using the orthogonal relations (5.5.10) and (5.5.12), the transfer matrix $t(u)$ can be expressed as

$$\begin{aligned}t(u)&=\mathrm{tr}\{K^+(m|u)\mathscr{U}(m|u)\}\\&=K^+_{11}(m|u)\mathscr{A}_m(u)+K^+_{12}(m|u)\mathscr{C}_m(u)\\&\quad+K^+_{21}(m|u)\mathscr{B}_m(u)+K^+_{22}(m|u)\mathscr{D}_m(u),\end{aligned} \tag{5.5.41}$$

where

$$
\begin{aligned}
K^{+}(m|u) &= \begin{pmatrix} K^{+}_{11}(m|u) & K^{+}_{12}(m|u) \\ K^{+}_{21}(m|u) & K^{+}_{22}(m|u) \end{pmatrix} \\
&= \begin{pmatrix} \bar{Y}_m(-u)K^{+}(u)X_m(u) & \bar{Y}_m(-u)K^{+}(u)Y_{m-2}(u) \\ \bar{X}_m(-u)K^{+}(u)X_{m+2}(u) & \bar{X}_m(-u)K^{+}(u)Y_m(u) \end{pmatrix}, \qquad (5.5.42) \\
\mathscr{U}(m|u) &= \begin{pmatrix} \mathscr{A}_m(u) & \mathscr{B}_m(u) \\ \mathscr{C}_m(u) & \mathscr{D}_m(u) \end{pmatrix} \\
&= \begin{pmatrix} \tilde{Y}_{m-2}(u)\mathscr{U}(u)X_m(-u) & \tilde{Y}_m(u)\mathscr{U}(u)Y_m(-u) \\ \tilde{X}_m(u)\mathscr{U}(u)X_m(-u) & \tilde{X}_{m+2}(u)\mathscr{U}(u)Y_m(-u) \end{pmatrix}. \qquad (5.5.43)
\end{aligned}
$$

The double-row monodromy matrix $\mathscr{U}(u)$ is defined as usual

$$
\mathscr{U}(u) = T(u)K^{-}(u)\hat{T}(u) = \begin{pmatrix} \mathscr{A}(u) & \mathscr{B}(u) \\ \mathscr{C}(u) & \mathscr{D}(u) \end{pmatrix}, \qquad (5.5.44)
$$

which satisfies RE

$$
R_{1,2}(u-v)\mathscr{U}_1(u)R_{2,1}(u+v)\mathscr{U}_2(v) = \mathscr{U}_2(v)R_{1,2}(u+v)\mathscr{U}_1(u)R_{2,1}(u-v). \qquad (5.5.45)
$$

The commutation relations among $\mathscr{A}_m(u)$, $\mathscr{B}_m(u)$, $\mathscr{C}_m(u)$ and $\mathscr{D}_m(u)$ can be derived from (5.5.45) and the commutation relations (5.5.13)–(5.5.40). Multiplying (5.5.45) with $\tilde{X}^1_{m+1}(u_1)\tilde{X}^2_m(u_2)$ from the left and $X^1_{m+1}(-u_1)X^2_{m+2}(-u_2)$ from the right, we have

$$
\mathscr{C}_m(u_1)\mathscr{C}_{m+2}(u_2) = \mathscr{C}_m(u_2)\mathscr{C}_{m+2}(u_1). \qquad (5.5.46)
$$

Multiplying (5.5.45) with $\tilde{Y}^1_{m-1}(u_1)\tilde{Y}^2_m(u_2)$ from the left and $X^1_{m+1}(-u_1)X^2_{m+2}(-u_2)$ from the right, we have

$$
\begin{aligned}
\mathscr{A}_{m+2}(u_1)\mathscr{A}_{m+2}(u_2) &= \mathscr{A}_{m+2}(u_2)\mathscr{A}_{m+2}(u_1) \\
&+ \frac{\sinh(m\eta - u_1 - u_2)\sinh\eta}{\sinh(m\eta)\sinh(u_1 + u_2 + \eta)}\mathscr{B}_m(u_2)\mathscr{C}_{m+2}(u_1) \\
&- \frac{\sinh(m\eta - u_1 - u_2)\sinh\eta}{\sinh(m\eta)\sinh(u_1 + u_2 + \eta)}\mathscr{B}_m(u_1)\mathscr{C}_{m+2}(u_2). \qquad (5.5.47)
\end{aligned}
$$

Multiplying (5.5.45) with $\tilde{X}^1_{m+1}(u_1)\tilde{Y}^2_{m-2}(u_2)$ from the left and $X^1_{m+1}(-u_1)X^2_{m+2}(-u_2)$ from the right, we have

$$\mathscr{C}_{m+2}(u_1)\mathscr{A}_{m+2}(u_2) = \frac{\sinh(u_1+u_2)\sinh(u_1-u_2+\eta)}{\sinh(u_1-u_2)\sinh(u_1+u_2+\eta)}\mathscr{A}_m(u_2)\mathscr{C}_{m+2}(u_1)$$
$$-\frac{\sinh(m\eta+u_1-u_2)\sinh(u_1+u_2)\sinh\eta}{\sinh(u_1+u_2+\eta)\sinh(u_1-u_2)\sinh(m\eta)}\mathscr{A}_m(u_1)\mathscr{C}_{m+2}(u_2)$$
$$-\frac{\sinh(m\eta-u_1-u_2)\sinh\eta}{\sinh(u_1+u_2+\eta)\sinh(m\eta)}\mathscr{D}_m(u_1)\mathscr{C}_{m+2}(u_2). \tag{5.5.48}$$

Multiplying (5.5.45) with $\tilde{X}^1_{m+1}(u_1)\tilde{Y}^2_{m-2}(u_2)$ from the left and $Y^1_{m-1}(-u_1)X^2_m(-u_2)$ from the right, we have

$$\mathscr{D}_m(u_1)\mathscr{A}_m(u_2) = \mathscr{A}_m(u_2)\mathscr{D}_m(u_1)$$
$$+\frac{\sinh(m\eta+u_1-u_2)\sinh(u_1+u_2+\eta)\sinh(m\eta)\sinh\eta}{\sinh(u_1+u_2)\sinh(u_1-u_2)\sinh(m+1)\eta\sinh(m-1)\eta}$$
$$\times[\mathscr{B}_{m-2}(u_2)\mathscr{C}_m(u_1)-\mathscr{B}_{m-2}(u_1)\mathscr{C}_m(u_2)]+\frac{\sinh(m\eta+u_1-u_2)}{\sinh(u_1-u_2)\sinh(u_1+u_2)}$$
$$\times\frac{\sinh(m\eta+u_1+u_2)\sinh^2\eta}{\sinh(m+1)\eta\sinh(m-1)\eta}[\mathscr{A}_m(u_2)\mathscr{A}_m(u_1)-\mathscr{A}_m(u_1)\mathscr{A}_m(u_2)]. \tag{5.5.49}$$

Using (5.5.47), the relation (5.5.49) can be rewritten as

$$\mathscr{D}_m(u_1)\mathscr{A}_m(u_2) = \mathscr{A}_m(u_2)\mathscr{D}_m(u_1)$$
$$+\frac{\sinh\eta\sinh(m-1)\eta\sinh(m\eta+u_1-u_2)\sinh(u_1+u_2+2\eta)}{\sinh(m+1)\eta\sinh(m-2)\eta\sinh(u_1-u_2)\sinh(u_1+u_2+\eta)}$$
$$\times[\mathscr{B}_{m-2}(u_2)\mathscr{C}_m(u_1)-\mathscr{B}_{m-2}(u_1)\mathscr{C}_m(u_2)]. \tag{5.5.50}$$

Note that $\mathscr{A}_m(u) = \mathscr{D}_{-m}(u)$ and $\mathscr{B}_m(u) = \mathscr{C}_{-m}(u)$. Similar commutation relations about $\mathscr{B}_m(u)$ can be constructed.

Using the orthogonal relations (5.5.10) and (5.5.12), the transfer matrix can also be written as

$$\begin{aligned} t(u) &= \mathrm{tr}\{\bar{\mathscr{U}}(m|u)\bar{K}^+(m|u)\} \\ &= \bar{K}^+_{11}(m|u)\bar{\mathscr{A}}_m(u) + \bar{K}^+_{21}(m|u)\bar{\mathscr{B}}_m(u) \\ &\quad +\bar{K}^+_{12}(m|u)\bar{\mathscr{C}}_m(u) + \bar{K}^+_{22}(m|u)\bar{\mathscr{D}}_m(u), \end{aligned} \tag{5.5.51}$$

where

$$\begin{aligned} \bar{K}^+(m|u) &= \begin{pmatrix} \bar{K}^+_{11}(m|u) & \bar{K}^+_{12}(m|u) \\ \bar{K}^+_{21}(m|u) & \bar{K}^+_{22}(m|u) \end{pmatrix} \\ &= \begin{pmatrix} \bar{Y}_m(-u)K^+(u)X_m(u) & \bar{Y}_{m+2}(-u)K^+(u)Y_m(u) \\ \bar{X}_{m-2}(-u)K^+(u)X_m(u) & \bar{X}_m(-u)K^+(u)Y_m(u) \end{pmatrix}, \end{aligned} \tag{5.5.52}$$

$$\bar{\mathscr{U}}(m|u) = \begin{pmatrix} \bar{\mathscr{A}}_m(u) & \bar{\mathscr{B}}_m(u) \\ \bar{\mathscr{C}}_m(u) & \bar{\mathscr{D}}_m(u) \end{pmatrix}$$
$$= \begin{pmatrix} \bar{Y}_m(u)\mathscr{U}(u)\hat{X}_{m-2}(-u) & \bar{Y}_m(u)\mathscr{U}(u)\hat{Y}_m(-u) \\ \bar{X}_m(u)\mathscr{U}(u)\hat{X}_m(-u) & \bar{X}_m(u)\mathscr{U}(u)\hat{Y}_{m+2}(-u) \end{pmatrix}. \quad (5.5.53)$$

Similarly, the commutation relations among $\bar{\mathscr{A}}_m(u)$, $\bar{\mathscr{B}}_m(u)$, $\bar{\mathscr{C}}_m(u)$ and $\bar{\mathscr{D}}_m(u)$ can be derived from (5.5.45) and the commutation relations (5.5.13)–(5.5.40). Multiplying (5.5.45) with $\bar{X}^1_{m+1}(u_1)\bar{X}^2_m(u_2)$ from the left and $\hat{X}^1_{m+1}(-u_1)\hat{X}^2_{m+2}(-u_2)$ from the right, we have

$$\bar{\mathscr{C}}_m(u_1)\bar{\mathscr{C}}_{m+2}(u_2) = \bar{\mathscr{C}}_m(u_2)\bar{\mathscr{C}}_{m+2}(u_1). \quad (5.5.54)$$

Multiplying (5.5.45) with $\bar{X}^1_{m-1}(u_1)\bar{X}^2_{m-2}(u_2)$ from the left and $\hat{Y}^1_{m+1}(-u_1)\hat{Y}^2_m(-u_2)$ from the right, we have

$$\bar{\mathscr{D}}_{m-2}(u_2)\bar{\mathscr{D}}_{m-2}(u_1) = \bar{\mathscr{D}}_{m-2}(u_1)\bar{\mathscr{D}}_{m-2}(u_2)$$
$$+\frac{\sinh(m\eta+u_1+u_2)\sinh\eta}{\sinh(m\eta)\sinh(u_1+u_2+\eta)}\bar{\mathscr{C}}_{m-2}(u_1)\bar{\mathscr{B}}_m(u_2)$$
$$-\frac{\sinh(m\eta+u_1+u_2)\sinh\eta}{\sinh m\eta\sinh(u_1+u_2+\eta)}\bar{\mathscr{C}}_{m-2}(u_2)\bar{\mathscr{B}}_m(u_1). \quad (5.5.55)$$

Multiplying (5.5.45) with $\bar{X}^1_{m-1}(u_1)\bar{X}^2_{m-2}(u_2)$ from the left and $\hat{X}^1_{m-1}(-u_1)\hat{Y}^2_{m+2}$ $(-u_2)$ from the right, we have

$$\bar{\mathscr{D}}_{m-2}(u_2)\bar{\mathscr{C}}_{m-2}(u_1) = \frac{\sinh(u_1-u_2+\eta)\sinh(u_1+u_2)}{\sinh(u_1+u_2+\eta)\sinh(u_1-u_2)}\bar{\mathscr{C}}_{m-2}(u_1)\bar{\mathscr{D}}_m(u_2)$$
$$-\frac{\sinh(m\eta-u_1+u_2)\sinh(u_1+u_2)\sinh\eta}{\sinh(m\eta)\sinh(u_1-u_2)\sinh(u_1+u_2+\eta)}\bar{\mathscr{C}}_{m-2}(u_2)\bar{\mathscr{D}}_m(u_1)$$
$$-\frac{\sinh(m\eta+u_1+u_2)\sinh\eta}{\sinh(m\eta)\sinh(u_1+u_2+\eta)}\bar{\mathscr{C}}_{m-2}(u_2)\bar{\mathscr{A}}_m(u_1). \quad (5.5.56)$$

Multiplying (5.5.45) with $\bar{Y}^1_{m+1}(u_1)\bar{X}^2_m(u_2)$ from the left and $\hat{X}^1_{m-1}(-u_1)\hat{Y}^2_{m+2}$ $(-u_2)$ from the right, we have

$$\bar{\mathscr{D}}_m(u_2)\bar{\mathscr{A}}_m(u_1) = \bar{\mathscr{A}}_m(u_1)\bar{\mathscr{D}}_m(u_2)$$
$$+\frac{\sinh(m\eta-u_1+u_2)\sinh(u_1+u_2+\eta)\sinh\eta\sinh(m\eta)}{\sinh(u_1-u_2)\sinh(u_1+u_2)\sinh(m-1)\eta\sinh(m+1)\eta}$$
$$\times[\bar{\mathscr{C}}_m(u_1)\bar{\mathscr{B}}_{m+2}(u_2)-\bar{\mathscr{C}}_m(u_2)\bar{\mathscr{B}}_{m+2}(u_1)]+\frac{\sinh(m\eta-u_1+u_2)}{\sinh(u_1-u_2)\sinh(u_1+u_2)}$$
$$\times\frac{\sinh(m\eta-u_1-u_2)\sinh^2\eta}{\sinh(m+1)\eta\sinh(m-1)\eta}[\bar{\mathscr{D}}_m(u_1)\bar{\mathscr{D}}_m(u_2)-\bar{\mathscr{D}}_m(u_2)\bar{\mathscr{D}}_m(u_1)]. \quad (5.5.57)$$

Using (5.5.55), (5.5.57) can be rewritten as

$$\begin{aligned}\bar{\mathscr{D}}_m(u_2)\bar{\mathscr{A}}_m(u_1) &= \bar{\mathscr{A}}_m(u_1)\bar{\mathscr{D}}_m(u_2)\\ &+\frac{\sinh(m+1)\eta\sinh\eta\sinh(m\eta-u_1+u_2)\sinh(u_1+u_2+2\eta)}{\sinh(m+2)\eta\sinh(m-1)\eta\sinh(u_1-u_2)\sinh(u_1+u_2+\eta)}\\ &\times[\bar{\mathscr{C}}_m(u_1)\bar{\mathscr{B}}_{m+2}(u_2)-\bar{\mathscr{C}}_m(u_2)\bar{\mathscr{B}}_{m+2}(u_1)]. \qquad (5.5.58)\end{aligned}$$

Note that the relations $\bar{\mathscr{A}}_m(u)=\bar{\mathscr{D}}_{-m}(u)$, $\bar{\mathscr{B}}_m(u)=\bar{\mathscr{C}}_{-m}(u)$, $\bar{\mathscr{B}}_m(u)\sim\mathscr{B}_m(u)$ and $\bar{\mathscr{C}}_m(u)\sim\mathscr{C}_m(u)$ hold.

5.5.3 Right SoV Basis

Before going further, we define a right local state $|m\rangle_n$ of nth site in the lattice:

$$|m\rangle_n = e^{-[\theta_n+(m+N-n+1+\alpha)\eta]}|\uparrow\rangle_n + |\downarrow\rangle_n. \qquad (5.5.59)$$

The gauged R-matrix is defined as

$$R_{0,n}(m|u-\theta_n) = \tilde{G}^{-1}_{m+N-n}(u)R_{0,n}(u-\theta_n)\tilde{G}_{m+N-n+1}(u), \qquad (5.5.60)$$

$$R_{0,n}(m|u+\theta_n) = \bar{G}^{-1}_{m+N-n+1}(-u)R_{0,n}(u+\theta_n)\bar{G}_{m+N-n}(-u). \qquad (5.5.61)$$

It is easy to check that the following relations hold:

$$R_{0,n}(m|u-\theta_n)^1_1|m\rangle_n = \frac{\sinh(u-\theta_n+\eta)}{\sinh\eta}|m+1\rangle_n, \qquad (5.5.62)$$

$$R_{0,n}(m|u-\theta_n)^1_2|m\rangle_n = 0, \qquad (5.5.63)$$

$$R_{0,n}(m|u-\theta_n)^2_2|m\rangle_n = \frac{\sinh(m+N-n+1)\eta}{\sinh(m+N-n)\eta}\frac{\sinh(u-\theta_n)}{\sinh\eta}|m-1\rangle_n, \qquad (5.5.64)$$

$$R_{0,n}(m|u+\theta_n)^1_1|m\rangle_n = \frac{\sinh(u+\theta_n+\eta)}{\sinh\eta}|m-1\rangle_n, \qquad (5.5.65)$$

$$R_{0,n}(m|u+\theta_n)^1_2|m\rangle_n = 0, \qquad (5.5.66)$$

$$R_{0,n}(m|u+\theta_n)^2_2|m\rangle_n = \frac{\sinh(m+N-n)\eta}{\sinh(m+N-n+1)\eta}\frac{\sinh(u+\theta_n)}{\sinh\eta}|m+1\rangle_n, \qquad (5.5.67)$$

where R^j_i means the ith row and jth column element of the R-matrix spanned in the auxiliary space. The gauged one-row monodromy matrices are defined as

$$T(m|u) = \tilde{G}_m^{-1}(u)T(u)\tilde{G}_{m+N}(u) = \begin{pmatrix} A_m(u) & B_m(u) \\ C_m(u) & D_m(u) \end{pmatrix} \tag{5.5.68}$$

$$= \begin{pmatrix} \tilde{Y}_{m-1}(u)T(u)X_{l+1}(u) & \tilde{Y}_{m-1}(u)T(u)Y_{l-1}(u) \\ \tilde{X}_{m+1}(u)T(u)X_{l+1}(u) & \tilde{X}_{m+1}(u)T(u)Y_{l-1}(u) \end{pmatrix}, \tag{5.5.69}$$

$$\hat{T}(m|u) = \bar{G}_{m+N}^{-1}(-u)\hat{T}(u)\bar{G}_m(-u) = \begin{pmatrix} \tilde{\mathcal{A}}_m(u) & \tilde{B}_m(u) \\ \tilde{C}_m(u) & \tilde{D}_m(u) \end{pmatrix} \tag{5.5.70}$$

$$= \begin{pmatrix} \bar{Y}_l(-u)\hat{T}(u)X_m(-u) & \bar{Y}_l(-u)\hat{T}(u)Y_m(-u) \\ \bar{X}_l(-u)\hat{T}(u)X_m(-u) & \bar{X}_l(-u)\hat{T}(u)Y_m(-u) \end{pmatrix}, \tag{5.5.71}$$

where $l = m + N$.

Let us define the following global state

$$|m\rangle = \otimes_{n=1}^N |m\rangle_n. \tag{5.5.72}$$

With the help of (5.5.62)–(5.5.67), we readily have

$$C_m(u)|m\rangle = 0, \tag{5.5.73}$$

$$A_m(u)|m\rangle = \prod_{j=1}^N \frac{\sinh(u - \theta_j + \eta)}{\sinh\eta}|m+1\rangle, \tag{5.5.74}$$

$$D_m(u)|m\rangle = \frac{\sinh(m+N)\eta}{\sinh(m\eta)}\prod_{j=1}^N \frac{\sinh(u - \theta_j)}{\sinh\eta}|m-1\rangle, \tag{5.5.75}$$

$$\tilde{C}_m(u)|m\rangle = 0, \tag{5.5.76}$$

$$\tilde{\mathcal{A}}_m(u)|m\rangle = \prod_{j=1}^N \frac{\sinh(u + \theta_j + \eta)}{\sinh\eta}|m-1\rangle, \tag{5.5.77}$$

$$\tilde{D}_m(u)|m\rangle = \frac{\sinh(m\eta)}{\sinh(m+N)\eta}\prod_{j=1}^N \frac{\sinh(u + \theta_j)}{\sinh\eta}|m+1\rangle. \tag{5.5.78}$$

Note that the elements of the gauged double-row monodromy matrix can be expressed in terms of those of the gauged one-row monodromy matrix by using the orthogonal relations (5.5.11) and (5.5.12)

$$\begin{aligned}
\mathscr{A}_m(u) = {} & \tilde{Y}_{m-2}(u)T(u)X_l(u)\tilde{Y}_{l-2}(u)K^-(u)X_l(-u)\bar{Y}_l(-u)\hat{T}(u)X_m(-u) \\
& + \tilde{Y}_{m-2}(u)T(u)Y_{l-2}(u)\tilde{X}_l(u)K^-(u)X_l(-u)\bar{Y}_l(-u)\hat{T}(u)X_m(-u) \\
& + \tilde{Y}_{m-2}(u)T(u)X_{l+2}(u)\tilde{Y}_l(u)K^-(u)Y_l(-u)\bar{X}_l(-u)\hat{T}(u)X_m(-u) \\
& + \tilde{Y}_{m-2}(u)T(u)Y_l(u)\tilde{X}_{l+2}(u)K^-(u)Y_l(-u)\bar{X}_l(-u)\hat{T}(u)X_m(-u),
\end{aligned} \tag{5.5.79}$$

$$\begin{aligned}
\mathscr{C}_m(u) = {} & \tilde{X}_m(u)T(u)X_l(u)\tilde{Y}_{l-2}(u)K^-(u)X_l(-u)\bar{Y}_l(-u)\hat{T}(u)X_m(-u) \\
& + \tilde{X}_m(u)T(u)Y_{l-2}(u)\tilde{X}_l(u)K^-(u)X_l(-u)\bar{Y}_l(-u)\hat{T}(u)X_m(-u)
\end{aligned}$$

$$+\tilde{X}_m(u)T(u)X_{l+2}(u)\tilde{Y}_l(u)K^-(u)Y_l(-u)\bar{X}_l(-u)\hat{T}(u)X_m(-u)$$
$$+\tilde{X}_m(u)T(u)Y_l(u)\tilde{X}_{l+2}(u)K^-(u)Y_l(-u)\bar{X}_l(-u)\hat{T}(u)X_m(-u), \quad (5.5.80)$$

$$\mathscr{D}_m(u) = \tilde{X}_{m+2}(u)T(u)X_l(u)\tilde{Y}_{l-2}(u)K^-(u)X_l(-u)\bar{Y}_l(-u)\hat{T}(u)Y_m(-u)$$
$$+\tilde{X}_{m+2}(u)T(u)Y_{l-2}(u)\tilde{X}_l(u)K^-(u)X_l(-u)\bar{Y}_l(-u)\hat{T}(u)Y_m(-u)$$
$$+\tilde{X}_{m+2}(u)T(u)X_{l+2}(u)\tilde{Y}_l(u)K^-(u)Y_l(-u)\bar{X}_l(-u)\hat{T}(u)Y_m(-u)$$
$$+\tilde{X}_{m+2}(u)T(u)Y_l(u)\tilde{X}_{l+2}(u)K^-(u)Y_l(-u)\bar{X}_l(-u)\hat{T}(u)Y_m(-u). \quad (5.5.81)$$

Define

$$K^-(l|u) = \begin{pmatrix} K^-_{11}(l|u) & K^-_{12}(l|u) \\ K^-_{21}(l|u) & K^-_{22}(l|u) \end{pmatrix}$$
$$= \begin{pmatrix} \tilde{Y}_{l-2}(u)K^-(u)X_l(-u) & \tilde{Y}_l(u)K^-(u)Y_l(-u) \\ \tilde{X}_l(u)K^-(u)X_l(-u) & \tilde{X}_{l+2}(u)K^-(u)Y_l(-u) \end{pmatrix}. \quad (5.5.82)$$

The relations (5.5.73)–(5.5.78) together with (5.5.79)–(5.5.80) and (5.5.82) imply that

$$\mathscr{C}_m(u)|m\rangle = K^-_{21}(l|u)\frac{\sinh(m+N-1)\eta}{\sinh(m-1)\eta}$$
$$\times\prod_{j=1}^{N}\frac{\sinh(u-\theta_j)\sinh(u+\theta_j+\eta)}{\sinh^2\eta}|m-2\rangle, \quad (5.5.83)$$

$$\mathscr{A}_m(u)|m\rangle = K^-_{11}(l|u)\prod_{j=1}^{N}\frac{\sinh(u-\theta_j+\eta)\sinh(u+\theta_j+\eta)}{\sinh^2\eta}|m\rangle$$
$$+K^-_{21}(l|u)\prod_{j=1}^{N}\frac{\sinh(u+\theta_j+\eta)}{\sinh\eta}B_{m-1}(u)|m-1\rangle. \quad (5.5.84)$$

The relation (5.5.83) implies that $|m\rangle$ is a pseudo eigenstate of $\mathscr{C}_m(u)$ and

$$\mathscr{C}_m(\theta_j)|m\rangle = \mathscr{C}_m(-\theta_j-\eta)|m\rangle = 0, \quad j=1,\ldots,N. \quad (5.5.85)$$

Let us introduce the following right states:

$$|\theta_{p_1},\ldots,\theta_{p_n};m\rangle = \mathscr{A}_m(\theta_{p_1})\cdots\mathscr{A}_m(\theta_{p_n})|m\rangle, \quad (5.5.86)$$

where $p_j \in \{1,\ldots,N\}$ and $p_1 < p_2 < \cdots < p_n$. The relation (5.5.47) shows that the operators $\mathscr{A}_m(u)$ with different spectral parameters are not commuting with each other. However, the relation (5.5.47) and (5.5.85) ensure that the state $|\theta_{p_1},\ldots,\theta_{p_n};m\rangle$ does not depend on the order of $\mathscr{A}_m(\theta_{p_j})$. Using the commutation relation (5.5.48) and the relations (5.5.83) and (5.5.85), we can obtain

$$\mathscr{C}_m(u)|\theta_{p_1},\ldots,\theta_{p_n};m\rangle = g_m(u,\{\theta_{p_1},\ldots,\theta_{p_n}\})|\theta_{p_1},\ldots,\theta_{p_n};m-2\rangle, \quad (5.5.87)$$

where

$$g_m(u,\{\theta_{p_1},\ldots,\theta_{p_n}\}) = K_{21}^{-}(l|u)\frac{\sinh(m+N-1)\eta}{\sinh(m-1)\eta}\prod_{l=1}^{N}\frac{\sinh(u-\theta_l)}{\sinh\eta}$$
$$\times\frac{\sinh(u+\theta_l+\eta)}{\sinh\eta}\prod_{j=1}^{n}\frac{\sinh(u+\theta_{p_j})\sinh(u-\theta_{p_j}+\eta)}{\sinh(u-\theta_{p_j})\sinh(u+\theta_{p_j}+\eta)}. \quad (5.5.88)$$

Obviously, the right states $\{|\theta_{p_1},\ldots,\theta_{p_n};m\rangle|n=0,\ldots,N\}$ given by (5.5.86) are pseudo eigenstates of $\mathscr{C}_m(u)$ and therefore form a basis [20] of the Hilbert space.

5.5.4 Left SoV Basis

A left local state of the nth site in the lattice can be defined as

$$_n\langle m| = \frac{e^{\theta_n+\alpha\eta}}{2\sinh(m+n-N-1)\eta}\left\{{}_n\langle\uparrow| - e^{-[\theta_n+(\alpha+m+n-N-1)\eta]}\,{}_n\langle\downarrow|\right\}. \quad (5.5.89)$$

Let us further define an alternative gauged R-matrix

$$\bar{R}_{0,n}(m|u-\theta_n) = \bar{G}^{-1}_{m+n-N}(u)R_{0,n}(u-\theta_n)\bar{G}_{m+n-N-1}(u), \quad (5.5.90)$$
$$\bar{R}_{0,n}(m|u+\theta_n) = \hat{G}^{-1}_{m+n-N-1}(-u)R_{0,n}(u+\theta_n)\hat{G}_{m+n-N}(-u). \quad (5.5.91)$$

Acting the elements of this gauged R-matrix on the left local state (5.5.89), we have

$$_n\langle m|\bar{R}_{0,n}(m|u-\theta_n)_2^1 = 0, \quad (5.5.92)$$
$$_n\langle m|\bar{R}_{0,n}(m|u-\theta_n)_2^2 = \frac{\sinh(u-\theta_n+\eta)}{\sinh\eta}\,{}_n\langle m+1|, \quad (5.5.93)$$
$$_n\langle m|\bar{R}_{0,n}(m|u-\theta_n)_1^1 = \frac{\sinh(m+n-N-2)\eta}{\sinh(m+n-N-1)\eta}\frac{\sinh(u-\theta_n)}{\sinh\eta}\,{}_n\langle m-1|, \quad (5.5.94)$$
$$_n\langle m|\bar{R}_{0,n}(m|u+\theta_n)_2^1 = 0, \quad (5.5.95)$$
$$_n\langle m|\bar{R}_{0,n}(m|u+0_n)_2^2 = \frac{\sinh(u+\theta_n+\eta)}{\sinh\eta}\,{}_n\langle m-1|, \quad (5.5.96)$$
$$_n\langle m|\bar{R}_{0,n}(m|u+\theta_n)_1^1 = \frac{\sinh(m+n-N+1)\eta}{\sinh(m+n-N)\eta}\frac{\sinh(u+\theta_n)}{\sinh\eta}\,{}_n\langle m+1|, \quad (5.5.97)$$

where $\bar{R}_i^j$ means the ith row and jth column element of the $\bar{R}$-matrix spanned in the auxiliary space. The corresponding gauged one-row monodromy matrices read

$$
\begin{aligned}
\bar{T}(m|u) &= \bar{G}_m^{-1}(u)T(u)\bar{G}_{m-N}(u) = \begin{pmatrix} \bar{A}_m(u) & \bar{B}_m(u) \\ \bar{C}_m(u) & \bar{D}_m(u) \end{pmatrix} \\
&= \begin{pmatrix} \bar{Y}_m(u)T(u)X_{\bar{l}}(u) & \bar{Y}_m(u)T(u)Y_{\bar{l}}(u) \\ \bar{X}_m(u)T(u)X_{\bar{l}}(u) & \bar{X}_m(u)T(u)Y_{\bar{l}}(u) \end{pmatrix},
\end{aligned} \tag{5.5.98}
$$

$$
\begin{aligned}
\bar{\hat{T}}(m|u) &= \hat{G}^{-1}_{m-N}(-u)\hat{T}(u)\hat{G}_m(-u) = \begin{pmatrix} \bar{\bar{A}}_m(u) & \bar{\bar{B}}_m(u) \\ \bar{\bar{C}}_m(u) & \bar{\bar{D}}_m(u) \end{pmatrix} \\
&= \begin{pmatrix} \bar{Y}_{\bar{l}+1}(-u)\hat{T}(u)\hat{X}_{m-1}(-u) & \bar{Y}_{\bar{l}+1}(-u)\hat{T}(u)\hat{Y}_{m+1}(-u) \\ \bar{X}_{\bar{l}-1}(-u)\hat{T}(u)\hat{X}_{m-1}(-u) & \bar{X}_{\bar{l}-1}(-u)\hat{T}(u)\hat{Y}_{m+1}(-u) \end{pmatrix},
\end{aligned} \tag{5.5.99}
$$

where $\bar{l} = m - N$.

Let us define the following left global state

$$
\langle m| = \otimes_{n=1}^{N}\, {}_n\langle m|. \tag{5.5.100}
$$

We readily have

$$
\langle m|\bar{C}_m(u) = 0, \tag{5.5.101}
$$

$$
\langle m|\bar{D}_m(u) = \prod_{j=1}^{N} \frac{\sinh(u-\theta_j+\eta)}{\sinh\eta}\langle m+1|, \tag{5.5.102}
$$

$$
\langle m|\bar{A}_m(u) = \frac{\sinh(m-N-1)\eta}{\sinh(m-1)\eta}\prod_{j=1}^{N}\frac{\sinh(u-\theta_j)}{\sinh\eta}\langle m-1|, \tag{5.5.103}
$$

$$
\langle m|\bar{\bar{C}}_m(u) = 0, \tag{5.5.104}
$$

$$
\langle m|\bar{\bar{D}}_m(u) = \prod_{j=1}^{N} \frac{\sinh(u+\theta_j+\eta)}{\sinh\eta}\langle m-1|, \tag{5.5.105}
$$

$$
\langle m|\bar{\bar{A}}_m(u) = \frac{\sinh(m+1)\eta}{\sinh(m+1-N)\eta}\prod_{j=1}^{N}\frac{\sinh(u+\theta_j)}{\sinh\eta}\langle m+1|. \tag{5.5.106}
$$

With the relations (5.5.10) and (5.5.12), we can rewrite $\bar{\mathscr{A}}_m(u)$, $\bar{\mathscr{C}}_m(u)$ and $\bar{\mathscr{D}}_m(u)$ as

$$
\begin{aligned}
\bar{\mathscr{A}}_m(u) &= \bar{Y}_m(u)T(u)X_{\bar{l}}(u)\bar{Y}_{\bar{l}}(u)K^-(u)\hat{X}_{\bar{l}-2}(-u)\bar{Y}_{\bar{l}}(-u)\hat{T}(u)\hat{X}_{m-2}(-u) \\
&+\bar{Y}_m(u)T(u)X_{\bar{l}}(u)\bar{Y}_{\bar{l}}(u)K^-(u)\hat{Y}_{\bar{l}}(-u)\bar{X}_{\bar{l}-2}(-u)\hat{T}(u)\hat{X}_{m-2}(-u) \\
&+\bar{Y}_m(u)T(u)Y_{\bar{l}}(u)\bar{X}_{\bar{l}}(u)K^-(u)\hat{X}_{\bar{l}}(-u)\bar{Y}_{\bar{l}+2}(-u)\hat{T}(u)\hat{X}_{m-2}(-u) \\
&+\bar{Y}_m(u)T(u)Y_{\bar{l}}(u)\bar{X}_{\bar{l}}(u)K^-(u)\hat{Y}_{\bar{l}+2}(-u)\bar{X}_{\bar{l}}(-u)\hat{T}(u)\hat{X}_{m-2}(-u),
\end{aligned} \tag{5.5.107}
$$

$$
\begin{aligned}
\bar{\mathscr{C}}_m(u) &= \bar{X}_m(u)T(u)X_{\bar{l}}(u)\bar{Y}_{\bar{l}}(u)K^-(u)\hat{X}_{\bar{l}-2}(-u)\bar{Y}_{\bar{l}}(-u)\hat{T}(u)\hat{X}_m(-u) \\
&+\bar{X}_m(u)T(u)X_{\bar{l}}(u)\bar{Y}_{\bar{l}}(u)K^-(u)\hat{Y}_{\bar{l}}(-u)\bar{X}_{\bar{l}-2}(-u)\hat{T}(u)\hat{X}_m(-u)
\end{aligned}
$$

$$+\bar{X}_m(u)T(u)Y_{\bar{l}}(u)\bar{X}_{\bar{l}}(u)K^-(u)\hat{X}_{\bar{l}}(-u)\bar{Y}_{\bar{l}+2}(-u)\hat{T}(u)\hat{X}_m(-u)$$
$$+\bar{X}_m(u)T(u)Y_{\bar{l}}(u)\bar{X}_{\bar{l}}(u)K^-(u)\hat{Y}_{\bar{l}+2}(-u)\bar{X}_{\bar{l}}(-u)\hat{T}(u)\hat{X}_m(-u), \qquad (5.5.108)$$

$$\bar{\mathscr{D}}_m(u) = \bar{X}_m(u)T(u)X_{\bar{l}}(u)\bar{Y}_{\bar{l}}(u)K^-(u)\hat{X}_{\bar{l}-2}(-u)\bar{Y}_{\bar{l}}(-u)\hat{T}(u)\hat{Y}_{m+2}(-u)$$
$$+\bar{X}_m(u)T(u)X_{\bar{l}}(u)\bar{Y}_{\bar{l}}(u)K^-(u)\hat{Y}_{\bar{l}}(-u)\bar{X}_{\bar{l}-2}(-u)\hat{T}(u)\hat{Y}_{m+2}(-u)$$
$$+\bar{X}_m(u)T(u)Y_{\bar{l}}(u)\bar{X}_{\bar{l}}(u)K^-(u)\hat{X}_{\bar{l}}(-u)\bar{Y}_{\bar{l}+2}(-u)\hat{T}(u)\hat{Y}_{m+2}(-u)$$
$$+\bar{X}_m(u)T(u)Y_{\bar{l}}(u)\bar{X}_{\bar{l}}(u)K^-(u)\hat{Y}_{\bar{l}+2}(-u)\bar{X}_{\bar{l}}(-u)\hat{T}(u)\hat{Y}_{m+2}(-u). \qquad (5.5.109)$$

Define

$$\bar{K}^-(\bar{l}|u) = \begin{pmatrix} \bar{K}^-_{11}(\bar{l}|u) & \bar{K}^-_{12}(\bar{l}|u) \\ \bar{K}^-_{21}(\bar{l}|u) & \bar{K}^-_{22}(\bar{l}|u) \end{pmatrix}$$
$$= \begin{pmatrix} \bar{Y}_{\bar{l}}(u)K^-(u)\hat{X}_{\bar{l}-2}(-u) & \bar{Y}_{\bar{l}}(u)K^-(u)\hat{Y}_{\bar{l}}(-u) \\ \bar{X}_{\bar{l}}(u)K^-(u)\hat{X}_{\bar{l}}(-u) & \bar{X}_{\bar{l}}(u)K^-(u)\hat{Y}_{\bar{l}+2}(-u) \end{pmatrix}. \qquad (5.5.110)$$

From (5.5.101)–(5.5.106), (5.5.108), (5.5.109) and (5.5.110), we have

$$\langle m|\bar{\mathscr{C}}_m(u) = \bar{K}^-_{21}(\bar{l}|u)\frac{\sinh(m+2)\eta}{\sinh(m+2-N)\eta}$$
$$\times\prod_{j=1}^{N}\frac{\sinh(u-\theta_j+\eta)\sinh(u+\theta_j)}{\sinh^2\eta}\langle m+2|, \qquad (5.5.111)$$

$$\langle m|\bar{\mathscr{D}}_m(u) = \bar{K}^-_{22}(\bar{l}|u)\prod_{j=1}^{N}\frac{\sinh(u-\theta_j+\eta)\sinh(u+\theta_j+\eta)}{\sinh^2\eta}\langle m|$$
$$+\bar{K}^-_{21}(\bar{l}|u)\prod_{j=1}^{N}\frac{\sinh(u-\theta_j+\eta)}{\sinh\eta}\langle m+1|\bar{\bar{B}}_{m+1}(u). \qquad (5.5.112)$$

From (5.5.111), we know that $\langle m|$ is a pseudo eigenstate of $\bar{\mathscr{C}}_m(u)$ and

$$\langle m|\bar{\mathscr{C}}_m(-\theta_j) = \langle m|\bar{\mathscr{C}}_m(\theta_j-\eta) = 0, \quad j = 1,\ldots,N. \qquad (5.5.113)$$

Introduce the following left states:

$$\langle\theta_{p_1}\cdots\theta_{p_n}; m| = \langle m|\bar{\mathscr{D}}_m(-\theta_{p_1})\cdots\bar{\mathscr{D}}_m(-\theta_{p_n}), \qquad (5.5.114)$$

where $p_j \in \{1,\ldots,N\}$, $\theta_{p_1} < \theta_{p_2}\cdots < \theta_{p_n}$. From (5.5.55) we know that the operators $\bar{\mathscr{D}}_m(u)$ with different spectral parameters are not commuting with each other. However, using the commutation relation (5.5.56) and (5.5.113), we find that the state $\langle\theta_{p_1},\ldots,\theta_{p_n}; m|$ is independent of the order of $\bar{\mathscr{D}}_m(-\theta_{p_j})$ in (5.5.114). Acting $\bar{\mathscr{C}}_m(u)$ on $\langle\theta_{p_1},\ldots,\theta_{p_n}; m|$ and using (5.5.111), (5.5.113) and (5.5.56), we have

$$\langle\theta_{p_1},\ldots,\theta_{p_n};m|\bar{\mathscr{C}}_m(u)=\bar{g}_m(u,\{\theta_{p_1},\ldots,\theta_{p_n}\})\langle\theta_{p_1},\ldots,\theta_{p_n};m+2|,\quad(5.5.115)$$

where

$$\bar{g}_m(u,\{\theta_{p_1},\ldots,\theta_{p_n}\})=\bar{K}^-_{21}(\bar{l}|u)\frac{\sinh(m+2)\eta}{\sinh(m+2-N)\eta}\times\prod_{l=1}^{N}\frac{\sinh(u-\theta_l+\eta)\sinh(u+\theta_l)}{\sinh^2\eta}\times\prod_{j=1}^{n}\frac{\sinh(u+\theta_{p_j}+\eta)\sinh(u-\theta_{p_j})}{\sinh(u-\theta_{p_j}+\eta)\sinh(u+\theta_{p_j})}.\quad(5.5.116)$$

The relation (5.5.115) indicates that the left states $\{\langle\theta_{p_1},\ldots,\theta_{p_n};m|\}$ are pseudo eigenstates of $\bar{\mathscr{C}}_m(u)$ and therefore form a left basis [26] of the Hilbert space.

5.5.5 The Scalar Product $\langle\theta_{p_1},\ldots,\theta_{p_n};m_0|\Psi\rangle$

An important fact is that with a proper choice of the parameters $\alpha=\alpha_0$ and $m=m_0$, $K^+(u)$ can be diagonalized with a corresponding gauge transformation. Thus the double-row transfer matrix becomes a linear combination of $\bar{\mathscr{A}}_{m_0}(u)$ and $\bar{\mathscr{D}}_{m_0}(u)$.

The operator $\bar{\mathscr{A}}_m(u)$ can be rewritten as

$$\begin{aligned}\bar{\mathscr{A}}_m(u)=&\bar{K}^-_{11}(\bar{l}|u)\bar{A}_m(u)\bar{\bar{A}}_{m-1}(u)+\bar{K}^-_{12}(\bar{l}|u)\bar{A}_m(u)\bar{\bar{C}}_{m-1}(u)\\&+\bar{K}^-_{21}(\bar{l}|u)\bar{Y}_m(u)T(u)Y_{\bar{l}}(u)\bar{Y}_{\bar{l}+2}(-u)\hat{T}(u)\hat{X}_{m-2}(-u)\\&+\bar{K}^-_{22}(\bar{l}|u)\bar{Y}_m(u)T(u)Y_{\bar{l}}(u)\bar{X}_{\bar{l}}(-u)\hat{T}(u)\hat{X}_{m-2}(-u).\end{aligned}\quad(5.5.117)$$

Using YBE, we have

$$T_2(u)R_{1,2}(2u)\hat{T}_1(u)=\hat{T}_1(u)R_{1,2}(2u)T_2(u).\quad(5.5.118)$$

Multiplying (5.5.118) with $\bar{Y}^1_m(u)\bar{X}^2_{\bar{l}-1}(-u)$ from the left and $Y^1_{\bar{l}-1}(u)\hat{X}^2_{m-2}(-u)$ from the right, we have

$$\begin{aligned}&\bar{Y}_m(u)T(u)Y_{\bar{l}}(u)\bar{X}_{\bar{l}}(-u)\hat{T}(u)\hat{X}_{m-2}(-u)\\&=\bar{\bar{C}}_m(u)\bar{Y}_{m+1}(u)T(u)Y_{\bar{l}-1}(u)\\&\quad+\frac{\sinh\eta\sinh((m-1)\eta-2u)}{\sinh(m-1)\eta\sinh(2u+\eta)}\bar{\bar{D}}_m(u)\bar{D}_{m-1}(u)\\&\quad-\frac{\sinh\eta\sinh((\bar{l}-1)\eta-2u)}{\sinh(\bar{l}-1)\eta\sinh(2u+\eta)}\bar{A}_m(u)\bar{\bar{A}}_{m-1}(u).\end{aligned}\quad(5.5.119)$$

Multiplying (5.5.118) with $\bar{Y}_m^1(u)\bar{Y}_{\bar{l}+1}^2(-u)$ from the left and $Y_{\bar{l}+1}^1(u)\hat{X}_{m-2}^2(-u)$ from the right, we have

$$\begin{aligned}
&\bar{Y}_m(u)T(u)Y_{\bar{l}}(u)\bar{Y}_{\bar{l}+2}(-u)\hat{T}(u)\hat{X}_{m-2}(-u)\\
&=\frac{\sinh((m-1)\eta-2u)\sinh\eta\sinh(\bar{l}+1)\eta}{\sinh(m-1)\eta\sinh(2u)\sinh(\bar{l}+2)\eta}\\
&\quad\times\bar{Y}_{\bar{l}+1}(-u)\hat{T}(u)\hat{Y}_{m+1}(-u)\bar{X}_{m-1}(u)T(u)Y_{\bar{l}+1}(u)\\
&\quad+\frac{\sinh(2u+\eta)\sinh(\bar{l}+1)\eta}{\sinh(2u)\sinh(\bar{l}+2)\eta}\bar{\bar{A}}_m(u)\bar{B}_{m+1}(u).
\end{aligned}\tag{5.5.120}$$

Multiplying (5.5.118) with $\bar{X}_m^1(u)\bar{Y}_{\bar{l}+1}^2(-u)$ from the left and $Y_{\bar{l}+1}^1(u)\hat{Y}_{m+2}^2(-u)$ from the right, we have

$$\begin{aligned}
&\bar{Y}_{\bar{l}+1}(-u)\hat{T}(u)\hat{Y}_{m+1}(-u)\bar{X}_{m-1}(u)T(u)Y_{\bar{l}+1}(u)\\
&=\frac{\sinh(2u)\sinh(\bar{l}+2)\eta}{\sinh(2u+\eta)\sinh(\bar{l}+1)\eta}\bar{D}_m(u)\bar{\bar{B}}_{m+1}(u)\\
&\quad-\frac{\sinh((m+1)\eta+2u)\sinh\eta}{\sinh(m+1)\eta\sinh(2u+\eta)}\bar{\bar{A}}_m(u)\bar{B}_{m+1}(u).
\end{aligned}\tag{5.5.121}$$

Combining (5.5.120) and (5.5.121), we obtain

$$\begin{aligned}
&\bar{Y}_m(u)T(u)Y_{\bar{l}}(u)\bar{Y}_{\bar{l}+2}(-u)\hat{T}(u)\hat{X}_{m-2}(-u)\\
&=\frac{\sinh((m-1)\eta-2u)\sinh\eta}{\sinh(m-1)\eta\sinh(2u+\eta)}\bar{D}_m(u)\bar{\bar{B}}_{m+1}(u)\\
&\quad+\frac{\sinh(2u+\eta)\sinh(\bar{l}+1)\eta}{\sinh(2u)\sinh(\bar{l}+2)\eta}\bar{\bar{A}}_m(u)\bar{B}_{m+1}(u)\\
&\quad-\frac{\sinh((m-1)\eta-2u)\sinh\eta\sinh(\bar{l}+1)\eta}{\sinh(m-1)\eta\sinh(2u)\sinh(\bar{l}+2)\eta}\\
&\quad\times\frac{\sinh((m+1)\eta+2u)\sinh\eta}{\sinh(m+1)\eta\sinh(2u+\eta)}\bar{\bar{A}}_m(u)\bar{B}_{m+1}(u).
\end{aligned}\tag{5.5.122}$$

It is easy to verify that $\langle m|\bar{\bar{A}}_m(-\theta_j)=0$. From (5.5.117), (5.5.119) and (5.5.122), we have

$$\langle m|\bar{\mathscr{A}}_m(-\theta_j)=-\frac{\sinh((m-1)\eta+2\theta_j)\sinh\eta}{\sinh(m-1)\eta\sinh(2\theta_j-\eta)}\langle m|\bar{\mathscr{D}}_m(-\theta_j).\tag{5.5.123}$$

From (5.5.113) and (5.5.58), we have

$$\langle \theta_{p_1}, \ldots, \theta_{p_n}; m | \bar{\mathscr{A}}_m(-\theta_{p_{n+1}}) = -\frac{\sinh((m-1)\eta + 2\theta_{p_{n+1}}) \sinh \eta}{\sinh(m-1)\eta \sinh(2\theta_{p_{n+1}} - \eta)} \times \langle \theta_{p_1}, \ldots, \theta_{p_n}, \theta_{p_{n+1}}; m |, \tag{5.5.124}$$

where $p_{n+1} \neq p_1, \ldots, p_n$.

Let $|\Psi\rangle$ be a common eigenstate of the transfer matrix $t(u)$, namely

$$t(u)|\Psi\rangle = \Lambda(u)|\Psi\rangle, \tag{5.5.125}$$

where $\Lambda(u)$ is given by the inhomogeneous $T-Q$ relation (5.3.41). As for the XXX case, we introduce the following scalar product

$$F_n(\theta_{p_1}, \ldots, \theta_{p_n}; m_0) = \langle \theta_{p_1}, \ldots, \theta_{p_n}; m_0 | \Psi \rangle, \quad n = 0, \ldots, N. \tag{5.5.126}$$

We consider the quantity $\langle \theta_{p_1}, \ldots, \theta_{p_n}; m_0 | t(-\theta_{p_{n+1}}) | \Psi \rangle$, where $p_{n+1} \neq p_1, \ldots, p_n$. By acting $t(-\theta_{p_{n+1}})$ to left and to right alternately with (5.5.123), we obtain the following recursive relations:

$$\begin{aligned} \Lambda(-\theta_{p_{n+1}}) F_n(\theta_{p_1}, \ldots, \theta_{p_n}; m_0) &= F_{n+1}(\theta_{p_1}, \ldots, \theta_{p_n}, \theta_{p_{n+1}}; m_0) \\ \times \Big\{ \bar{K}^+_{22}(m_0| -\theta_{p_{n+1}}) &- \frac{\sinh((m_0-1)\eta + 2\theta_{p_{n+1}}) \sinh \eta}{\sinh(m_0-1)\eta \sinh(2\theta_{p_{n+1}} - \eta)} \\ \times \bar{K}^+_{11}(m_0| -\theta_{p_{n+1}}) \Big\}, \end{aligned} \tag{5.5.127}$$

which directly induces

$$\begin{aligned} F_n(\theta_{p_1}, \ldots, \theta_{p_n}; m_0) = \prod_{j=1}^{n} \Bigg[\bar{K}^+_{22}(m_0| -\theta_{p_j}) &- \frac{\sinh((m_0-1)\eta + 2\theta_{p_j}) \sinh \eta}{\sinh(m_0-1)\eta \sinh(2\theta_{p_j} - \eta)} \\ \times \bar{K}^+_{11}(m_0| -\theta_{p_j}) \Bigg]^{-1} \Lambda(-\theta_{p_j}) F_0(m_0), \end{aligned} \tag{5.5.128}$$

where $F_0(m_0) = \langle m_0 | \Psi \rangle$ is an overall scalar factor.

5.5.6 The Inner Product $\langle \theta_{p_1}, \ldots, \theta_{p_n}; m_0 | \bar{m}_0 \rangle$

Similarly, we can always chose $m = \bar{m}_0$ and $\alpha = \alpha_0$ to make

$$K^-_{21}(\bar{m}_0 + N|u) = 0. \tag{5.5.129}$$

In such a case, from (5.5.83) and (5.5.84) we have

$$\mathscr{C}_{\bar{m}_0}(u)|\bar{m}_0\rangle = 0, \tag{5.5.130}$$

$$\mathscr{A}_{\bar{m}_0}(u)|\bar{m}_0\rangle = K^-_{11}(\bar{m}_0 + N|u) \times \prod_{j=1}^{N} \frac{\sinh(u-\theta_j+\eta)\sinh(u+\theta_j+\eta)}{\sinh^2\eta}|\bar{m}_0\rangle. \tag{5.5.131}$$

From the definitions of $\mathscr{U}(m|u)$ and $\bar{\mathscr{U}}(m|u)$, we can derive the relations

$$\begin{aligned}\bar{\mathscr{C}}_m(u) = {} & \frac{e^{-\bar{m}\eta}-e^{-m\eta}}{2\sinh(m\eta)} \frac{\sinh(m+2)\eta[e^{(\bar{m}-1)\eta}-e^{-(m+1)\eta}]}{2\sinh(m+1)\eta\sinh(\bar{m}\eta)}\mathscr{A}_{\bar{m}}(u) \\ & + \frac{e^{(\bar{m}-2)\eta}-e^{-m\eta}}{2\sinh(m\eta)} \frac{\sinh(m+2)\eta[e^{(\bar{m}-1)\eta}-e^{-(m+1)\eta}]}{2\sinh(m+1)\eta\sinh(\bar{m}\eta)}\mathscr{C}_{\bar{m}}(u) \\ & + \frac{e^{-(\bar{m}+2)\eta}-e^{-m\eta}}{2\sinh(m\eta)} \frac{\sinh(m+2)\eta[e^{-(m+1)\eta}-e^{-(\bar{m}+1)\eta}]}{2\sinh(m+1)\eta\sinh(\bar{m}\eta)}\mathscr{B}_{\bar{m}}(u) \\ & + \frac{e^{\bar{m}\eta}-e^{-m\eta}}{2\sinh(m\eta)} \frac{\sinh(m+2)\eta[e^{-(m+1)\eta}-e^{-(\bar{m}+1)\eta}]}{2\sinh(m+1)\eta\sinh(\bar{m}\eta)}\mathscr{D}_{\bar{m}}(u),\end{aligned} \tag{5.5.132}$$

$$\begin{aligned}\bar{\mathscr{D}}_m(u) = {} & \frac{[e^{-\bar{m}\eta}-e^{-m\eta}][e^{(\bar{m}-1)\eta}-e^{(m+1)\eta}]}{4\sinh(m+1)\eta\sinh(\bar{m}\eta)}\mathscr{A}_{\bar{m}}(u) \\ & + \frac{[e^{(\bar{m}-2)\eta}-e^{-m\eta}][e^{(\bar{m}-1)\eta}-e^{(m+1)\eta}]}{4\sinh(m+1)\eta\sinh(\bar{m}\eta)}\mathscr{C}_{\bar{m}}(u) \\ & + \frac{[e^{-(\bar{m}+2)\eta}-e^{-m\eta}][e^{(m+1)\eta}-e^{-(\bar{m}+1)\eta}]}{4\sinh(m+1)\eta\sinh(\bar{m}\eta)}\mathscr{B}_{\bar{m}}(u) \\ & + \frac{[e^{\bar{m}\eta}-e^{-m\eta}][e^{(m+1)\eta}-e^{-(\bar{m}+1)\eta}]}{4\sinh(m+1)\eta\sinh(\bar{m}\eta)}\mathscr{D}_{\bar{m}}(u),\end{aligned} \tag{5.5.133}$$

where m and $\bar{m}$ are arbitrary.

Let us consider the quantity of $\langle\theta_{p_1},\ldots,\theta_{p_n};m_0|\bar{\mathscr{C}}_{m_0}(-\theta_j)|\bar{m}_0\rangle$ for $j \neq p_l, l = 1,\ldots,n$. Acting $\bar{\mathscr{C}}_{m_0}(-\theta_j)$ to the left, from the relations (5.5.115) and (5.5.116), we know

$$\langle\theta_{p_1},\ldots,\theta_{p_n};m_0|\bar{\mathscr{C}}_{m_0}(-\theta_j)|\bar{m}_0\rangle = 0. \tag{5.5.134}$$

From the definition (5.5.114), we have

$$\langle\theta_{p_1},\ldots,\theta_{p_n},\theta_{p_{n+1}};m_0|\bar{m}_0\rangle = \langle\theta_{p_1},\ldots,\theta_{p_n};m_0|\bar{\mathscr{D}}_{m_0}(-\theta_{p_{n+1}})|\bar{m}_0\rangle. \tag{5.5.135}$$

Acting $\bar{\mathscr{D}}_{m_0}(-\theta_{p_{n+1}})$ to the right and using (5.5.130)–(5.5.134), we obtain the recursive relations

$$\langle\theta_{p_1},\ldots,\theta_{p_n},\theta_{p_{n+1}};m_0|\bar{m}_0\rangle = \frac{\sinh(m_0\eta)}{\sinh(m_0+2)\eta}\frac{e^{(m_0+1)\eta}-e^{-(\bar{m}_0+1)\eta}}{e^{-(m_0+1)\eta}-e^{-(\bar{m}_0+1)\eta}}$$
$$\times\langle\theta_{p_1},\ldots,\theta_{p_n};m_0|\bar{\mathscr{C}}_{m_0}(-\theta_{p_{n+1}})|\bar{m}_0\rangle + \langle\theta_{p_1},\ldots,\theta_{p_n};m_0|\bar{\mathscr{A}}_{\bar{m}_0}(-\theta_{p_{n+1}})|\bar{m}_0\rangle$$
$$= K^{-}_{\bar{1}\bar{1}}(\bar{m}_0+N|-\theta_{p_{n+1}})\prod_{j=1}^{N}\frac{\sinh(-\theta_{p_{n+1}}-\theta_j+\eta)\sinh(-\theta_{p_{n+1}}+\theta_j+\eta)}{\sinh^2\eta}$$
$$\times\langle\theta_{p_1},\ldots,\theta_{p_n};m_0|\bar{m}_0\rangle,\quad n=0,\ldots,N-1. \tag{5.5.136}$$

5.5.7 Bethe States

Define

$$|\mu_1,\ldots,\mu_N;m;\bar{m}\rangle = \bar{\mathscr{C}}_m(\mu_1)\bar{\mathscr{C}}_{m+2}(\mu_2)\cdots\bar{\mathscr{C}}_{m+2(N-1)}(\mu_N)|\bar{m}\rangle, \tag{5.5.137}$$

where $\{\mu_1,\ldots,\mu_N\}$ is a set of Bethe roots. From (5.5.115), (5.5.116) and (5.5.136), we have

$$\langle\theta_{p_1},\ldots,\theta_{p_n};m_0|\mu_1,\ldots,\mu_N;m_0;\bar{m}_0\rangle = \prod_{k=1}^{n}K^{-}_{\bar{1}\bar{1}}(\bar{m}_0+N|-\theta_{p_k})$$
$$\times\prod_{j=1}^{N}\frac{\sinh(-\theta_{p_k}-\theta_j+\eta)\sinh(-\theta_{p_k}+\theta_j+\eta)}{\sinh^2\eta}\prod_{l=1}^{N}\frac{\sinh(\mu_l+\theta_{p_k}+\eta)}{\sinh(\mu_l-\theta_{p_k}+\eta)}$$
$$\times\frac{\sinh(\mu_l-\theta_{p_k})}{\sinh(\mu_l+\theta_{p_k})}G_0(m_0,\bar{m}_0,\{\mu_1,\ldots,\mu_N\}), \tag{5.5.138}$$

where the overall scalar factor G_0 reads

$$G_0(m_0,\bar{m}_0,\{\mu_1,\ldots,\mu_N\}) = \prod_{j=1}^{N}\bar{K}^{-}_{21}(m_0+2j-2-N|\mu_j)\frac{\sinh(m_0+2j)\eta}{\sinh(m_0+2j-N)\eta}$$
$$\times\prod_{k=1}^{N}\frac{\sinh(\mu_j-\theta_k+\eta)\sinh(\mu_j+\theta_k)}{\sinh^2\eta}\langle m_0+2N|\bar{m}_0\rangle. \tag{5.5.139}$$

To make $\bar{K}^{+}_{21}(m_0|u)=0$ and $\bar{K}^{+}_{12}(m_0|u)=0$, we have

$$\sinh(\alpha_+ + \beta_+) = \sinh(\theta_+ + (\alpha_0-1)\eta + m_0\eta), \tag{5.5.140}$$
$$\sinh(\alpha_+ + \beta_+) = \sinh(\theta_+ + (\alpha_0-1)\eta - m_0\eta). \tag{5.5.141}$$

To make $K^-_{21}(\bar{m}_0 + N|u) = 0$ under the condition that $\alpha = \alpha_0$, we have

$$\sinh(\alpha_- + \beta_-) + \sinh(\theta_- + \alpha_0\eta + (\bar{m}_0 + N)\eta) = 0. \tag{5.5.142}$$

There are several solutions to the above equations. We shall take one set of the solutions to derive the right Bethe-type eigenstates, namely

$$\alpha_0\eta = -\theta_+ + \eta + i\frac{\pi}{2}, \tag{5.5.143}$$

$$m_0\eta = \alpha_+ + \beta_+ - i\frac{\pi}{2}, \tag{5.5.144}$$

$$(\bar{m}_0 + N)\eta = -\theta_- - \alpha_0\eta + \alpha_- + \beta_- + i\pi. \tag{5.5.145}$$

In such a case, we have

$$\bar{K}^+_{11}(m_0|u) = \frac{-2e^{-u}}{\cosh(\alpha_+ + \beta_+)}\sinh(u + \alpha_+ + \eta) \times \cosh(u + \beta_+ + \eta)\cosh(\alpha_+ + \beta_+ - \eta), \tag{5.5.146}$$

$$\bar{K}^+_{22}(m_0|u) = \frac{2e^{-u}}{\cosh(\alpha_+ + \beta_+)}\sinh(u - \alpha_+ + \eta) \times \cosh(u - \beta_+ + \eta)\cosh(\alpha_+ + \beta_+ + \eta), \tag{5.5.147}$$

$$K^-_{11}(\bar{m}_0 + N|u) = -2e^{u}\sinh(u - \alpha_-)\cosh(u - \beta_-), \tag{5.5.148}$$

and

$$\bar{K}^+_{22}(m_0|u) + \frac{\sinh\eta\sinh((m_0 - 1)\eta - 2u)}{\sinh(2u + \eta)\sinh(m_0 - 1)\eta}\bar{K}^+_{11}(m_0|u) = 2e^{-u}\frac{\sinh(2u + 2\eta)}{\sinh(2u + \eta)}\sinh(u - \alpha_+)\cosh(u - \beta_+). \tag{5.5.149}$$

With the help of (5.5.149), the $T - Q$ relation (5.3.41) and the Q-function defined in (5.3.42), it is easy to rewrite $F_n(\theta_{p_1}, \ldots, \theta_{p_n}; m_0)$ as follows

$$F_n(\theta_{p_1}, \ldots, \theta_{p_n}; m_0) = \prod_{j=1}^{n} 2e^{-\theta_{p_j}}\sinh(\theta_{p_j} + \alpha_-)\cosh(\theta_{p_j} + \beta_-)\frac{Q(-\theta_{p_j} - \eta)}{Q(-\theta_{p_j})} \times \prod_{l=1}^{N}\frac{\sinh(\theta_{p_j} + \theta_l - \eta)\sinh(\theta_{p_j} - \theta_l - \eta)}{\sinh^2\eta}F_0(m_0). \tag{5.5.150}$$

Similarly, we also have

$$\begin{aligned}\langle\theta_{p_1},\ldots,\theta_{p_n};m_0|\mu_1,\ldots,\mu_N;m_0;\bar{m}_0\rangle &= \prod_{j=1}^{n}2e^{-\theta_{p_j}}\sinh(\theta_{p_j}+\alpha_-)\\ &\times\cosh(\theta_{p_j}+\beta_-)\frac{Q(-\theta_{p_j}-\eta)}{Q(-\theta_{p_j})}\prod_{l=1}^{N}\frac{\sinh(\theta_{p_j}+\theta_l-\eta)}{\sinh\eta}\\ &\times\frac{\sinh(\theta_{p_j}-\theta_l-\eta)}{\sinh\eta}G_0(m_0,\bar{m}_0,\{\mu_1,\ldots,\mu_N\}).\end{aligned} \tag{5.5.151}$$

Comparing (5.5.150) and (5.5.151), we conclude that $|\mu_1,\ldots,\mu_N;m_0;\bar{m}_0\rangle$ is exactly proportional to the state $|\Psi\rangle$ and is an eigenstate of the transfer matrix $t(u)$ with the corresponding eigenvalue (5.3.41), provided that the Bethe roots $\{\mu_j|j=1,\ldots,N\}$ satisfy the BAEs (5.3.44).

With the same procedure, we can also construct the left Bethe states as

$$\langle\mu_1,\ldots,\mu_N;m_0;\bar{m}_0| = \langle\bar{m}_0|\mathscr{C}_{m_0-2(N-1)}(\mu_1)\cdots\mathscr{C}_{m_0}(\mu_N), \tag{5.5.152}$$

with

$$\begin{aligned}&\alpha_0\eta=-\theta_+-\eta+i\frac{\pi}{2},\quad m_0\eta=-\alpha_+-\beta_++i\frac{\pi}{2},\\ &(\bar{m}_0-N)\eta=-\theta_--\alpha_0\eta-\alpha_--\beta_-.\end{aligned} \tag{5.5.153}$$

It is easy to check that the two reference states $|\bar{m}_0\rangle$ and $\langle\bar{m}_0|$ and all the elements of the gauged monodromy matrices have well-defined homogeneous limits.

5.5.8 Degenerate Case

When $\bar{m}_0 = m_0 - 2\bar{M}$ and $\bar{M}$ is an integer, the boundary parameters satisfy the constraint conditions (5.3.35). For this degenerate case, the method introduced in the above subsection fails and instead we can use the algebraic Bethe Ansatz method. Without losing generality, we use the following parametrization to reach the constraint (5.3.35)

$$\alpha_++\beta_+=i\frac{\pi}{2}+m_0\eta, \tag{5.5.154}$$

$$\theta_++(\alpha_0+1)\eta=i\frac{\pi}{2}, \tag{5.5.155}$$

$$\alpha_-+\beta_-=-\theta_--(\bar{m}_0+N+\alpha_0)\eta. \tag{5.5.156}$$

In this case, due to the conditions

$$K^+_{12}(m_0|u) = \overline{Y}_{m_0}(-u)K^+(u)Y_{m_0-2}(u) = 0, \tag{5.5.157}$$

$$K^+_{21}(m_0|u) = \overline{X}_{m_0}(-u)K^+(u)X_{m_0+2}(u) = 0, \tag{5.5.158}$$

the transfer matrix becomes

$$t(u) = K^+_{11}(m_0|u)\mathscr{A}_{m_0}(u) + K^+_{22}(m_0|u)\mathscr{D}_{m_0}(u). \tag{5.5.159}$$

Further, from the condition

$$K^-_{21}(\bar{m}_0 + N|u) = \widetilde{X}_{\bar{m}_0+N}(u)K^-(u)X_{\bar{m}_0+N}(-u) = 0, \tag{5.5.160}$$

we know that and

$$\mathscr{A}_{\bar{m}_0}(u)|\bar{m}_0\rangle = K^-_{11}(\bar{m}_0 + N|u)\prod_{j=1}^{N}\frac{\sinh(u-\theta_j+\eta)\sinh(u+\theta_j+\eta)}{\sinh^2\eta}|\bar{m}_0\rangle,$$

$$\mathscr{C}_{\bar{m}_0}(u)|\bar{m}_0\rangle = 0,$$

$$\begin{aligned}\mathscr{D}_{\bar{m}_0}(u)|\bar{m}_0\rangle &= K^-_{11}(\bar{m}_0 + N|u)\frac{\sinh(2u+(m_0+1)\eta)\sinh\eta}{\sinh(\bar{m}_0+1)\eta\sinh(2u+\eta)}\\ &\times\prod_{j=1}^{N}\frac{\sinh(u-\theta_j+\eta)\sinh(u+\theta_j+\eta)}{\sinh^2\eta}|\bar{m}_0\rangle\\ &+\left[K^-_{22}(\bar{m}_0+N|u) - K^-_{11}(\bar{m}_0+N|u)\frac{\sinh(2u+(\bar{m}_0+N+1)\eta)\sinh\eta}{\sinh(\bar{m}_0+N+1)\eta\sinh(2u+\eta)}\right]\\ &\times\frac{\sinh(\bar{m}_0\eta)\sinh(\bar{m}_0+N+1)\eta}{\sinh(\bar{m}_0+N)\eta\sinh(\bar{m}_0+1)\eta}\prod_{j=1}^{N}\frac{\sinh(u-\theta_j)\sinh(u+\theta_j)}{\sinh^2\eta}|\bar{m}_0\rangle.\end{aligned}$$

With the same procedure introduced in Sect. 5.5.2, multiplying the reflection equation with $\widetilde{Y}^1_{m-1}(u_1)\widetilde{Y}^2_m(u_2)$ from the left and $Y^1_{m-1}(-u_1)X^2_m(-u_2)$ from the right, we have

$$\begin{aligned}\mathscr{A}_{m+2}(u_1)\mathscr{B}_m(u_2) &= \frac{\sinh(u_1+u_2)\sinh(u_1-u_2-\eta)}{\sinh(u_1-u_2)\sinh(u_1+u_2+\eta)}\mathscr{B}_m(u_2)\mathscr{A}_m(u_1)\\ &-\frac{\sinh(-(m+1)\eta+u_1-u_2)\sinh(2u_2)\sinh\eta}{\sinh(2u_2+\eta)\sinh(u_1-u_2)\sinh(m+1)\eta}\mathscr{B}_m(u_1)\mathscr{A}_m(u_2)\\ &-\frac{\sinh(m\eta-u_1-u_2)\sinh\eta\sinh\eta}{\sinh(u_1+u_2+\eta)\sinh(m+1)\eta\sinh(2u_2+\eta)}\mathscr{B}_m(u_1)\tilde{\mathscr{D}}_m(u_2),\end{aligned} \tag{5.5.161}$$

while multiplying the reflection equation with $\widetilde{X}^1_{m+3}(u_1)\widetilde{Y}^2_m(u_2)$ from the left and $Y^1_{m+1}(-u_1)Y^2_m(-u_2)$ from the right, we have

$$\begin{aligned}\tilde{\mathscr{D}}_{m+2}(u_1)\mathscr{B}_m(u_2) &= \frac{\sinh(u_1+u_2+2\eta)\sinh(u_1-u_2+\eta)}{\sinh(u_1-u_2)\sinh(u_1+u_2+\eta)}\mathscr{B}_m(u_2)\tilde{\mathscr{D}}_m(u_1)\\ &+\frac{\sinh((m+2)\eta+u_1+u_2)\sinh(2u_2)\sinh(2u_1+2\eta)}{\sinh(u_1+u_2+\eta)\sinh(m+1)\eta\sinh(2u_2+\eta)}\mathscr{B}_m(u_1)\mathscr{A}_m(u_2)\\ &-\frac{\sinh((m+1)\eta+u_1-u_2)\sinh(2u_1+2\eta)\sinh\eta}{\sinh(m+1)\eta\sinh(u_1-u_2)\sinh(2u_2+\eta)}\mathscr{B}_m(u_1)\tilde{\mathscr{D}}_m(u_2), \end{aligned}\quad (5.5.162)$$

with

$$\begin{aligned}\tilde{\mathscr{D}}_m(u) = &\frac{\sinh(2u+\eta)\sinh(m+1)\eta}{\sinh\eta\sinh(m\eta)}\mathscr{D}_m(u)\\ &-\frac{\sinh(m\eta+2u+\eta)}{\sinh(m\eta)}\mathscr{A}_m(u). \end{aligned}\quad (5.5.163)$$

The eigenvalue of $\tilde{\mathscr{D}}_{\bar{m}_0}(u)$ acting on the reference state $|\bar{m}_0\rangle$ is

$$\begin{aligned}\tilde{\mathscr{D}}_{\bar{m}_0}(u)|\bar{m}_0\rangle = &\Bigg[K^-_{22}(\bar{m}_0+N|u)\frac{\sinh(2u+\eta)\sinh(\bar{m}_0+N+1)\eta}{\sinh\eta\sinh(\bar{m}_0+N)\eta}\\ &-K^-_{11}(\bar{m}_0+N|u)\frac{\sinh(2u+(\bar{m}_0+N+1)\eta)}{\sinh(\bar{m}_0+N)\eta}\Bigg]\\ &\times\prod_{j=1}^{N}\frac{\sinh(u-\theta_j)\sinh(u+\theta_j)}{\sinh^2\eta}|\bar{m}_0\rangle. \end{aligned}\quad (5.5.164)$$

Let us introduce further the notation

$$\begin{aligned}\mathscr{B}_{\bar{M}} &= \mathscr{B}_{m_0-2}(\mu_1)\cdots\mathscr{B}_{m_0-2\bar{M}}(\mu_{\bar{M}}),\\ \mathscr{B}^j_{\bar{M}} &= \mathscr{B}_{m_0-2}(\mu_1)\cdots\mathscr{B}_{m_0-2j}(u)\cdots\mathscr{B}_{m_0-2\bar{M}}(\mu_{\bar{M}}). \end{aligned}\quad (5.5.165)$$

By using the commutation relations (5.5.161) and (5.5.162), with a similar procedure used to demonstrate (2.2.26) we can prove that

$$\begin{aligned}\mathscr{A}_{m_0}(u)\mathscr{B}_{\bar{M}} = &\prod_{j=1}^{\bar{M}}\frac{\sinh(u+\mu_j)\sinh(u-\mu_j-\eta)}{\sinh(u-\mu_j)\sinh(u+\mu_j+\eta)}\mathscr{B}_{\bar{M}}\mathscr{A}_{m_0-2\bar{M}}(u)\\ &-\sum_{j=1}^{\bar{M}}\frac{\sinh(2\mu_j)\sinh(u-\mu_j-(m_0-2j+1)\eta)\sinh\eta}{\sinh(u-\mu_j)\sinh(2\mu_j+\eta)\sinh(m_0-2j+1)\eta}\end{aligned}$$

$$\times \prod_{l \neq j}^{\bar{M}} \frac{\sinh(\mu_j - \mu_l - \eta)\sinh(\mu_j + \mu_l)}{\sinh(\mu_j - \mu_l)\sinh(\mu_j + \mu_l + \eta)} \mathscr{B}^j_{\bar{M}} \mathscr{A}_{m_0 - 2\bar{M}}(\mu_j)$$

$$- \sum_{j=1}^{\bar{M}} \frac{\sinh\eta \sinh(-u - \mu_j + (m_0 - 2j)\eta)\sinh\eta}{\sinh(u + \mu_j + \eta)\sinh(2\mu_j + \eta)\sinh(m_0 - 2j + 1)\eta}$$

$$\times \prod_{l \neq j}^{\bar{M}} \frac{\sinh(\mu_j - \mu_l + \eta)\sinh(\mu_j + \mu_l + 2\eta)}{\sinh(\mu_j - \mu_l)\sinh(\mu_j + \mu_l + \eta)} \mathscr{B}^j_{\bar{M}} \tilde{\mathscr{D}}_{m_0 - 2\bar{M}}(\mu_j), \quad (5.5.166)$$

$$\tilde{\mathscr{D}}_{m_0}(u)\mathscr{B}_{\bar{M}} = \prod_{j=1}^{\bar{M}} \frac{\sinh(u + \mu_j + 2\eta)\sinh(u - \mu_j + \eta)}{\sinh(u - \mu_j)\sinh(u + \mu_j + \eta)} \mathscr{B}_{\bar{M}} \tilde{\mathscr{D}}_{m_0 - 2\bar{M}}(u)$$

$$- \sum_{j=1}^{\bar{M}} \frac{\sinh(2u + 2\eta)\sinh(u - \mu_j + (m_0 - 2j + 1)\eta)\sinh\eta}{\sinh(u - \mu_j)\sinh(2\mu_j + \eta)\sinh(m_0 - 2j + 1)\eta}$$

$$\times \prod_{l \neq j}^{\bar{M}} \frac{\sinh(\mu_j - \mu_l + \eta)\sinh(\mu_j + \mu_l + 2\eta)}{\sinh(\mu_j - \mu_l)\sinh(\mu_j + \mu_l + \eta)} \mathscr{B}^j_{\bar{M}} \tilde{\mathscr{D}}_{m_0 - 2\bar{M}}(\mu_j)$$

$$+ \sum_{j=1}^{\bar{M}} \frac{\sinh(2\mu_j)\sinh(u + \mu_j + (m_0 - 2j + 2)\eta)\sinh(2u + 2\eta)}{\sinh(u + \mu_j + \eta)\sinh(2\mu_j + \eta)\sinh(m_0 - 2j + 1)\eta}$$

$$\times \prod_{l \neq j}^{\bar{M}} \frac{\sinh(\mu_j - \mu_l - \eta)\sinh(\mu_j + \mu_l)}{\sinh(\mu_j - \mu_l)\sinh(\mu_j + \mu_l + \eta)} \mathscr{B}^j_{\bar{M}} \mathscr{A}_{m_0 - 2\bar{M}}(\mu_j). \quad (5.5.167)$$

Assume the eigenstate of the transfer matrix takes the form

$$|\mu_1, \ldots, \mu_{\bar{M}}\rangle = \mathscr{B}_{m_0 - 2}(\mu_1) \cdots \mathscr{B}_{m_0 - 2\bar{M}}(\mu_{\bar{M}})|\bar{m}_0\rangle, \quad (5.5.168)$$

with $m_0 - 2\bar{M} = \bar{m}_0$. The transfer matrix (5.5.159) acting on the states (5.5.168) gives

$$t(u)|\mu_1, \ldots, \mu_{\bar{M}}\rangle - \Lambda(u)|\mu_1, \ldots, \mu_{\bar{M}}\rangle + \sum_{j=1}^{\bar{M}} \Lambda_j(u)\mathscr{B}^j_M|\bar{m}_0\rangle, \quad (5.5.169)$$

where $\Lambda(u)$ is the eigenvalue

$$\Lambda(u) = \mathscr{K}^+_{11}(m_0|u)\mathscr{K}^-_{11}(\bar{m}_0 \mid N|u) \prod_{j=1}^{\bar{M}} \frac{\sinh(u + \mu_j)\sinh(u - \mu_j - \eta)}{\sinh(u - \mu_j)\sinh(u + \mu_j + \eta)}$$

$$\times \prod_{l=1}^{N} \frac{\sinh(u + \theta_l + \eta)\sinh(u - \theta_l + \eta)}{\sinh^2\eta}$$

$$+\mathcal{K}^{+}_{22}(m_0|u)\mathcal{K}^{-}_{22}(\bar{m}_0+N|u)\prod_{j=1}^{\bar{M}}\frac{\sinh(u+\mu_j+2\eta)\sinh(u-\mu_j+\eta)}{\sinh(u-\mu_j)\sinh(u+\mu_j+\eta)}$$
$$\times\prod_{l=1}^{N}\frac{\sinh(u+\theta_l)\sinh(u-\theta_l)}{\sinh^2\eta}, \tag{5.5.170}$$

and the unwanted coefficients $\Lambda_j(u)$ read

$$\Lambda_j(u)=\frac{\sinh(2\mu_j)}{\sinh(2\mu_j+\eta)\sinh(m_0-2j+1)\eta}$$
$$\times\left\{\left[\frac{\sinh(u+\mu_j+(m_0-2j+2)\eta)\sinh(2u+2\eta)}{\sinh(u+\mu_j+\eta)}\mathcal{K}^{+}_{22}(m_0|u)\right.\right.$$
$$\left.-\frac{\sinh(u-\mu_j-(m_0-2j+1)\eta)\sinh\eta}{\sinh(u-\mu_j)}\mathcal{K}^{+}_{11}(m_0|u)\right]\mathcal{K}^{-}_{11}(\bar{m}_0+N|\mu_j)$$
$$\times\prod_{l=1}^{N}\frac{\sinh(\mu_j-\theta_l+\eta)\sinh(\mu_j+\theta_l+\eta)}{\sinh^2\eta}\prod_{k\neq j}^{\bar{M}}\frac{\sinh(\mu_j-\mu_k-\eta)\sinh(\mu_j+\mu_k)}{\sinh(\mu_j-\mu_k)\sinh(\mu_j+\mu_k+\eta)}$$
$$-\left[\frac{\sinh(u-\mu_j+(m_0-2j+1)\eta)\sinh(2u+2\eta)}{\sinh(u-\mu_j)}\mathcal{K}^{+}_{22}(m_0|u)\right.$$
$$\left.+\frac{\sinh(-u-\mu_j+(m_0-2j)\eta)\sinh\eta}{\sinh(u+\mu_j+\eta)}\mathcal{K}^{+}_{11}(m_0|u)\right]\mathcal{K}^{-}_{22}(\bar{m}_0+N|\mu_j)$$
$$\left.\times\prod_{l=1}^{N}\frac{\sinh(\mu_j-\theta_l)\sinh(\mu_j+\theta_l)}{\sinh^2\eta}\prod_{k\neq j}^{\bar{M}}\frac{\sinh(\mu_j-\mu_k+\eta)\sinh(\mu_j+\mu_k+2\eta)}{\sinh(\mu_j-\mu_k)\sinh(\mu_j+\mu_k+\eta)}\right\},$$

where

$$\mathcal{K}^{+}_{11}(m_0|u)=K^{+}_{11}(m_0|u)+\frac{\sinh\eta\sinh((m_0+1)\eta+2u)}{\sinh(2u+\eta)\sinh(m_0+1)\eta}K^{+}_{22}(m_0|u),$$
$$\mathcal{K}^{-}_{11}(\bar{m}_0+N|u)=K^{-}_{11}(\bar{m}_0+N|u),$$
$$\mathcal{K}^{+}_{22}(m_0|u)=\frac{\sinh\eta\sinh(m_0\eta)}{\sinh(2u+\eta)\sinh(m_0+1)\eta}K^{+}_{22}(m_0|u),$$
$$\mathcal{K}^{-}_{22}(\bar{m}_0+N|u)=\frac{\sinh(2u+\eta)\sinh(\bar{m}_0+N+1)\eta}{\sinh\eta\sinh(\bar{m}_0+N)\eta}K^{-}_{22}(\bar{m}_0+N|u)$$
$$-\frac{\sinh(2u+(\bar{m}_0+N+1)\eta)}{\sinh(\bar{m}_0+N)\eta}K^{-}_{11}(\bar{m}_0+N|u). \tag{5.5.171}$$

Putting $\Lambda_j(u) = 0$, we obtain the following BAEs:

$$\begin{aligned}1 = &\frac{\mathcal{K}^+_{11}(m_0|\mu_j)\mathcal{K}^-_{11}(\bar{m}_0 + N|\mu_j)}{\mathcal{K}^+_{22}(m_0|\mu_j)\mathcal{K}^-_{22}(\bar{m}_0 + N|\mu_j)} \frac{\sinh(2\mu_j)}{\sinh(2\mu_j + 2\eta)} \\ &\times \prod_{l=1}^{N} \frac{\sinh(\mu_j - \theta_l + \eta)\sinh(\mu_j + \theta_l + \eta)}{\sinh(\mu_j - \theta_l)\sinh(\mu_j + \theta_l)} \\ &\times \prod_{l \neq j}^{\bar{M}} \frac{\sinh(\mu_j + \mu_l)\sinh(\mu_j - \mu_l - \eta)}{\sinh(\mu_j + \mu_l + 2\eta)\sinh(\mu_j - \mu_l + \eta)}, \quad j = 1, \ldots, \bar{M}.\end{aligned} \tag{5.5.172}$$

By using the parameterization (5.5.156), we obtain

$$\begin{aligned}\mathcal{K}^+_{11}(m_0|u) &= \frac{-2e^{-u}}{\sinh(2u+\eta)} \sinh(2u + 2\eta)\sinh(u + \alpha_+)\cosh(u + \beta_+), \\ \mathcal{K}^-_{11}(\bar{m}_0 + N|u) &= 2e^u \sinh(u + \alpha_-)\cosh(u + \beta_-), \\ \mathcal{K}^+_{22}(m_0|u) &= \frac{2e^{-u}}{\sinh(2u+\eta)} \sinh\eta \sinh(u - \alpha_+ + \eta)\cosh(u - \beta_+ + \eta), \\ \mathcal{K}^-_{22}(\bar{m}_0 + N|u) &= \frac{-2e^u}{\sinh\eta} \sinh(2u)\sinh(u - \alpha_- + \eta)\cosh(u - \beta_- + \eta).\end{aligned}$$

Taking the homogeneous limit, the BAEs become

$$\begin{aligned}1 = &\frac{\sinh(\mu_j + \alpha_+)\cosh(\mu_j + \beta_+)\sinh(\mu_j + \alpha_-)\cosh(\mu_j + \beta_-)}{\sinh(\mu_j - \alpha_+ + \eta)\cosh(\mu_j - \beta_+ + \eta)\sinh(\mu_j - \alpha_- + \eta)\cosh(\mu_j - \beta_- + \eta)} \\ &\times \frac{\sinh^{2N}(\mu_j + \eta)}{\sinh^{2N}(\mu_j)} \prod_{l \neq j}^{\bar{M}} \frac{\sinh(\mu_j + \mu_l)\sinh(\mu_j - \mu_l - \eta)}{\sinh(\mu_j + \mu_l + 2\eta)\sinh(\mu_j - \mu_l + \eta)}, \\ &j = 1, \ldots, \bar{M}.\end{aligned} \tag{5.5.173}$$

By replacing $\bar{M}$ with $M = N - \bar{M} - 1$, $\alpha^\pm$ with $-\alpha^\pm$ and $\beta^\pm$ with $-\beta^\pm$, we can get another set of solutions. These two solutions constitute the complete set of eigenstates in the degenerate case.

5.6 The Open XYZ Spin-$\frac{1}{2}$ Chain

5.6.1 The Model Hamiltonian

The model Hamiltonian of the XYZ spin chain with generic boundaries in terms of the transfer matrix $t(u)$ is

$$
\begin{aligned}
H &= \frac{\sigma(\eta)}{\sigma'(0)}\left\{\frac{\partial}{\partial u}\ln t(u)\Big|_{u=0} - (N-1)\zeta(\eta) - 2\zeta(2\eta)\right\} \\
&= \sum_{j=1}^{N-1}\left(J_x\sigma_j^x\sigma_{j+1}^x + J_y\sigma_j^y\sigma_{j+1}^y + J_z\sigma_j^z\sigma_{j+1}^z\right) + h_x^{(-)}\sigma_1^x \\
&\quad + h_y^{(-)}\sigma_1^y + h_z^{(-)}\sigma_1^z + h_x^{(+)}\sigma_N^x + h_y^{(+)}\sigma_N^y + h_z^{(+)}\sigma_N^z,
\end{aligned} \tag{5.6.1}
$$

where the coupling constants are parameterized as

$$
J_x = \frac{e^{i\pi\eta}\sigma(\eta+\frac{\tau}{2})}{\sigma(\frac{\tau}{2})}, \quad J_y = \frac{e^{i\pi\eta}\sigma(\eta+\frac{1}{2}+\frac{\tau}{2})}{\sigma(\frac{1}{2}+\frac{\tau}{2})}, \quad J_z = \frac{\sigma(\eta+\frac{1}{2})}{\sigma(\frac{1}{2})},
$$

and the boundary magnetic fields are parameterized as

$$
\begin{aligned}
h_z^{(\mp)} &= \pm\frac{\sigma(\eta)}{\sigma(\frac{1}{2})}\prod_{l=1}^{3}\frac{\sigma(\alpha_l^{(\mp)}-\frac{1}{2})}{\sigma(\alpha_l^{(\mp)})}, \\
h_x^{(\mp)} &= \pm e^{-i\pi\left(\sum_{l=1}^{3}\alpha_l^{(\mp)}-\frac{\tau}{2}\right)}\frac{\sigma(\eta)}{\sigma(\frac{\tau}{2})}\prod_{l=1}^{3}\frac{\sigma(\alpha_l^{(\mp)}-\frac{\tau}{2})}{\sigma(\alpha_l^{(\mp)})}, \\
h_y^{(\mp)} &= \pm e^{-i\pi\left(\sum_{l=1}^{3}\alpha_l^{(\mp)}-\frac{1}{2}-\frac{\tau}{2}\right)}\frac{\sigma(\eta)}{\sigma(\frac{1}{2}+\frac{\tau}{2})}\prod_{l=1}^{3}\frac{\sigma(\alpha_l^{(\mp)}-\frac{1}{2}-\frac{\tau}{2})}{\sigma(\alpha_l^{(\mp)})}.
\end{aligned} \tag{5.6.2}
$$

Here $\sigma(u)$ is the σ-function defined by (3.2.3) and $\{\alpha_l^{(\mp)}\}$ are parameters contained in the most general K-matrices [42, 43]

$$
\begin{aligned}
K^-(u) &= \frac{\sigma(2u)}{2\sigma(u)}\left\{\mathrm{id} + \frac{c_x^{(-)}\sigma(u)e^{-i\pi u}}{\sigma(u+\frac{\tau}{2})}\sigma^x \right. \\
&\quad \left. + \frac{c_y^{(-)}\sigma(u)e^{-i\pi u}}{\sigma(u+\frac{1+\tau}{2})}\sigma^y + \frac{c_z^{(-)}\sigma(u)}{\sigma(u+\frac{1}{2})}\sigma^z\right\},
\end{aligned} \tag{5.6.3}
$$

$$
K^+(u) = K^-(-u-\eta)\big|_{\{c_l^{(-)}\}\to\{c_l^{(+)}\}}, \tag{5.6.4}
$$

where the constants $\{c_l^{(\mp)}\}$ are expressed in terms of boundary parameters $\{\alpha_l^{(\mp)}\}$ as

$$
\begin{aligned}
c_x^{(\mp)} &= e^{-i\pi(\sum_l\alpha_l^{(\mp)}-\frac{\tau}{2})}\prod_{l=1}^{3}\frac{\sigma(\alpha_l^{(\mp)}-\frac{\tau}{2})}{\sigma(\alpha_l^{(\mp)})}, \quad c_z^{(\mp)} = \prod_{l=1}^{3}\frac{\sigma(\alpha_l^{(\mp)}-\frac{1}{2})}{\sigma(\alpha_l^{(\mp)})}, \\
c_y^{(\mp)} &= e^{-i\pi(\sum_l\alpha_l^{(\mp)}-\frac{1}{2}-\frac{\tau}{2})}\prod_{l=1}^{3}\frac{\sigma(\alpha_l^{(\mp)}-\frac{1}{2}-\frac{\tau}{2})}{\sigma(\alpha_l^{(\mp)})}.
\end{aligned} \tag{5.6.5}
$$

The integrability of the above Hamiltonian is associated with the transfer matrix defined by the R-matrix (3.2.10) and the K-matrices (5.6.3)–(5.6.4) respectively, which satisfy RE (5.1.5) and the dual RE (5.1.6).

5.6.2 Operator Product Identities

Direct calculation [14] shows that

$$\mathrm{Det}_q\{K^-(u)\} = \frac{\sigma(2u-2\eta)}{\sigma(\eta)}\prod_{l=1}^{3}\frac{\sigma(\alpha_l^{(-)}+u)\sigma(\alpha_l^{(-)}-u)}{\sigma(\alpha_l^{(-)})\sigma(\alpha_l^{(-)})}, \tag{5.6.6}$$

$$\mathrm{Det}_q\{K^+(u)\} = -\frac{\sigma(2u+2\eta)}{\sigma(\eta)}\prod_{l=1}^{3}\frac{\sigma(\alpha_l^{(+)}+u)\sigma(\alpha_l^{(+)}-u)}{\sigma(\alpha_l^{(+)})\sigma(\alpha_l^{(+)})}. \tag{5.6.7}$$

This leads to the explicit value of the quantum determinant $\Delta_q(u)$

$$\Delta_q(u) = -\frac{\sigma(2u+2\eta)\sigma(2u-2\eta)}{\sigma(\eta)\,\sigma(\eta)}\prod_{\gamma=\pm}\prod_{l=1}^{3}\frac{\sigma(u+\alpha_l^{(\gamma)})\sigma(u-\alpha_l^{(\gamma)})}{\sigma(\alpha_l^{(\gamma)})\sigma(\alpha_l^{(\gamma)})}$$
$$\times\prod_{l=1}^{N}\frac{\sigma(u+\theta_l+\eta)\sigma(u+\theta_l-\eta)\sigma(u-\theta_l+\eta)\sigma(u-\theta_l-\eta)}{\sigma(\eta)\sigma(\eta)\sigma(\eta)\sigma(\eta)}. \tag{5.6.8}$$

The operator product identities thus read

$$t(\theta_j)t(\theta_j-\eta) = \frac{\Delta_q(\theta_j)\,\sigma(\eta)\,\sigma(\eta)}{\sigma(\eta-2\theta_j)\,\sigma(\eta+2\theta_j)}\times \mathrm{id},\quad j=1,\ldots,N. \tag{5.6.9}$$

Similarly, the corresponding transfer matrix $t(u)$ satisfies the crossing relation

$$t(-u-\eta) = t(u). \tag{5.6.10}$$

From the periodic properties of σ-function we can easily derive the relations

$$R_{1,2}(u+1) = -\sigma_1^z R_{1,2}(u)\sigma_1^z = -\sigma_2^z R_{1,2}(u)\sigma_2^z,$$
$$K^{\mp}(u+1) = -\sigma^z K^{\mp}(u)\sigma^z,$$
$$R_{1,2}(u+\tau) = -e^{-2i\pi(u+\frac{\eta}{2}+\frac{\tau}{2})}\sigma_1^x R_{1,2}(u)\sigma_1^x = -e^{-2i\pi(u+\frac{\eta}{2}+\frac{\tau}{2})}\sigma_2^x R_{1,2}(u)\sigma_2^x,$$
$$R_{1,2}(u+1+\tau) = e^{-2i\pi(u+\frac{\eta}{2}+\frac{\tau}{2})}\sigma_1^y R_{1,2}(u)\sigma_1^y = e^{-2i\pi(u+\frac{\eta}{2}+\frac{\tau}{2})}\sigma_2^y R_{1,2}(u)\sigma_2^y,$$
$$K^-(u+\tau) = -e^{-2i\pi(3u+\frac{3}{2}\tau)}\sigma^x K^-(u)\sigma^x,$$

$$K^-(u+1+\tau) = e^{-2i\pi(3u+\frac{3}{2}\tau)}\sigma^y K^-(u)\sigma^y,$$
$$K^+(u+\tau) = -e^{-2i\pi(3u+3\eta+\frac{3}{2}\tau)}\sigma^x K^+(u)\sigma^x,$$
$$K^+(u+1+\tau) = e^{-2i\pi(3u+3\eta+\frac{3}{2}\tau)}\sigma^y K^+(u)\sigma^y. \quad (5.6.11)$$

These relations lead to the following quasi-periodic properties of the transfer matrix $t(u)$:

$$t(u+1) = t(u), \quad t(u+\tau) = e^{-2i\pi(N+3)(2u+\eta+\tau)}\,t(u). \quad (5.6.12)$$

From the definition of $K^-(u)$ in (5.6.3), we have

$$K^-(0) = \frac{1}{2}tr\left[K^-(0)\right] \times \text{id}, \quad (5.6.13)$$
$$K^-\left(\frac{1}{2}\right) = \frac{1}{2}tr\left[K^-\left(\frac{1}{2}\right)\sigma^z\right] \times \sigma^z, \quad (5.6.14)$$
$$K^-\left(\frac{\tau}{2}\right) = \frac{1}{2}tr\left[K^-\left(\frac{\tau}{2}\right)\sigma^x\right] \times \sigma^x, \quad (5.6.15)$$
$$K^-\left(\frac{1+\tau}{2}\right) = \frac{1}{2}tr\left[K^-\left(\frac{1+\tau}{2}\right)\sigma^y\right] \times \sigma^y. \quad (5.6.16)$$

The above relations allow us to derive

$$t(0) = tr\left[K^+(0)\right]tr\left[K^-(0)\right]\prod_{l=1}^{N}\frac{\sigma(\eta+\theta_l)\sigma(\eta-\theta_l)}{\sigma(\eta)\sigma(\eta)} \times \text{id}, \quad (5.6.17)$$

$$t\left(\frac{1}{2}\right) = (-1)^N tr\left[K^+\left(\frac{1}{2}\right)\sigma^z\right]tr\left[K^-\left(\frac{1}{2}\right)\sigma^z\right] \times\prod_{l=1}^{N}\frac{\sigma(\eta+\frac{1}{2}+\theta_l)\sigma(\eta-\frac{1}{2}-\theta_l)}{\sigma(\eta)\sigma(\eta)} \times \text{id}, \quad (5.6.18)$$

$$t\left(\frac{\tau}{2}\right) = (-1)^N e^{-2\pi i\left\{\frac{N}{2}\eta-\sum_{j=1}^{N}\theta_j\right\}} tr\left[K^+\left(\frac{\tau}{2}\right)\sigma^x\right]tr\left[K^-\left(\frac{\tau}{2}\right)\sigma^x\right] \times\prod_{l=1}^{N}\frac{\sigma(\eta+\frac{\tau}{2}+\theta_l)\sigma(\eta-\frac{\tau}{2}-\theta_l)}{\sigma(\eta)\sigma(\eta)} \times \text{id}, \quad (5.6.19)$$

$$t\left(\frac{1+\tau}{2}\right) = e^{-2\pi i\left\{\frac{N}{2}\eta-\sum_{j=1}^{N}\theta_j\right\}} tr\left[K^+\left(\frac{1+\tau}{2}\right)\sigma^y\right]tr\left[K^-\left(\frac{1+\tau}{2}\right)\sigma^y\right] \times(-1)^N\prod_{l=1}^{N}\frac{\sigma(\eta+\frac{1+\tau}{2}+\theta_l)\sigma(\eta-\frac{1+\tau}{2}-\theta_l)}{\sigma(\eta)\sigma(\eta)} \times \text{id}. \quad (5.6.20)$$

As we discussed above, $\Lambda(u)$ should satisfy the same relations of $t(u)$, namely, the functional relations

$$\Lambda(\theta_j)\Lambda(\theta_j-\eta)=\frac{\Delta_q(\theta_j)\,\sigma(\eta)\,\sigma(\eta)}{\sigma(\eta-2\theta_j)\,\sigma(\eta+2\theta_j)},\quad j=1,\ldots,N, \tag{5.6.21}$$

the crossing symmetry and the initial values

$$\Lambda(-u-\eta)=\Lambda(u), \tag{5.6.22}$$

$$\Lambda(0)=tr(K^+(0))tr(K^-(0))\prod_{l=1}^{N}\frac{\sigma(\eta+\theta_l)\sigma(\eta-\theta_l)}{\sigma(\eta)\sigma(\eta)}, \tag{5.6.23}$$

$$\Lambda\left(\frac{1}{2}\right)=(-1)^N tr\left[K^+\left(\frac{1}{2}\right)\sigma^z\right]tr\left[K^-\left(\frac{1}{2}\right)\sigma^z\right]\times\prod_{l=1}^{N}\frac{\sigma(\eta+\frac{1}{2}+\theta_l)\sigma(\eta-\frac{1}{2}-\theta_l)}{\sigma(\eta)\sigma(\eta)}, \tag{5.6.24}$$

$$\Lambda\left(\frac{\tau}{2}\right)=(-1)^N e^{-2\pi i\left\{\frac{N}{2}\eta-\sum_{j=1}^{N}\theta_j\right\}}tr\left[K^+\left(\frac{\tau}{2}\right)\sigma^x\right]tr\left[K^-\left(\frac{\tau}{2}\right)\sigma^x\right]\times\prod_{l=1}^{N}\frac{\sigma(\eta+\frac{\tau}{2}+\theta_l)\sigma(\eta-\frac{\tau}{2}-\theta_l)}{\sigma(\eta)\sigma(\eta)}, \tag{5.6.25}$$

$$\Lambda\left(\frac{1+\tau}{2}\right)=(-1)^N e^{-2\pi i\left\{\frac{N}{2}\eta-\sum_{j=1}^{N}\theta_j\right\}}tr\left[K^+\left(\frac{1+\tau}{2}\right)\sigma^y\right]\times tr\left[K^-\left(\frac{1+\tau}{2}\right)\sigma^y\right]\prod_{l=1}^{N}\frac{\sigma(\eta+\frac{1+\tau}{2}+\theta_l)\sigma(\eta-\frac{1+\tau}{2}-\theta_l)}{\sigma(\eta)\sigma(\eta)}, \tag{5.6.26}$$

the quasi-periodic properties

$$\Lambda(u+1)=\Lambda(u),\quad \Lambda(u+\tau)=e^{-2i\pi(N+3)(2u+\eta+\tau)}\,\Lambda(u), \tag{5.6.27}$$

and the analyticity

$$\Lambda(u),\text{ as an entire function of }u\text{, is an elliptic polynomial of degree }2N+6. \tag{5.6.28}$$

Therefore the values of $\Lambda(u)$ at generic $2N+6$ points in the fundamental domain of the elliptic functions together with its quasi-periodicity, i.e., Eqs. (5.6.21)–(5.6.28) suffice to determine the function completely.

5.6.3 The Inhomogeneous $T - Q$ Relation

For convenience, let us introduce the notation

$$A(u) = \prod_{j=1}^{N} \frac{\sigma(u+\theta_j+\eta)\,\sigma(u-\theta_j+\eta)}{\sigma(\eta)\,\sigma(\eta)}, \tag{5.6.29}$$

$$a(u) = -e^{-2i\pi l_1 u}\frac{\sigma(2u+2\eta)}{\sigma(2u+\eta)} \prod_{\gamma=\pm}\prod_{l=1}^{3} \frac{\sigma(u-\varepsilon_l^{(\gamma)}\alpha_l^{(\gamma)})}{\sigma(\varepsilon_l^{(\gamma)}\alpha_l^{(\gamma)})} A(u), \tag{5.6.30}$$

$$d(u) = a(-u-\eta), \tag{5.6.31}$$

where l_1 is an even integer and $\varepsilon_l^{(\gamma)} = \pm 1$. The solutions of (5.6.21)–(5.6.28) can thus be constructed by the $T - Q$ relation

$$\begin{aligned} \Lambda(u) &= a(u)\frac{Q_1(u-\eta)}{Q_2(u)} + d(u)\frac{Q_2(u+\eta)}{Q_1(u)} \\ &\quad + \frac{c\,\sigma(2u)\sigma(2u+2\eta)\sigma^m(u)\sigma^m(u+\eta)}{Q_1(u)Q_2(u)\sigma^m(\eta)\sigma^m(\eta)} A(u)A(-u-\eta), \end{aligned} \tag{5.6.32}$$

where the Q-functions parameterized by $M = N + 1 + m$ Bethe roots $\{\mu_j | j = 1, \ldots, M\}$ are defined as

$$Q_1(u) = \prod_{j=1}^{M} \frac{\sigma(u-\mu_j)}{\sigma(\eta)}, \quad Q_2(u) = \prod_{j=1}^{M} \frac{\sigma(u+\mu_j+\eta)}{\sigma(\eta)}, \tag{5.6.33}$$

and $m = 0$ ($m = 1$) for an even N (an odd N). The M Bethe roots μ_j and the constant c should satisfy the following $M + 1$ equations

$$\sum_{\gamma=\pm}\sum_{l=1}^{3} \varepsilon_l^{(\gamma)}\alpha_l^{(\gamma)} + (M+2+n)\eta + 2\sum_{j=1}^{M}\mu_j = l_1\tau \mod(2), \tag{5.6.34}$$

$$\frac{c\,\sigma(2\mu_j)\sigma(2\mu_j+2\eta)\sigma^m(\mu_j)\sigma^m(\mu_j+\eta)A(\mu_j)A(-\mu_j-\eta)}{d(\mu_j)\sigma^m(\eta)\sigma^m(\eta)Q_2(\mu_j)Q_2(\mu_j+\eta)} = -1, \tag{5.6.35}$$

with the selection rule (5.3.29) for the Bethe roots. In the homogeneous limit, the BAEs (5.6.35) read

$$\begin{aligned} &c\,e^{-2i\pi l_1(\mu_j+\eta)}\,\sigma^2(\eta)\sigma^m(\mu_j)\sigma^{2N+n}(\mu_j+\eta)\sigma(2\mu_j+\eta)\sigma(2\mu_j+2\eta) \\ &= \prod_{\gamma=\pm}\prod_{l=1}^{3} \frac{\sigma(\mu_j+\varepsilon_l^{(\gamma)}\alpha_l^{(\gamma)}+\eta)}{\sigma(\varepsilon_l^{(\gamma)}\alpha_l^{(\gamma)})} \prod_{l=1}^{M}\sigma(\mu_j+\mu_l+\eta)\sigma(\mu_j+\mu_l+2\eta), \end{aligned} \tag{5.6.36}$$

Table 5.11 The numerical solutions of (5.6.34) and (5.6.36) for $N = 2$, $\eta = -\frac{2}{3}$, $\tau = i$, $\alpha_1^{(-)} = 0.7$, $\alpha_2^{(-)} = 0.4$, $\alpha_3^{(-)} = 0.6$, $\alpha_1^{(+)} = 0.3$, $\alpha_2^{(+)} = 0.54$, $\alpha_3^{(+)} = 0.8$ and $l_1 = 0$

μ_1	μ_2	μ_3	c	E_n	n
$0.32451 - 0.10312i$	$0.32451 + 0.10312i$	$0.34764 - 0.00000i$	-0.00592	-4.62857	1
$-0.16841 - 0.15355i$	$-0.16841 + 0.15355i$	$0.33348 - 0.00000i$	-0.04571	0.20911	2
$-0.13394 + 0.00000i$	$-0.02024 - 0.00000i$	$0.15085 + 0.00000i$	0.00377	1.85348	3
$-0.16408 - 0.05393i$	$-0.16408 + 0.05393i$	$0.32482 + 0.00000i$	-0.02092	2.56597	4

The eigenvalues E_n calculated from (5.6.37) are the same as those from the exact diagonalization of the Hamiltonian (5.6.1)

Table 5.12 The numerical solutions of (5.6.34) and (5.6.36) for $N = 3, \eta = -\frac{2}{3}, \tau = i, \alpha_1^{(-)} = 0.7$, $\alpha_2^{(-)} = 0.4$, $\alpha_3^{(-)} = 0.6$, $\alpha_1^{(+)} = 0.3$, $\alpha_2^{(+)} = 0.54$, $\alpha_3^{(+)} = 0.8$ and $l_1 = 0$

μ_1	μ_2	μ_3
$-0.14193 - 0.00000i$	$0.07809 + 0.00000i$	$0.21920 - 0.00000i$
$-0.03293 - 0.00000i$	$0.07930 + 0.00000i$	$0.28247 - 0.02571i$
$-0.33332 - 0.16659i$	$-0.33332 + 0.16659i$	$-0.01760 + 0.00000i$
$-0.13331 + 0.00000i$	$0.10195 - 0.00000i$	$0.33228 - 0.11710i$
$-0.33332 - 0.09623i$	$-0.33332 + 0.09623i$	$-0.16457 - 0.15392i$
$-0.16671 - 0.16577i$	$-0.16671 + 0.16577i$	$-0.13317 + 0.00000i$
$-0.37731 + 0.00000i$	$-0.28935 - 0.00000i$	$-0.16917 - 0.05505i$
$-0.16659 - 0.06662i$	$-0.16659 + 0.06662i$	$-0.13271 + 0.00000i$

μ_4	μ_5	c	E_n	n
$0.36608 - 0.00000i$	$0.47522 + 0.00000i$	0.00004	-3.39556	1
$0.28247 + 0.02571i$	$0.38537 + 0.00000i$	-0.00002	-3.33751	2
$0.34046 - 0.10824i$	$0.34046 + 0.10824i$	0.01360	-3.02894	3
$0.33228 + 0.11710i$	$0.36346 + 0.00000i$	-0.00035	-2.51981	4
$-0.16457 + 0.15392i$	$-0.00754 + 0.00000i$	0.05665	1.76506	5
$0.10246 - 0.00000i$	$0.36080 + 0.00000i$	-0.00255	2.00149	6
$-0.16917 + 0.05505i$	$0.00167 - 0.00000i$	0.02178	4.16110	7
$0.10441 + 0.00000i$	$0.35815 - 0.00000i$	-0.00135	4.35417	8

The eigenvalues E_n calculated from (5.6.37) are the same as those from the exact diagonalization of the Hamiltonian (5.6.1)

and the eigenvalue of the Hamiltonian in terms of the Bethe roots is

$$E = \frac{\sigma(\eta)}{\sigma'(0)} \left\{ -2\sum_{j=1}^{M} \zeta(u_j + \eta) + (N-1)\zeta(\eta) \right.$$
$$\left. -2i\pi l_1 - \sum_{\gamma=\pm} \sum_{l=1}^{3} \zeta(\varepsilon_l^{(\gamma)} \alpha_l^{(\gamma)}) \right\}. \tag{5.6.37}$$

We note that any choice of $\varepsilon_l^{(\gamma)}$ should give a complete set of solutions. Numerical solutions of the BAEs given in Table 5.11 and in Table 5.12 indicate that the $T-Q$ relation and the BAEs indeed give the correct and complete spectrum of the Hamiltonian even with a fixed $l_1 = 0$.

Finally, we should point out that if the boundary parameters $\{\alpha_l^{(\gamma)}\}$ satisfy the constraint [14]

$$\sum_{\gamma=\pm}\sum_{l=1}^{3}\varepsilon_l^{(\gamma)}\alpha_l^{(\gamma)} - l_1\tau - k\eta = 0 \mod(2), \tag{5.6.38}$$

where k is an integer, then $c = 0$ and the $T - Q$ relation (5.6.32) becomes a usual one [8, 14]. It should be remarked that, similar to the situation of the periodic XYZ model with an odd N, a $T - Q$ relation of the form (5.1.38) with a single Q-function cannot express the spectrum of the open XYZ spin chain with generic boundaries for both even and odd N.

References

1. M. Gaudin, Boundary energy of a Bose gas in one dimension. Phys. Rev. A **4**, 386 (1971)
2. F.C. Alcaraz, M.N. Barber, M.T. Batchelor, R.J. Baxter, G.R.W. Quispel, Surface exponents of the quantum XXZ, Ashkin-Teller and Potts models. J. Phys. A **20**, 6397 (1987)
3. E.K. Sklyanin, Boundary conditions for integrable quantum systems. J. Phys. A **21**, 2375 (1988)
4. I. Cherednik, Factorizing particles on a half line and root sytsems. Theor. Math. Phys. **61**, 977 (1984)
5. H.J. de Vega, A. González-Ruiz, Boundary K-matrices for the six vertex and the $n(2n-1)A_{n-1}$ vertex models. J. Phys. A **26**, L519 (1993)
6. H.J. de Vega, A. González-Ruiz, Boundary K-matrices for the XYZ, XXZ and XXX spin chains. J. Phys. A **27**, 6129 (1994)
7. S. Ghoshal, A.B. Zamolodchikov, Boundary S-Matrix and boundary state in two-dimensional integrable quantum field theory. Int. J. Mod. Phys. A **9**, 3841 (1994)
8. H. Fan, B.-Y. Hou, K.-J. Shi, Z.-X. Yang, Algebraic Bethe Ansatz for the eight-vertex model with general open boundary conditions. Nucl. Phys. B **478**, 723 (1996)
9. J. Cao, H.-Q. Lin, K.-J. Shi, Y. Wang, Exact solution of XXZ spin chain with unparallel boundary fields. Nucl. Phys. B **663**, 487 (2003)
10. R.I. Nepomechie, Solving the open XXZ spin chain with nondiagonal boundary terms at roots of unity. Nucl. Phys. B **622**, 615 (2002)
11. R.I. Nepomechie, Functional relations and Bethe Ansatz for the XXZ chain. J. Stat. Phys. **111**, 1363 (2003)
12. R.I. Nepomechie, Bethe Ansatz solution of the open XXZ chain with nondiagonal boundary terms. J. Phys. A **37**, 433 (2004)
13. W.-L. Yang, R.I. Nepomechie, Y.-Z. Zhang, Q-operator and T-Q relation from the fusion hierarchy. Phys. Lett. B **633**, 664 (2006)
14. W.-L. Yang, Y.-Z. Zhang, T-Q relation and exact solution for the XYZ chain with general nondiagonal boundary terms. Nucl. Phys. B **744**, 312 (2006)
15. P. Baseilhac, K. Koizumi, Exact spectrum of the XXZ open spin chain from the q-Onsager algebra representation theory. J. Stat. Mech. P09006 (2007)
16. H. Frahm, A. Seel, T. Wirth, Separation of variables in the open XXX chain. Nucl. Phys. B **802**, 351 (2008)
17. G. Niccoli, Non-diagonal open spin-1/2 XXZ quantum chains by separation of variables: complete spectrum and matrix elements of some quasi-local operators. J. Stat. Mech. P10025 (2012)

18. G. Niccoli, Antiperiodic spin-1/2 XXZ quantum chains by separation of variables: complete spectrum and form factors. Nucl. Phys. B **870**, 397 (2013)
19. G. Niccoli, Antiperiodic dynamical 6-vertex model I: complete spectrum by SOV, matrix elements of the identity on separate states and connections to the periodic 8-vertex model. J. Phys. A **46**, 075003 (2013)
20. S. Faldella, N. Kitanine, G. Niccoli, The complete spectrum and scalar products for the open spin-1/2 XXZ quantum chains with non-diagonal boundary terms. J. Stat. Mech. P01011 (2014)
21. N. Kitanine, J.-M. Maillet, G. Niccoli, Open spin chains with generic integrable boundaries: Baxter equation and Bethe Ansatz completeness from SOV, J. Stat. Mech. P05015 (2014)
22. J. Cao, W.-L. Yang, K. Shi, Y. Wang, Off-diagonal Bethe Ansatz solution of the XXX spin chain with arbitrary boundary conditions. Nucl. Phys. B **875**, 152 (2013)
23. J. Cao, W.-L. Yang, K. Shi, Y. Wang, Off-diagonal Bethe Ansatz solutions of the anisotropic spin-1/2 chains with arbitrary boundary fields. Nucl. Phys. B **877**, 152 (2013)
24. S. Belliard, N. Crampé, Heisenberg XXX model with general boundaries: eigenvectors from algebraic Bethe Ansatz. SIGMA **9**, 072 (2013)
25. X. Zhang, Y.-Y. Li, J. Cao, W. -L. Yang, K. Shi, Y. Wang, Retrive the Bethe states of quantum integrable models solved via off-diagonal Bethe Ansatz.arXiv:1407.5294
26. X. Zhang, Y.-Y. Li, J. Cao, W.-L. Yang, K. Shi, Y. Wang, Bethe states of the XXZ spin-1/2 chian with arbitrary boundary fields. Nucl. Phys. B **893**, 70 (2015)
27. Y.-Y. Li, J. Cao, W.-L. Yang, K. Shi, Y. Wang, Thermodynamic limit and surface energy of the XXZ spin chain with arbitrary boundary fields. Nucl. Phys. B **884**, 17 (2014)
28. R.I. Nepomechie, Inhomogeneous T-Q equation for the open XXX chain with general boundary terms: completeness and arbitrary spin. J. Phys. A **46**, 442002 (2013)
29. W.-L. Yang, Y.-Z. Zhang, On the second reference state and complete eigenstates of the open XXZ chain. J. High Energy Phys. **04**, 044 (2007)
30. W.-L. Yang, Y.-Z. Zhang, Multiple reference states and complete spectrum of the Z_n Belavin model with open boundaries. Nucl. Phys. B **789**, 591 (2008)
31. R.I. Nepomechie, F. Ravanini, Completeness of the Bethe Ansatz solution of the open XXZ chain with nondiagonal boundary terms. J. Phys. A **36**, 11391 (2003)
32. M. Takahashi, *Thermodynamics of One-Dimensional Solvable Models* (Cambridge University Press, Cambridge, England, 1999)
33. C.J. Hamer, G.R.W. Quispel, M.T. Batchelor, Conformal anomaly and surface energy for Potts and Ashkin-Teller quantum chains. J. Phys. A **20**, 5677 (1987)
34. M.T. Batchelor, C.J. Hamer, Surface energy of integrable quantum spin chains. J. Phys. A **23**, 761 (1990)
35. A. Kapustin, S. Skorik, Surface excitations and surface energy of the antiferromagnetic XXZ chain by the Bethe Ansatz approach. J. Phys. A **29**, 1629 (1996)
36. R. Murgan, R.I. Nepomechie, C. Shi, Boundary energy of the general open XXZ chain at roots of unity. J. High Energy Phys. **01**, 038 (2007)
37. C. Ahn, R.I. Nepomechie, Finite size effects in the XXZ and sine-Gordon models with two boundaries. Nucl. Phys. B **676**, 637 (2004)
38. C.N. Yang, C.P. Yang, One-dimensional chain of anisotropic spin-spin interactions II. Properties of the ground state energy per lattice site for an infinite system. Phys. Rev. **150**, 327 (1966)
39. H.J. de Vega, F. Woynarovich, Method for calculating finite size corrections in Bethe Ansatz systems—Heisenberg chains and 6-vertex model. Nucl. Phys. B **251**, 439 (1985)
40. F. Woynarovich, H.P. Eckle, Finite-size corrections and numerical calculations for long spin-1/2 Heisenberg chains in the critical region. J. Phys. A **20**, L97 (1987)
41. J. de Gier, F.H.L. Essler, Bethe Ansatz solution of the asymmetric exclusion process with open boundaries. Phys. Rev. Lett. **95**, 240601 (2005)
42. T. Inami, H. Konno, Integrable XYZ spin chain with boundaries. J. Phys. A **27**, L913 (1994)
43. B.-Y. Hou, K.-J. Shi, H. Fan, Z.-X. Yang, Solution of reflection equation. Commun. Theor. Phys. **23**, 163 (1995)

Chapter 6
The One-Dimensional Hubbard Model

As one of the minimal models for strongly correlated electron systems, the Hubbard model plays a central role in modern condensed matter physics [1–3]. Interestingly, this model in one spatial dimension is exactly solvable [4], which provides an important benchmark for understanding the Mott insulators. In the past several decades, numerous efforts have been made to study the integrability and the physical properties of this model [5–10]. A remarkable result was obtained by Shastry [11–13] who constructed the corresponding R-matrix and the Lax matrix of the one-dimensional Hubbard model, thus demonstrating its integrability under the framework of YBE. The graded version of the R-matrix was then obtained by Wadati et al. [14–18]. With Shastry's findings, the model was resolved [19, 20] via the algebraic Bethe Ansatz method. On the other hand, the reason that the open-boundary problem of this model has attracted a lot of attention is because of its relationship to the impurity problem in a Luttinger liquid [21]. The exact solution of the open Hubbard chain was first obtained by Shulz [22], whose result was subsequently generalized to the non-zero boundary potential case [23–25]. The integrability of the one-dimensional Hubbard model with diagonal open boundaries was demonstrated in [26] by the construction of the Lax representation, while the generic integrable boundary conditions were obtained in [27] by solving YBE and RE. It was found that applying either scalar potentials or magnetic fields on the two end sites does not break the integrability of this model. However, the conventional Bethe Ansatz methods could only construct the exact solutions for parallel boundary conditions. The Bethe Ansatz solution of this model with generic non-diagonal boundaries was given by the ODBA method [28].

The Hubbard model in the strongly repulsive limit becomes the $t - J$ model where double occupancy of two electrons on a single site is forbidden. This model in two spatial dimensions may provide a non-phonon mechanism for high-T_c superconductivity [29]. Interestingly, the $t - J$ model in one spatial dimension at the super-symmetric points $2t = \pm J$ is also exactly solvable [30–34]. Based on this observation, some important physical properties including the elementary excitations [35], the correlation functions [36] and the thermodynamics [37, 38] were extensively studied. It was found that special cases of the model with certain open boundaries are also integrable [39–45].

Y. Wang et al., *Off-Diagonal Bethe Ansatz for Exactly Solvable Models*,
DOI 10.1007/978-3-662-46756-5_6

This chapter is devoted to the exact solutions of both the one-dimensional Hubbard model and the super-symmetric $t-J$ model with generic integrable boundary conditions by the combination of the coordinate Bethe Ansatz and the ODBA [28, 46].

6.1 The Periodic Hubbard Model

The Hamiltonian of the one-dimensional Hubbard model reads

$$H=-t\sum_{\alpha,j=1}^{N}[c^{\dagger}_{j,\alpha}c_{j+1,\alpha}+c^{\dagger}_{j+1,\alpha}c_{j,\alpha}]+U\sum_{j=1}^{N}n_{j,\uparrow}n_{j,\downarrow}, \tag{6.1.1}$$

where $c^{\dagger}_{j,\alpha}$ and $c_{j,\alpha}$ are the creation and annihilation operators of electrons on site j with spin component $\alpha=\uparrow,\downarrow$; N is the number of sites; t and U are the hopping constant and the on-site repulsive interaction constant, respectively. Note that the periodic boundary condition $c_{N+1,\alpha}=c_{1,\alpha}$ is assumed. Obviously, the model possesses $U(1)\times SU(2)_{spin}\times SU(2)_{charge}$ symmetry. In fact, a hidden $SO(4)$ symmetry [47, 48] also holds.

6.1.1 Coordinate Bethe Ansatz

The $U(1)$ symmetry in the charge sector allows us to construct the eigenstates of the Hamiltonian (6.1.1) in the form

$$|\Psi\rangle=\sum_{j=1}^{M}\sum_{\alpha_j=\uparrow,\downarrow}\sum_{x_j=1}^{N}\Psi^{\{\alpha\}}(x_1,\ldots,x_M)c^{\dagger}_{x_1,\alpha_1}\cdots c^{\dagger}_{x_M,\alpha_M}|0\rangle, \tag{6.1.2}$$

where M is the number of electrons and $\{\alpha\}=\{\alpha_1,\ldots,\alpha_M\}$. The eigenvalue equation of the Hamiltonian (6.1.1) thus reads

$$\begin{aligned}&-t\sum_{j=1}^{M}[\Psi^{\{\alpha\}}(\ldots,x_j+1,\ldots)+\Psi^{\{\alpha\}}(\ldots,x_j-1,\ldots)]\\&+U\sum_{i<j}^{M}\delta_{x_i,x_j}\Psi^{\{\alpha\}}(x_1,\ldots,x_M)=E\,\Psi^{\{\alpha\}}(x_1,\ldots,x_M).\end{aligned} \tag{6.1.3}$$

The wave function takes the following Bethe Ansatz form [49]

$$\Psi^{\{\alpha\}}(x_1, \ldots, x_M) = \sum_{p,q} A_p^{\{\alpha\}}(q) \exp\left[i \sum_{j=1}^{M} k_{p_j} x_{q_j}\right] \times\theta(x_{q_1} \leq x_{q_2} \leq \cdots \leq x_{q_M}), \tag{6.1.4}$$

where $p = \{p_1, \ldots, p_M\}$ and $q = \{q_1, \ldots, q_M\}$ are the permutations of $\{1, \ldots, M\}$ and $\theta(x_1 \leq \cdots \leq x_M)$ is the generalized step function, which is one in the noted variables' region and zero otherwise.

For all $x_j \neq x_l$ cases, Eq. (6.1.3) is automatically satisfied and the corresponding eigenvalue is

$$E = -2t \sum_{j=1}^{M} \cos k_j. \tag{6.1.5}$$

For the case of two electrons occupying the same site, we should consider the continuity of the wave function $\Psi^{\{\alpha\}}(x_1, \ldots, x_M)$. Considering the sector

$$I: \quad x_{q_1} \leq x_{q_2} \leq \cdots \leq x_{q_j} \leq x_{q_{j+1}} \leq \cdots \leq x_{q_M}, \tag{6.1.6}$$

and the sector

$$II: \quad x_{q_1} \leq x_{q_2} \leq \cdots \leq x_{q_{j+1}} \leq x_{q_j} \leq \cdots \leq x_{q_M}, \tag{6.1.7}$$

when $x_{q_j} = x_{q_{j+1}} = x$, the continuity of the wave function $\Psi^{\{\alpha\}}(x_1, \ldots, x_M)$ requires that

$$\Psi_I^{\{\alpha\}}(\ldots, x, x, \ldots) = \Psi_{II}^{\{\alpha\}}(\ldots, x, x, \ldots). \tag{6.1.8}$$

For convenience, we omit the superscript $\{\alpha\}$ and treat $A_p(q)$ as a column vector in the spin space. The continuity condition (6.1.8) implies

$$A_p(q) + A_{p'}(q) = A_p(q') + A_{p'}(q'), \tag{6.1.9}$$

where $q' = \{\ldots, q_{j+1}, q_j, \ldots\}$, and $p' = \{\ldots, p_{j+1}, p_j, \ldots\}$. For $x_{q_j} = x_{q_{j+1}}$, the Schrödinger equation (6.1.3) gives rise to

$$\begin{aligned} -t\Big[&A_p(q')e^{ik_{p_{j+1}}} + A_{p'}(q')e^{ik_{p_j}} + A_p(q)e^{-ik_{p_j}} + A_{p'}(q)e^{-ik_{p_{j+1}}} \\ &+A_p(q)e^{ik_{p_{j+1}}} + A_{p'}(q)e^{ik_{p_j}} + A_p(q')e^{-ik_{p_j}} + A_{p'}(q')e^{-ik_{p_{j+1}}}\Big] \\ &+U[A_p(q) + A_{p'}(q)] = -2t[\cos k_{p_j} + \cos k_{p_{j+1}}][A_p(q) + A_{p'}(q)]. \end{aligned} \tag{6.1.10}$$

Substituting Eqs. (6.1.9) into (6.1.10), we have

$$\begin{aligned}&\left[\sin(k_{p_j}) - \sin(k_{p_{j+1}})\right] A_p(q) - i\frac{U}{2t}A_p(q')\\&= \left[\sin(k_{p_j}) - \sin(k_{p_{j+1}}) + i\frac{U}{2t}\right] A_{p'}(q').\end{aligned} \tag{6.1.11}$$

Now let us define the coordinate permutation operator $\bar{P}_{i,j}$,

$$\bar{P}_{i,j} A_p(\ldots, q_i, \ldots, q_j, \ldots) = A_p(\ldots, q_j, \ldots, q_i, \ldots). \tag{6.1.12}$$

Since the fermionic wave function is completely antisymmetric with both the coordinates and spins of two particles exchanged, if we denote further $P_{i,j}$ as the spin permutation operator, we have

$$P_{i,j}\bar{P}_{i,j} = -1, \quad P_{i,j}^2 = \bar{P}_{i,j}^2 = 1, \tag{6.1.13}$$

which implies the relation

$$-P_{j,j+1}A_p(q) = A_p(q'). \tag{6.1.14}$$

Substituting this relation into Eq. (6.1.11), we readily obtain

$$A_p(q) = S_{p_j,p_{j+1}}(k_{p_j}, k_{p_{j+1}})A_{p'}(q'), \tag{6.1.15}$$

with the S-matrix given by

$$S_{j,l}(k_j, k_l) = \frac{\sin k_j - \sin k_l - i\frac{U}{2t}P_{j,l}}{\sin k_j - \sin k_l - i\frac{U}{2t}}. \tag{6.1.16}$$

Now let us consider the following process: The jth particle moves from the left end to the right end by scattering with all the other particles to its right. This process can be described by the relations

$$\begin{aligned}A_p(q) = {} & S_{j,j+1}(k_j, k_{j+1})S_{j,j+2}(k_j, k_{j+2}) \cdots S_{j,M}(k_j, k_M)\\&\times S_{j,1}(k_j, k_1) \cdots S_{j,j-1}(k_j, k_{j-1})e^{-ik_jN}A_p(q).\end{aligned} \tag{6.1.17}$$

Note above we have used the periodic boundary condition

$$\Psi(\ldots, x_j + N, \ldots) = \Psi(\ldots, x_j, \ldots). \tag{6.1.18}$$

The relation (6.1.17) was first obtained and solved by Yang [49] with a very opaque Ansatz. Here we introduce an alternative approach to this second eigenvalue problem.

6.1.2 Solution of the Second Eigenvalue Problem

For convenience, we introduce the notations

$$R_{j,l}(u) = u + \eta P_{j,l}, \tag{6.1.19}$$

with $\eta = -i\frac{U}{2t}$ and

$$\begin{aligned} T_0(u) = & R_{0,j}(u - \sin k_j)R_{0,j+1}(u - \sin k_{j+1}) \cdots R_{0,M}(u - \sin k_M) \\ & \times R_{0,1}(u - \sin k_1) \cdots R_{0,j-1}(u - \sin k_{j-1}). \end{aligned} \tag{6.1.20}$$

$T_0(u)$ is nothing but the monodromy matrix of the XXX spin chain with the inhomogeneity parameters $\{\theta_j = \sin k_j\}$. This allows us to transform the eigenvalue problem (6.1.17) to that of the transfer matrix of the XXX spin chain model with the correspondence

$$\begin{aligned} & S_{j,j+1}(k_j, k_{j+1})S_{j,j+2}(k_j, k_{j+2}) \cdots S_{j,M}(k_j, k_M) \\ & \times S_{j,1}(k_j, k_1) \cdots S_{j,j-1}(k_j, k_{j-1}) = \frac{\tau(\sin k_j)}{\eta \prod_{l \neq j}^{M} (\sin k_j - \sin k_l + \eta)}, \end{aligned} \tag{6.1.21}$$

where $\tau(u) = tr_0 T_0(u)$ is the transfer matrix. As the operator identities

$$\begin{aligned} \tau(\sin k_j)\tau(\sin k_j - \eta) = & \prod_{l=1}^{M} (\sin k_j - \sin k_l + \eta)(\sin k_j - \sin k_l - \eta) \times \mathrm{id}, \\ & j = 1, \ldots, M, \end{aligned} \tag{6.1.22}$$

hold and

$$\lim_{u \to \infty} \tau(u) = 2u^M + \cdots, \tag{6.1.23}$$

by the same procedure introduced in Chap. 1 we conclude that the eigenvalue of the transfer matrix takes the form

$$\Lambda(u) = \prod_{j=1}^{M} (u - \sin k_j + \eta) \frac{Q(u - \eta)}{Q(u)} + \prod_{j=1}^{M} (u - \sin k_j) \frac{Q(u + \eta)}{Q(u)}, \tag{6.1.24}$$

with

$$Q(u) = \prod_{\alpha=1}^{\bar{M}} (u - \lambda_\alpha), \tag{6.1.25}$$

and $\bar{M} \leq M$ is a non-negative integer. Because $\Lambda(u)$ must be a polynomial of degree M as required, its regularity induces

$$\prod_{j=1}^{M} \frac{\lambda_\alpha - \sin k_j}{\lambda_\alpha - \sin k_j + \eta} = -\prod_{\beta=1}^{\bar{M}} \frac{\lambda_\alpha - \lambda_\beta - \eta}{\lambda_\alpha - \lambda_\beta + \eta}, \quad \alpha = 1, \ldots, \bar{M}. \tag{6.1.26}$$

From (6.1.17) and (6.1.21) we have

$$e^{ik_jN} = \frac{\Lambda(\sin k_j)}{\eta \prod_{l \neq j}(\sin k_j - \sin k_l + \eta)}, \quad j = 1, \ldots, M. \tag{6.1.27}$$

Substituting (6.1.24) into (6.1.27) and putting $\lambda_\alpha = \mu_\alpha - \eta/2$ we readily obtain the BAEs of the present model, namely

$$e^{ik_jN} = \prod_{\alpha=1}^{\bar{M}} \frac{\sin k_j - \mu_\alpha - \frac{\eta}{2}}{\sin k_j - \mu_\alpha + \frac{\eta}{2}}, \quad j = 1, \ldots, M, \tag{6.1.28}$$

$$\prod_{j=1}^{M} \frac{\mu_\alpha - \sin k_j - \frac{\eta}{2}}{\mu_\alpha - \sin k_j + \frac{\eta}{2}} = -\prod_{\beta=1}^{\bar{M}} \frac{\mu_\alpha - \mu_\beta - \eta}{\mu_\alpha - \mu_\beta + \eta}, \quad \alpha = 1, \ldots, \bar{M}. \tag{6.1.29}$$

These equations should determine the spectrum of the Hamiltonian (6.1.1) completely.

6.1.3 Ground State Energy and Mott Gap at Half Filling

Substituting $\eta = -i\frac{U}{2t}$ into (6.1.28) and (6.1.29) and taking the logarithm, we obtain

$$Nk_j = 2\pi I_j - \sum_{\alpha=1}^{\bar{M}} \theta_1(\sin k_j - \mu_\alpha), \tag{6.1.30}$$

$$\sum_{j=1}^{M} \theta_1(\mu_\alpha - \sin k_j) = 2\pi J_\alpha + \sum_{\beta=1}^{\bar{M}} \theta_2(\mu_\alpha - \mu_\beta), \tag{6.1.31}$$

where $\theta_n(x) = 2\arctan(\frac{4xt}{Un})$, I_j are integers (half odd integers) for even (odd) $\bar{M}$ and J_α are integers (half odd integers) for odd (even) $M - \bar{M}$. As for the Heisenberg chain model, let us define the counting functions

$$Z_c(k) = \frac{k}{2\pi} + \frac{1}{2N\pi} \sum_{\alpha=1}^{\bar{M}} \theta_1(\sin k - \mu_\alpha), \tag{6.1.32}$$

$$Z_s(\mu) = \frac{1}{2N\pi}\left\{\sum_{j=1}^{M}\theta_1(\mu - \sin k_j) - \sum_{\beta=1}^{\bar{M}}\theta_2(\mu - \mu_\beta)\right\}. \tag{6.1.33}$$

The corresponding density functions read

$$\rho_c(k) + \rho_c^h(k) = \frac{dZ_c(k)}{dk}, \tag{6.1.34}$$

$$\rho_s(\mu) + \rho_s^h(\mu) = \frac{dZ_s(\mu)}{d\mu}. \tag{6.1.35}$$

In the ground state, all k_j and μ_α take real values ($M \leq N$ is assumed). There is no μ hole in the whole real axis and no k hole in the interval $[-B, B]$. We thus have the equations

$$\rho_c(k) = \frac{1}{2\pi} + \cos k \int_{-\infty}^{\infty} a_1(\sin k - \mu)\rho_s(\mu)d\mu, \tag{6.1.36}$$

$$\rho_s(\mu) = \int_{-B}^{B} a_1(\mu - \sin k)\rho_c(k)dk - \int_{-\infty}^{\infty} a_2(\mu - \lambda)\rho_s(\lambda)d\lambda, \tag{6.1.37}$$

where

$$a_n(x) = \frac{1}{\pi}\frac{4tnU}{n^2U^2 + 16t^2x^2}, \tag{6.1.38}$$

and B is determined by

$$\int_{-B}^{B} \rho_c(k)dk = \frac{M}{N}. \tag{6.1.39}$$

We consider the half-filling case, i.e., $M = N$. In this case, $B = \pi$, the integral equations can be solved exactly via Fourier transformation. From equation (6.1.37) we obtain

$$\rho_s(\mu) = \frac{t}{U}\int_{-\pi}^{\pi} \frac{\rho_c(k)}{\cosh \frac{2\pi t(\mu - \sin k)}{U}}dk. \tag{6.1.40}$$

Substituting equation (6.1.36) into the above equation we have

$$\rho_s(\mu) = \frac{1}{2\pi}\int_{-\infty}^{\infty} \frac{e^{-i\omega\mu} J_0(\omega)}{2\cosh\left(\frac{U\omega}{4t}\right)}d\omega. \tag{6.1.41}$$

Substituting equation (6.1.41) into equation (6.1.36) we get

$$\rho_c(k) = \frac{1}{2\pi} + \frac{\cos k}{\pi}\int_0^{\infty} \frac{J_0(\omega)\cos(\omega \sin k)}{1 + e^{\frac{U\omega}{2t}}}d\omega. \tag{6.1.42}$$

The ground state energy per site thus reads

$$e_g = -2t \int_{-\pi}^{\pi} \cos k\rho_c(k)dk = -4t \int_0^{\infty} \frac{J_0(\omega)J_1(\omega)}{\omega(1+e^{\frac{U\omega}{2t}})} d\omega, \tag{6.1.43}$$

where

$$J_0(\omega) = \frac{1}{2\pi} \int_{-\pi}^{\pi} e^{i\omega \sin k} dk, \tag{6.1.44}$$

$$J_1(\omega) = \frac{\omega}{2\pi} \int_{-\pi}^{\pi} \cos^2 k \cos(\omega \sin k) dk. \tag{6.1.45}$$

The Mott gap is defined as (even N is assumed)

$$\Delta E = \frac{1}{2}\left[E\left(\frac{N}{2}, \frac{N}{2}+1\right) + E\left(\frac{N}{2}, \frac{N}{2}-1\right) - 2E\left(\frac{N}{2}, \frac{N}{2}\right)\right], \tag{6.1.46}$$

where $E(N_\uparrow, N_\downarrow)$ is the ground state energy of the system with $N_\uparrow$ spin-up electrons and $N_\downarrow$ spin-down electrons, $M = N_\uparrow + N_\downarrow$. With the particle-hole transformation, we can easily deduce that [4]

$$E(N_\uparrow, N_\downarrow) = E(N - N_\uparrow, N - N_\downarrow) + (M - N)U. \tag{6.1.47}$$

Therefore, the Mott gap can be expressed as

$$\Delta E = \frac{1}{2}U - \left[E\left(\frac{N}{2}, \frac{N}{2}\right) - E\left(\frac{N}{2}, \frac{N}{2}-1\right)\right]. \tag{6.1.48}$$

Suppose the charge density and the spin density of the system with $N = N_\uparrow + N_\downarrow$ electrons are $\rho_c^0(k)$ and $\rho_s^0(\mu)$, respectively. We also denote the charge density and spin density of the system with $N-1$ electrons as ρ_c^- and ρ_s^-. Thus

$$\rho_c^0(k) - \rho_c^-(k) = \frac{1}{2N}[\delta(k-\pi) + \delta(k+\pi)] + \delta\rho_c(k). \tag{6.1.49}$$

From (6.1.36) and (6.1.37), we know that the deviation of charge density induced with one electron taken away should satisfy

$$\begin{aligned}\delta\rho_c(k) &+ \frac{1}{2N}[\delta(k+\pi) + \delta(k-\pi)] \\ &= \cos k \int_{-\infty}^{\infty} a_1(\sin k - \mu)\delta\rho_s(\mu)d\mu,\end{aligned} \tag{6.1.50}$$

$$\frac{1}{N}a_1(\mu) + \int_{-\pi^-}^{\pi^-} a_1(\sin k - \mu)\delta\rho_c(k)dk$$
$$= \delta\rho_s(\mu) + \int_{-\infty}^{\infty} a_2(\mu - \lambda)\delta\rho_s(\lambda)d\lambda, \tag{6.1.51}$$

where $\delta\rho_s(\mu) = \rho_s^0(\mu) - \rho_s^-(\mu)$. With Fourier transformation, we readily have

$$\delta\rho_c(k) = -\frac{1}{2N}[\delta(k-\pi) + \delta(k+\pi)] + \frac{\cos k}{N\pi}\int_0^\infty \frac{\cos(\omega \sin k)d\omega}{1 + e^{\frac{U\omega}{2t}}}. \tag{6.1.52}$$

Therefore, the Mott gap reads

$$\Delta E = \frac{U}{2} - 2t + 4t\int_0^\infty \frac{J_1(\omega)}{\omega(e^{\frac{U\omega}{2t}} + 1)}d\omega. \tag{6.1.53}$$

By using the formula

$$\int_0^\infty \frac{J_1(\omega)}{\omega(e^{\frac{U\omega}{2t}} + 1)}d\omega = \sum_{n=1}^\infty (-1)^{n+1}\left[\left(1 + \frac{n^2U^2}{4t^2}\right)^{\frac{1}{2}} - \frac{nU}{2t}\right]$$
$$= -\frac{U}{8t} + \frac{1}{2} + \frac{2t}{U}\int_1^\infty \frac{(x^2-1)^{\frac{1}{2}}}{\sinh\frac{2\pi x t}{U}}dx, \tag{6.1.54}$$

we conclude that the energy gap ΔE is always positive as long as U is positive.

6.2 Hubbard Model with Open Boundaries

6.2.1 Coordinate Bethe Ansatz

The model Hamiltonian of the open Hubbard chain with unparallel boundary fields reads

$$H = -t\sum_{\alpha,j=1}^{N-1}[c_{j,\alpha}^\dagger c_{j+1,\alpha} + c_{j+1,\alpha}^\dagger c_{j,\alpha}] + U\sum_{j=1}^{N} n_{j,\uparrow}n_{j,\downarrow}$$
$$+ \mathbf{h}_1 \cdot \sigma_1 + \mathbf{h}_N \cdot \sigma_N, \tag{6.2.1}$$

where $\sigma_j = \sum_{\alpha,\beta} c_{j,\alpha}^\dagger \sigma_{\alpha,\beta} c_{j,\beta}$ is the spin operator on site j, and $\mathbf{h}_1$ and $\mathbf{h}_N$ indicate the boundary fields. This Hamiltonian has also been demonstrated [27] to be integrable.

We note that the unparallel boundary fields break the $U(1)$ symmetry in the spin sector. However, the $U(1)$ symmetry in the charge sector still exists. The left $U(1)$-symmetry allows us to construct the eigenstates of the Hamiltonian (6.2.1) as

$$|\Psi\rangle = \sum_{j=1}^{M} \sum_{\alpha_j=\uparrow,\downarrow} \sum_{x_j=1}^{N} \Psi^{\{\alpha\}}(x_1, \ldots, x_M) c^{\dagger}_{x_1,\alpha_1} \cdots c^{\dagger}_{x_M,\alpha_M} |0\rangle. \tag{6.2.2}$$

The associated Schrödinger equation thus reads

$$\begin{aligned}
&-t \sum_{j=1}^{M} \Big[(1-\delta_{x_j,N}) \Psi^{\{\alpha\}}(\ldots, x_j+1, \ldots) + (1-\delta_{x_j,1}) \Psi^{\{\alpha\}}(\ldots, x_j-1, \ldots) \Big] \\
&+ \sum_{j=1}^{M} \sum_{\beta_j=\uparrow,\downarrow} \left[\delta_{x_j,1} \mathbf{h}_1 \cdot \boldsymbol{\sigma}_{\alpha_j,\beta_j} + \delta_{x_j,N} \mathbf{h}_N \cdot \sigma_{\alpha_j,\beta_j} \right] \Psi^{\{\alpha\}_j}(x_1, \ldots, x_M) \\
&+ U \sum_{i<j} \delta_{x_i,x_j} \Psi^{\{\alpha\}}(x_1, \ldots, x_M) = E\, \Psi^{\{\alpha\}}(x_1, \ldots, x_M),
\end{aligned} \tag{6.2.3}$$

where $\{\alpha\}_j$ means that α_j is replaced by β_j in the set $\{\alpha\}$. The wave function takes the Bethe Ansatz form

$$\begin{aligned}
\Psi^{\{\alpha\}}(x_1, \ldots, x_M) = &\sum_{p,q,r} A_p^{\{\alpha\},r}(q) \exp\left[i \sum_{j=1}^{M} r_{p_j} k_{p_j} x_{q_j} \right] \\
&\times \theta(x_{q_1} \le x_{q_2} \le \cdots \le x_{q_M}),
\end{aligned} \tag{6.2.4}$$

where $r = \{r_1, \ldots, r_M\}$ with $r_j = \pm$. The eigenvalue E can be easily derived from (6.2.3) with the case of all $x_j \neq 1, N$ and $x_j \neq x_l$ considered

$$E = -2t \sum_{j=1}^{M} \cos k_j. \tag{6.2.5}$$

The wave function must be continuous when two electrons occupy the same site. This indicates that

$$A_p^r(q) + A_{p'}^{r'}(q) = A_p^r(q') + A_{p'}^{r'}(q'), \tag{6.2.6}$$

where $q' = \{\ldots, q_{j+1}, q_j, \ldots\}$, $p' = \{\ldots, p_{j+1}, p_j, \ldots\}$ and $r' = \{\ldots, r_{j+1}, r_j, \ldots\}$. As for the periodic boundary case, we can deduce

$$A_p^r(q) = S_{p_j,p_{j+1}}(r_{p_j} k_{p_j}, r_{p_{j+1}} k_{p_{j+1}}) A_{p'}^{r'}(q'), \tag{6.2.7}$$

with the S-matrix given by (6.1.19).

Now let us turn to the case of $x_{q_1} = 1$, $x_{q_i} \neq x_{q_j} (i \neq j)$ and $x_{q_M} \neq N$. In this case, the eigenvalue equation (6.2.3) becomes

$$-t\Psi^{\{\alpha\}}(2,\ldots) + \sum_{\beta_1} \mathbf{h}_1 \cdot \sigma_{\alpha_1,\beta_1} \Psi^{(\beta_1,\ldots)}(1,\ldots)$$
$$= -2t\cos k_{p_1} \Psi^{\{\alpha\}}(1,\ldots). \quad (6.2.8)$$

This induces

$$\sum_{\beta_1} \mathbf{h}_1 \cdot \sigma_{\alpha_1,\beta_1} \Psi^{(\beta_1,\ldots)}(1,\ldots) = -t\Psi^{\{\alpha\}}(0,\ldots), \quad (6.2.9)$$

and

$$A_p^{(+,\cdots)}(q) = \bar{K}_1^+(k_{p_1}) A_p^{(-,\cdots)}(q), \quad (6.2.10)$$

with

$$\bar{K}_j^+(k) = -\frac{t^2 - \mathbf{h}_1^2 - 2it\sin k\mathbf{h}_1 \cdot \boldsymbol{\sigma}_j}{t^2 - \mathbf{h}_1^2 e^{2ik}}, \quad (6.2.11)$$

by employing the identity $(\mathbf{h}_1 \cdot \boldsymbol{\sigma})^2 = \mathbf{h}_1^2$. With the same procedure we have

$$\sum_{\beta_M} \mathbf{h}_N \cdot \sigma_{\alpha_M,\beta_M} \Psi^{(\ldots,\beta_M)}(\ldots,N) = -t\Psi^{\{\alpha\}}(\ldots,N+1), \quad (6.2.12)$$

and

$$e^{-2ik_{p_M}N} A_p^{(\cdots,-)}(q) = \bar{K}_M^-(k_{p_M}) A_p^{(\cdots,+)}(q), \quad (6.2.13)$$

with

$$\bar{K}_j^-(k) = -\frac{t^2 - \mathbf{h}_N^2 - 2it\sin k\mathbf{h}_N \cdot \boldsymbol{\sigma}_j}{t^2 e^{-2ik} - \mathbf{h}_N^2}. \quad (6.2.14)$$

With the help of (6.2.9) and (6.2.12), we can easily deduce that (6.2.3) also holds with the Ansatz (6.2.4) for $x_{q_1} = x_{q_2} = 1$ or $x_{q_{M-1}} = x_{q_M} = N$.

Following the coordinate Bethe Ansatz procedure introduced in Chap. 1, we obtain the second eigenvalue equation

$$\bar{\tau}(k_j) A^{(\cdots,+,\cdots)} = e^{-2ik_jN} A^{(\cdots,+,\cdots)}, \quad (6.2.15)$$

where

$$\begin{aligned}\bar{\tau}(k_j) = &S_{j-1,j}(k_{j-1},k_j)\cdots S_{1,j}(k_1,k_j)\bar{K}_j^+(k_j)S_{j,1}(-k_j,k_1)\cdots\\ &\times S_{j,j-1}(-k_j,k_{j-1})S_{j,j+1}(-k_j,k_{j+1})\cdots S_{j,M}(-k_j,k_M)\bar{K}_j^-(k_j)\\ &\times S_{M,j}(k_M,k_j)\cdots S_{j+1,j}(k_{j+1},k_j).\end{aligned} \tag{6.2.16}$$

6.2.2 *Off-Diagonal Bethe Ansatz*

Let us introduce the K-matrices

$$K^-(u) = \bar{p} + u\,\mathbf{h}_N\cdot\boldsymbol{\sigma}, \tag{6.2.17}$$

$$K^+(u) = \bar{q} - (u+\eta)\mathbf{h}_1\cdot\boldsymbol{\sigma}, \tag{6.2.18}$$

with

$$\bar{p} = i\frac{\mathbf{h}_N^2 - t^2}{2t}, \quad \bar{q} = i\frac{t^2 - \mathbf{h}_1^2}{2t}.$$

The following RE and its dual equation hold:

$$\begin{aligned}&R_{0,\bar{0}}(u-v)K_0^-(u)R_{\bar{0},0}(u+v)K_{\bar{0}}^-(v)\\ &\quad= K_{\bar{0}}^-(v)R_{0,\bar{0}}(u+v)K_0^-(u)R_{\bar{0},0}(u-v),\end{aligned} \tag{6.2.19}$$

$$\begin{aligned}&R_{0,\bar{0}}(v-u)K_0^+(u)R_{\bar{0},0}(-u-v-2\eta)K_{\bar{0}}^+(v)\\ &\quad= K_{\bar{0}}^+(v)R_{\bar{0},0}(-u-v-2\eta)K_0^+(u)R_{\bar{0},0}(v-u),\end{aligned} \tag{6.2.20}$$

where the R-matrix is given by (6.1.20). To solve the eigenvalue problem (6.2.16), let us introduce the inhomogeneous double-row monodromy matrix

$$\begin{aligned}\mathscr{U}_0(u) = &R_{0,1}(u-\sin k_1)\cdots R_{0,M}(u-\sin k_M)K_0^-(u)\\ &\times R_{M,0}(u+\sin k_M)\cdots R_{1,0}(u+\sin k_1),\end{aligned} \tag{6.2.21}$$

and the transfer matrix $\tau(u)$,

$$\tau(u) = tr_0\{K_0^+(u)\mathscr{U}_0(u)\}. \tag{6.2.22}$$

Noting that

$$\bar{K}_j^-(k_j) = it\frac{2K_j^-(-\sin k_j)}{\mathbf{h}_N^2 - t^2e^{-2ik_j}}, \tag{6.2.23}$$

$$\bar{K}_j^+(k_j) = it\frac{tr_0\{K_0^+(-\sin k_j)R_{0,j}(-2\sin k_j)P_{0,j}\}}{(\sin k_j - \eta)(\mathbf{h}_1^2 e^{2ik_j} - t^2)}, \tag{6.2.24}$$

we have the important identification between the operator $\{\bar{\tau}(k_j)\}$ and $\tau(u)$:

$$\bar{\tau}(k_j) = \prod_{l\neq j}^{M}(\sin k_j - \sin k_l - \eta)^{-1}(\sin k_j + \sin k_l - \eta)^{-1}$$
$$\times \frac{-2t^2\,\tau(-\sin k_j)}{\eta(\sin k_j - \eta)(t^2 - \mathbf{h}_1^2 e^{2ik_j})(t^2 e^{-2ik_j} - \mathbf{h}_N^2)}. \tag{6.2.25}$$

The eigenvalue problem (6.2.15) is thus transformed to the eigenvalue problem of the transfer matrix associated with the inhomogeneous open XXX chain model introduced in Chap. 5. In the present case, the "inhomogeneity" parameters are $\theta_j = \sin k_j$ and the crossing parameter is $\eta = -i\frac{U}{2t}$.

Let us introduce the notation:

$$\bar{A}(u) = \prod_{l=1}^{M}(u - \sin k_l + \eta)(u + \sin k_l + \eta), \tag{6.2.26}$$

$$a(u) = \frac{2u + 2\eta}{2u + \eta}(\bar{p} + u\,\varepsilon|\mathbf{h}_N|)(\bar{q} - u\,|\mathbf{h}_1|)\bar{A}(u), \tag{6.2.27}$$

$$d(u) = a(-u - \eta), \tag{6.2.28}$$

$$c = 2\left[\varepsilon|\mathbf{h}_1||\mathbf{h}_N| - \mathbf{h}_1 \cdot \mathbf{h}_N)\right], \tag{6.2.29}$$

with

$$\varepsilon = \frac{\mathbf{h}_1 \cdot \mathbf{h}_N}{|\mathbf{h}_1 \cdot \mathbf{h}_N|}. \tag{6.2.30}$$

The eigenvalue $\Lambda(u)$ of the transfer matrix $\tau(u)$ can thus be expressed as [28]

$$\Lambda(u) = a(u)\frac{Q_1(u-\eta)}{Q_2(u)} + d(u)\frac{Q_2(u+\eta)}{Q_1(u)}$$
$$+(-1)^M cu(u+\eta)\frac{\bar{A}(u)\bar{A}(-u-\eta)}{Q_1(u)Q_2(u)}, \tag{6.2.31}$$

in which the functions $Q_1(u)$ and $Q_2(u)$ are parameterized by M Bethe roots $\{\mu_j | j = 1, \ldots, M\}$ for a generic non-vanishing c as follows

$$Q_1(u) = \prod_{j=1}^{M}(u - \mu_j), \tag{6.2.32}$$

$$Q_2(u) = (-1)^M \prod_{j=1}^{M} (u + \mu_j + \eta) = Q_1(-u - \eta). \tag{6.2.33}$$

We note that the regularity of $\Lambda(u)$ results in the BAEs

$$\frac{c\,(\mu_j+\eta)(\mu_j+\frac{\eta}{2})}{(\bar{p}-(\mu_j+\eta)\,\varepsilon|\mathbf{h}_N|)(\bar{q}+(\mu_j+\eta)|\mathbf{h}_1|)} = (-1)^{M+1}$$
$$\times \prod_{l=1}^{M} \frac{(\mu_j+\mu_l+\eta)(\mu_j+\mu_l+2\eta)}{(\mu_j-\sin k_l+\eta)(\mu_j+\sin k_l+\eta)}, \quad j = 1, \ldots, M, \tag{6.2.34}$$

with the selection rules $\mu_j \neq \mu_l$, and $\mu_j \neq -\mu_l - \eta$. Based on the $T-Q$ relation (6.2.31) and the correspondence (6.2.25), the eigenvalue problem (6.2.15) implies the following BAEs for the quasi-momenta $\{k_j\}$:

$$e^{-2ik_jN} = \prod_{l \neq j} (\sin k_j - \sin k_l - \eta)^{-1} (\sin k_j + \sin k_l - \eta)^{-1}$$
$$\times \frac{-2t^2 \Lambda(-\sin k_j)}{\eta(\sin k_j - \eta)(t^2 - \mathbf{h}_1^2 e^{2ik_j})(t^2 e^{-2ik_j} - \mathbf{h}_N^2)}. \tag{6.2.35}$$

Noting that $d(-\sin k_j) = \bar{A}(\sin k_j - \eta) = 0$, the above BAEs become

$$\frac{4t^2(\bar{p} - \sin k_j\,\varepsilon|\mathbf{h}_N|)(\bar{q} + \sin k_j\,|\mathbf{h}_1|)}{(t^2 - \mathbf{h}_1^2 e^{2ik_j})(t^2 e^{-2ik_j} - \mathbf{h}_N^2)}$$
$$= e^{-2ik_jN} \prod_{l=1}^{M} \frac{(\sin k_j - \mu_l - \eta)}{(\sin k_j + \mu_l + \eta)}, \quad j = 1, \ldots, M. \tag{6.2.36}$$

Equations (6.2.34) and (6.2.36) determine the spectrum of the Hamiltonian completely.

It should be remarked that if the two boundary fields $\mathbf{h}_1$ and $\mathbf{h}_N$ are parallel or anti-parallel, the associated $K^{\pm}(u)$-matrices can be diagonalized simultaneously. In this case, the $U(1)$ symmetry in the spin sector is recovered and the constant c given by (6.2.29) is zero. The $T-Q$ Ansatz is thus reduced to the conventional form no matter whether M is even or odd

$$\Lambda(u) = a(u)\frac{Q(u-\eta)}{Q(u)} + d(u)\frac{Q(u+\eta)}{Q(u)}, \tag{6.2.37}$$

where the functions $Q(u)$ are parameterized by $\bar{M}$ unequal Bethe roots $\{\lambda_j | j = 1, \ldots, \bar{M}\}$ as follows

$$Q(u) = \prod_{l=1}^{\bar{M}} (u - \lambda_l)(u + \lambda_l + \eta) = Q(-u - \eta), \quad \bar{M} = 0, \ldots, M. \tag{6.2.38}$$

These $\bar{M}$ parameters $\{\lambda_j\}$ and M quasi-momenta $\{k_j\}$ satisfy the BAEs

$$\frac{4t^2(\bar{p} - \sin k_j \varepsilon |\mathbf{h}_N|)(\bar{q} + \sin k_j |\mathbf{h}_1|)}{(t^2 - \mathbf{h}_1^2 e^{2ik_j})(t^2 e^{-2ik_j} - \mathbf{h}_N^2)} = e^{-2ik_j N}$$
$$\times \prod_{l=1}^{\bar{M}} \frac{(\sin k_j + \lambda_l)(\sin k_j - \lambda_l - \eta)}{(\sin k_j - \lambda_l)(\sin k_j + \lambda_l + \eta)}, \quad j = 1, \ldots, M, \tag{6.2.39}$$

$$\frac{\lambda_j(\bar{p} - (\lambda_j + \eta)|\mathbf{h}_N|)(\bar{q} + (\lambda_j + \eta)|\mathbf{h}_1|)}{(\lambda_j + \eta)(\bar{p} + \lambda_j |\mathbf{h}_N|)(\bar{q} - \lambda_j |\mathbf{h}_1|)} \prod_{l=1}^{M} \frac{(\lambda_j + \sin k_l)(\lambda_j - \sin k_l)}{(\lambda_j - \sin k_l + \eta)(\lambda_j + \sin k_l + \eta)}$$
$$= -\prod_{l=1}^{\bar{M}} \frac{(\lambda_j - \lambda_l - \eta)(\lambda_j + \lambda_l)}{(\lambda_j - \lambda_l + \eta)(\lambda_j + \lambda_l + 2\eta)}, \quad j = 1, \ldots, \bar{M}. \tag{6.2.40}$$

6.3 The Super-symmetric $t - J$ Model with Non-diagonal Boundaries

In this section, we consider the exact solution of the super-symmetric $t - J$ model with generic non-diagonal boundaries [46]. The model Hamiltonian is

$$H = -t \sum_{\alpha, j=1}^{N-1} \mathscr{P}[c_{j,\alpha}^\dagger c_{j+1,\alpha} + c_{j+1,\alpha}^\dagger c_{j,\alpha}]\mathscr{P} + 2t \sum_{j=1}^{N-1} [\mathbf{S}_j \cdot \mathbf{S}_{j+1} - \frac{1}{4} n_j n_{j+1}]$$
$$+ \xi_1 n_1 + 2\mathbf{h}_1 \cdot \mathbf{S}_1 + \xi_N n_N + 2\mathbf{h}_N \cdot \mathbf{S}_N, \tag{6.3.1}$$

where N is the site number; t is the hopping constant; $\mathscr{P}$ projects out double occupancies; $\mathbf{S}_j = \frac{1}{2} \sum_{\alpha,\beta} c_{j,\alpha}^\dagger \boldsymbol{\sigma}_{\alpha,\beta} c_{j,\beta}$ are the spin operators; and ξ_1 and ξ_N are the boundary chemical potentials.

6.3.1 Coordinate Bethe Ansatz

As for the Hubbard model, we construct the eigenstate of the Hamiltonian (6.3.1) as

$$|\Psi\rangle = \sum_{j=1}^{M} \sum_{\alpha_j = \uparrow,\downarrow} \sum_{x_j=1}^{N} \Psi^{\{\alpha\}}(x_1, \ldots, x_M) c_{x_1,\alpha_1}^\dagger \cdots c_{x_M,\alpha_M}^\dagger |0\rangle. \tag{6.3.2}$$

To exclude double occupancy, we need

$$\Psi^{\{\alpha\}}(\ldots, x, \ldots, x, \ldots) \equiv 0. \tag{6.3.3}$$

The eigenvalue equation can be written as

$$\begin{aligned}
&-t\sum_{j=1}^{M}\Big[(1-\delta_{x_j,N})\Psi^{\{\alpha\}}(\ldots, x_j+1, \ldots) + (1-\delta_{x_j,1})\Psi^{\{\alpha\}}(\ldots, x_j-1, \ldots)\Big] \\
&+\sum_{j=1}^{M}\sum_{\beta_j=\uparrow,\downarrow}\left[\delta_{x_j,1}(\xi_1+\mathbf{h}_1\cdot\boldsymbol{\sigma}_{\alpha_j,\beta_j}) + \delta_{x_j,N}(\xi_N+\mathbf{h}_N\cdot\sigma_{\alpha_j,\beta_j})\right]\Psi^{\{\alpha\}_j}(x_1, \ldots, x_M) \\
&-t\sum_{j=1}^{N-1}\sum_{l\neq k}^{M}\delta_{x_l,j}\delta_{x_k,j+1}(1-P_{j,j+1})\Psi^{\{\alpha\}}(\ldots, x_l, \ldots, x_k, \ldots) \\
&= E\,\Psi^{\{\alpha\}}(x_1, \ldots, x_M)
\end{aligned} \tag{6.3.4}$$

The wave function takes the following Bethe Ansatz form:

$$\begin{aligned}
\Psi^{\{\alpha\}}(x_1, \ldots, x_M) = &\sum_{p,q,r} A_p^{\{\alpha\},r}(q)\exp\left[i\sum_{j=1}^{M} r_{p_j}k_{p_j}x_{q_j}\right] \\
&\times\theta(x_{q_1} < x_{q_2} < \cdots < x_{q_M}).
\end{aligned} \tag{6.3.5}$$

For $x_j \neq 1, N$, $x_i \neq 1, N$, and $|x_i - x_j| > 1$, the corresponding eigenvalue is

$$E = -2t\sum_{j=1}^{M}\cos k_j. \tag{6.3.6}$$

When two electrons occupy two adjacent sites, namely, $x_{q_j} = x_{q_{j+1}} - 1 = x$ and $x \neq 1, N$, the Schrödinger equation (6.3.4) induces the relation

$$A_p^r(q) = S(r_{p_j}k_{p_j}, r_{p_{j+1}}k_{p_{j+1}})A_{p'}^{r'}(q'), \tag{6.3.7}$$

with $p' = \{\ldots, p_{j+1}, p_j, \ldots\}$ and $q' = \{\ldots, q_{j+1}, q_j, \ldots\}$. After introducing a new parametrization

$$e^{ik_j} = \frac{\lambda_j - \frac{i}{2}}{\lambda_j + \frac{i}{2}}, \tag{6.3.8}$$

we obtain the S-matrix

$$S(\lambda_j-\lambda_k)=\frac{\lambda_j-\lambda_k+iP_{i,j}}{\lambda_j-\lambda_k+i}. \tag{6.3.9}$$

The Schrödinger equation (6.3.4) for $x_{q1}=1, x_{q2}\neq 2$ induces

$$A_p^{(+,\cdots)}(q)=\bar{K}^+_{p1}A_p^{(-,\cdots)}(q), \tag{6.3.10}$$

with

$$\bar{K}^+(\lambda)=-\frac{\lambda+\frac{i}{2}}{\lambda-\frac{i}{2}}\frac{(t^2+\xi_1^2-\mathbf{h}_1^2)(\lambda^2+\frac{1}{4})+2\xi_1t(\lambda^2-\frac{1}{4})+2i\lambda t\,\mathbf{h}_1\cdot\boldsymbol{\sigma}}{[(t+\xi_1)\lambda+\frac{i}{2}(t-\xi_1)]^2-(\lambda-\frac{i}{2})^2\mathbf{h}_1^2}.$$

Similarly, we have

$$e^{-2ik_{p_M}N}A_p^{(\cdots,-)}(q)=\bar{K}^-_{p_M}A_p^{(\cdots,+)}(q), \tag{6.3.11}$$

with

$$\bar{K}^-(\lambda)=-\frac{\lambda-\frac{i}{2}}{\lambda+\frac{i}{2}}\frac{(t^2+\xi_N^2-\mathbf{h}_N^2)(\lambda^2+\frac{1}{4})+2\xi_Nt(\lambda^2-\frac{1}{4})+2i\lambda t\,\mathbf{h}_N\cdot\boldsymbol{\sigma}}{[(t+\xi_N)\lambda+\frac{i}{2}(t-\xi_N)]^2-(\lambda-\frac{i}{2})^2\mathbf{h}_N^2}.$$

To ensure the integrability of the model, $\bar{K}^{\pm}(u)$ must satisfy RE

$$\begin{aligned}&S_{1,2}(u_1-u_2)\bar{K}_1^{\pm}(u_1)S_{2,1}(u_1+u_2)\bar{K}_2^{\pm}(u_2)\\&\quad=\bar{K}_2^{\pm}(u_2)S_{1,2}(u_1+u_2)\bar{K}_1^{\pm}(u_1)S_{2,1}(u_1-u_2),\end{aligned} \tag{6.3.12}$$

which implies the integrable conditions of the model

$$(t+\xi_1)^2=\mathbf{h}_1^2,\quad (t+\xi_N)^2=\mathbf{h}_N^2. \tag{6.3.13}$$

Under this restriction, the reflection matrices become

$$\bar{K}^-(\lambda)=\frac{2\lambda-i}{2\lambda+i}\frac{\xi_N-2i\lambda\,\mathbf{h}_N\cdot\boldsymbol{\sigma}}{\xi_N+2i\lambda(t+\xi_N)}, \tag{6.3.14}$$

$$\bar{K}^+(\lambda)=\frac{2\lambda+i}{2\lambda-i}\frac{\xi_1-2i\lambda\,\mathbf{h}_1\cdot\boldsymbol{\sigma}}{\xi_1+2i\lambda(t+\xi_1)}. \tag{6.3.15}$$

As for the Hubbard model, the eigenvalue problem of the Hamiltonian is related to the eigenvalue problem

$$\bar{\tau}(\lambda_j)A^{(\cdots,+,\cdots)}=\left(\frac{2\lambda_j-i}{2\lambda_j+i}\right)^{-2N}A^{(\cdots,+,\cdots)}, \tag{6.3.16}$$

with the resulting operator

$$\begin{aligned}\bar{\tau}(u) = {} & S_{j-1,j}(\lambda_{j-1}-u)\cdots S_{1,j}(\lambda_1-u)\bar{K}_j^+(u)S_{j,1}(-u-\lambda_1)\cdots \\ & \times S_{j,j-1}(-u-\lambda_{j-1})S_{j,j+1}(-u-\lambda_{j+1})\cdots S_{j,M}(-u-\lambda_M)\bar{K}_j^-(u) \\ & \times S_{M,j}(\lambda_M-u)\cdots S_{j+1,j}(\lambda_{j+1}-u). \end{aligned} \tag{6.3.17}$$

6.3.2 Off-Diagonal Bethe Ansatz

Let us introduce the following R-matrix and K-matrices:

$$R_{0,j}(u) = u + \eta P_{0,j}, \tag{6.3.18}$$

$$K_0^-(u) = \bar{p} + u\mathbf{h}_N \cdot \boldsymbol{\sigma}_0, \tag{6.3.19}$$

$$K_0^+(u) = \bar{q} - (u + \eta)\mathbf{h}_1 \cdot \boldsymbol{\sigma}_0, \tag{6.3.20}$$

where

$$\eta = i, \quad \bar{p} = \frac{\xi_N}{2i}, \quad \bar{q} = -\frac{\xi_1}{2i}.$$

With the same notation as for the Hubbard model, i.e.,

$$\begin{aligned}\mathscr{U}_0(u) = {} & R_{0,1}(u-\lambda_1)\cdots R_{0,M}(u-\lambda_M)K_0^-(u) \\ & \times R_{M,0}(u+\lambda_M)\cdots R_{1,0}(u+\lambda_1), \end{aligned} \tag{6.3.21}$$

and

$$\tau(u) = tr_0\left\{K_0^+(u)\mathscr{U}_0(u)\right\}, \tag{6.3.22}$$

we may derive the following important identification between $\{\bar{\tau}(\lambda_j)\}$ given by (6.3.17) and $\{\tau(\lambda_j)\}$:

$$\begin{aligned}\bar{\tau}(\lambda_j) = {} & \prod_{l\neq j}^{M}(\lambda_j-\lambda_l-\eta)^{-1}(\lambda_j+\lambda_l-\eta)^{-1} \\ & \times\frac{\tau(-\lambda_j)}{2\eta(\lambda_j-\eta)[\bar{p}+\lambda_j(t+\xi_N)][-\bar{q}+\lambda_j(t+\xi_1)]}. \end{aligned} \tag{6.3.23}$$

For convenience, we introduce the notation

$$\bar{A}(u) = \prod_{l=1}^{M}(u-\lambda_l+\eta)(u+\lambda_l+\eta), \tag{6.3.24}$$

$$a(u) = \frac{2u+2\eta}{2u+\eta}[\bar{p}+u\varepsilon|\mathbf{h}_N|](\bar{q}-u|\mathbf{h}_1|)\bar{A}(u), \tag{6.3.25}$$

$$d(u)=a(-u-\eta), \tag{6.3.26}$$
$$c=2[\varepsilon|\mathbf{h}_1||\mathbf{h}_N|-\mathbf{h}_1\cdot\mathbf{h}_N]. \tag{6.3.27}$$

The $T-Q$ equation for the eigenvalue $\Lambda(u)$ of $\tau(u)$ thus reads

$$\begin{aligned}\Lambda(u)&=a(u)\frac{Q_1(u-\eta)}{Q_2(u)}+d(u)\frac{Q_2(u+\eta)}{Q_1(u)}\\&+(-1)^M cu(u+\eta)\frac{\bar{A}(u)\bar{A}(-u-\eta)}{Q_1(u)Q_2(u)}.\end{aligned} \tag{6.3.28}$$

The Q-functions are given by

$$Q_1(u)=\prod_{j=1}^{M}(u-\mu_j),\quad Q_2(u)=Q_1(-u-\eta). \tag{6.3.29}$$

With the same procedure introduced in the previous section, we obtain the BAEs

$$\begin{aligned}&\frac{[\bar{p}-\lambda_j\varepsilon|\mathbf{h}_N|](\bar{q}+\lambda_j|\mathbf{h}_1|)}{[\bar{p}+\lambda_j\varepsilon|\mathbf{h}_N|][\bar{q}-\lambda_j|\mathbf{h}_1|]}\left(\frac{2\lambda_j-\eta}{2\lambda_j+\eta}\right)^{2N}=\prod_{l=1}^{M}\frac{\lambda_j-\mu_l-\eta}{\lambda_j+\mu_l+\eta},\\&j=1,\ldots,M,\end{aligned} \tag{6.3.30}$$

$$\begin{aligned}&\frac{c\,(\mu_j+\eta)(2\mu_j+\eta)}{2[\bar{p}-(\mu_j+\eta)\,\varepsilon|\mathbf{h}_N|][\bar{q}+(\mu_j+\eta)|\mathbf{h}_1|]}=(-1)^{M+1}\\&\times\prod_{l=1}^{M}\frac{(\mu_j+\mu_l+\eta)(\mu_j+\mu_l+2\eta)}{(\mu_j-\lambda_l+\eta)(\mu_j+\lambda_l+\eta)},\quad j=1,\ldots,M.\end{aligned} \tag{6.3.31}$$

Equations (6.3.30) and (6.3.31) determine the spectrum of the Hamiltonian (6.3.1) completely.

References

1. F.H.L. Essler, H. Frahm, F. Göhmann, A. Klümper, V.E. Korepin, *The One-Dimensional Hubbard Model* (Cambridge University Press, Cambridge, 2005)
2. N. Andrei, Integrable models in condensed matter physics, in *Low-Dimensional Quantum Field Theories for Condensed Matter Physicists*, ed. by S. Lundqvist, G. Morandi, Y. Lu. Series in Modern Condensed Matter Physics, vol. 6 (World Scientific Publishing Co., Pte. Ltd., Singapore, 1995), p. 457
3. T. Deguchi, F.H.L. Essler, F. Göhmann, A. Klümper, V.E. Korepin, K. Kusakabe, Thermodynamics and excitations of the one-dimensional Hubbard model. Phys. Rep. **331**, 197 (2000)
4. E.H. Lieb, F.Y. Wu, Absence of Mott transition in an exact solution of the short-range one-band model in one dimension. Phys. Rev. Lett. **20**, 1445 (1968)
5. F.H.L. Essler, V.E. Korepin, K. Schoutens, Complete solution of the one-dimensional Hubbard model. Phys. Rev. Lett. **67**, 3848 (1991)

6. F.H.L. Essler, V.E. Korepin, K. Schoutens, New eigenstates of the 1-dimensional Hubbard model. Nucl. Phys. B **372**, 559 (1992)
7. M. Takahashi, Magnetization curve for the half-filled Hubbard model. Prog. Theor. Phys. **42**, 1098 (1969)
8. M. Takahashi, Magnetization curve for the half-filled Hubbard model. Prog. Theor. Phys. **43**, 860 (1970)
9. M. Takahashi, Magnetic susceptibility for the half-filled Hubbard model. Prog. Theor. Phys. **43**, 1619 (1970)
10. E. Olmedilla, M. Wadati, Conserved quantities of the one-dimensional Hubbard model. Phys. Rev. Lett. **60**, 1595 (1988)
11. B.S. Shastry, Infinite conservation laws in the one-dimensional Hubbard model. Phys. Rev. Lett. **56**, 1529 (1986)
12. B.S. Shastry, Exact integrability of the one-dimensional Hubbard model. Phys. Rev. Lett. **56**, 2453 (1986)
13. B.S. Shastry, Decorated star-triangle relations and exact integrability of the one-dimensional Hubbard model. J. Stat. Phys. **30**, 57 (1987)
14. M. Wadati, E. Olmedilla, Y. Akutsu, Lax pair for the one-dimensional Hubbard model. J. Phys. Soc. Jpn. **36**, 1340 (1987)
15. E. Olmedilla, M. Wadati, Y. Akutsu, Yang-Baxter relations for spin models and fermion models. J. Phys. Soc. Jpn. **36**, 2298 (1987)
16. M. Shiroishi, M. Wadati, Decorated star-triangle relations for the free-fermion model and a new solvable bilayer vertex model. J. Phys. Soc. Jpn. **64**, 2795 (1995)
17. M. Shiroishi, M. Wadati, Tetrahedral Zamolodchikov algebra related to the six-vertex free-fermion model and a new solution of the Yang-Baxter equation. J. Phys. Soc. Jpn. **64**, 4598 (1995)
18. Y. Umeno, M. Shiroishi, M. Wadati, Fermionic R-operator and integrability of the one-dimensional Hubbard model. J. Phys. Soc. Jpn. **67**, 2242 (1998)
19. M.J. Matins, P.B. Ramos, Exact solution of the lattice vertex model analogue of the coupled Bariev XY chains. J. Phys. A **30**, L465 (1997)
20. M.J. Matins, P.B. Ramos, The quantum inverse scattering method for Hubbard-like models. Nucl. Phys. B **522**, 413 (1998)
21. Y. Wang, J. Voit, F.-C. Pu, Exact boundary critical exponents and tunneling effects in integrable models for quantum wires. Phys. Rev. B **54**, 8491 (1996)
22. H. Shulz, Hubbard chain with reflecting ends. J. Phys. C **18**, 581 (1985)
23. H. Asakawa, M. Suzuki, Finite-size corrections in the XXZ model and the Hubbard model with boundary fields. J. Phys. A **29**, 225 (1996)
24. X.-W. Guan, Algebraic Bethe Ansatz for the one-dimensional Hubbard model with open boundaries. J. Phys. A **33**, 5391 (2000)
25. M. Shiroishi, M. Wadati, Bethe Ansatz equation for the Hubbard model with boundary fields. J. Phys. Soc. Jpn **66**, 1 (1997)
26. X.-W. Guan, M.S. Wang, S.D. Yang, Lax pair and boundary K-matrices for the one-dimensional Hubbard model. Nucl. Phys. B **485**, 685 (1997)
27. M. Shiroishi, M. Wadati, Integrable boundary conditions for the one-dimensional Hubbard model. J. Phys. Soc. Jpn. **66**, 2288 (1997)
28. Y.-Y. Li, J. Cao, W.-L. Yang, K. Shi, Y. Wang, Exact solution of the one-dimensional Hubbard model with arbitrary boundary magnetic fields. Nucl. Phys. B **879**, 98 (2014)
29. F.C. Zhang, T.M. Rice, Effective hamiltonian for the superconducting Cu oxides. Phys. Rev. B **37**, 3759 (1988)
30. P. Schlottmann, Integrable narrow-band model with possible relevance to heavy-fermion systems. Phys. Rev. B **36**, 5177 (1987)
31. P.B. Wiegmann, Superconductivity in strongly correlated electronic systems and confinement versus deconfinement phenomenon. Phys. Rev. Lett. **60**, 821 (1988)
32. F.H.L. Essler, V.E. Korepin, K. Schoutens, New exactly solvable model of strongly correlated electrons motivated by high-Tc superconductivity. Phys. Revs. Lett. **68**, 2960 (1992)

33. A. Foerster, M. Karowski, Algebraic properties of the Bethe Ansatz for an $spl(2, 1)$-supersymmetric t-J model. Nucl. Phys. B **396**, 611 (1993)
34. F.H.L. Essler, V.E. Korepin, Higher conservation laws and algebraic Bethe Ansatze for the supersymmetric t-J model. Phys. Rev. B **46**, 9147 (1992)
35. P.-A. Bares, G. Blatter, M. Ogata, Exact solution of the t-J model in one dimension at $2t = \pm J$: ground state and excitation spectrum. Phys. Rev. B **44**, 130 (1991)
36. N. Kawakami, S.-K. Yang, Correlation functions in the one-dimensional t-J model. Phys. Rev. Lett. **65**, 2309 (1990)
37. E.D. Williams, Thermodynamics and excitations of the supersymmetric t-J model. Int. J. Mod. Phys B **09**, 3607 (1995)
38. G. Jüttner, A. Klümper, J. Suzuki, Exact thermodynamics and Luttinger liquid properties of the integrable t-J model. Nucl. Phys. B **487**, 650 (1997)
39. A. Foerster, M. Karowski, The supersymmetric t-J model with quantum group invariance. Nucl. Phys. B **408**, 512 (1993)
40. A. González-Ruiz, Integrable open-boundary conditions for the supersymmetric t-J model the quantum-group-invariant case. Nucl. Phys. B **424**, 468 (1994)
41. F.H.L. Essler, The supersymmetric t-J model with a boundary. J. Phys. A **29**, 6183 (1996)
42. Y. Wang, J. Dai, Z. Hu, F.-C. Pu, Exact results for a Kondo problem in a one-dimensional t-J model. Phys. Rev. Lett. **79**, 1901 (1997)
43. Y. K. Zhou, M.T. Batchelor, Spin excitations in the integrable open quantum group invariant supersymmetric t-J model. Nucl. Phys. B **490**, 576 (1997)
44. G. Bedürftig, H. Frahm, Open t-J chain with boundary impurities. J. Phys. A **32**, 4585 (1999)
45. W. Galleas, Spectrum of the supersymmetric t-J model with non-diagonal open boundaries. Nucl. Phys. B **777**, 352 (2007)
46. X. Zhang, J. Cao, W.-L. Yang, K. Shi, Y. Wang, Exact solution of the one-dimensional supersymmetric t-J model with unparallel boundary fields. J. Stat. Mech. **P04031** (2014)
47. C.N. Yang, S.C. Zhang, $SO(4)$ symmetry in the Hubbard model. Mod. Phys. Lett. B **4**, 759 (1990)
48. F.H.L. Essler, V.E. Korepin, K. Schoutens, Completeness of the SO(4) extended Bethe Ansatz for the one-dimensional Hubbard model. Nucl. Phys. B **384**, 431 (1992)
49. C.N. Yang, Some exact results for the many-body problem in one dimension with repulsive delta-function interaction. Phys. Rev. Lett. **19**, 1312 (1967)

Chapter 7
The Nested Off-Diagonal Bethe Ansatz

In Chap. 2, we introduced how the nested algebraic Bethe Ansatz method was used in the exact solution of the periodic $SU(n)$-invariant spin chain. This method can also solve the open chain with diagonal boundaries [1–5]. However, as for the $SU(2)$ case [6], it is difficult to use this method to solve the $SU(n)$-invariant spin chain with generic integrable boundaries. In this chapter, we introduce the nested ODBA method to diagonalize the $SU(n)$ invariant spin chain with both periodic and generic integrable open boundary conditions [7]. The central points of the nested generalization of the ODBA are: (1) based on the intrinsic properties of the R-matrix, the recursive operator product identities of the transfer matrix can be constructed with the fusion technique [8–12]; (2) from the definition of the transfer matrix, its asymptotic behavior for $u \to \infty$ and its values at some special points can be derived explicitly; (3) based on the relations in (1) and (2) the nested inhomogeneous $T - Q$ relation can be constructed with the self-consistency of the final BAEs checked. This nested generalization to ODBA is useful and general to the eigenvalue problems of integrable models associated with high-rank algebras.

7.1 The Fusion Procedure

7.1.1 Fundamental Fusion Relations

Generally, the transfer matrices of the integrable models defined in high-rank algebras cannot form closed operator product identities by themselves as in Eq. (1.5.31). Instead, a recursive set of operator product identities is needed in order to obtain the eigenvalues of the transfer matrices. The situation is similar to that of the $n \times n$ c-number matrix where n invariants, i.e., its n principal minors (with the first one and the last one being the trace and the determinant, respectively), are needed to characterize the n eigenvalues. The main tool to construct such recursive operator identities is the fusion technique [8–12] which was initially employed to construct

Y. Wang et al., *Off-Diagonal Bethe Ansatz for Exactly Solvable Models*,
DOI 10.1007/978-3-662-46756-5_7

new solutions of YBE based on the known R-matrices. The fusion procedure that we shall use in this chapter is the generalization of that introduced in Sect. 2.4 in multiple tensor space. In the framework of nested ODBA, this method is used to construct a set of fused transfer matrices, with which a nested analogue of Eq. (1.5.31) can be derived.

Throughout this chapter, we denote $\mathbf{V}$ as an n-dimensional linear vector space with an orthonormal basis $\{|i\rangle|, i=1,\ldots,n\}$. Let us consider the R-matrix $R_{1,2}(u) = u+\eta P_{1,2}$ which is related to the Hamiltonian (2.3.1). It possesses the following properties:

$$\text{Initial condition}: R_{1,2}(0)=\eta P_{1,2}, \tag{7.1.1}$$

$$\text{Unitarity}: R_{1,2}(u)R_{2,1}(-u)=\varphi_1(u)\times \text{id},$$
$$\varphi_1(u)=-(u+\eta)(u-\eta), \tag{7.1.2}$$

$$\text{Crossing-unitarity}: R^{t_1}_{1,2}(u)R^{t_1}_{2,1}(-u-n\eta)=\varphi_2(u)\times \text{id},$$
$$\varphi_2(u)=-u(u+n\eta), \tag{7.1.3}$$

$$\text{Fusion condition}: R_{1,2}(-\eta)=-\eta+\eta P_{1,2}=-2\eta P^{(-)}_{1,2}. \tag{7.1.4}$$

The anti-symmetric projectors $\{P^{(-)}_{1,\ldots,m}|m=2,\ldots,n\}$ in a tensor space of $\mathbf{V}$ are defined by the induction relations

$$P^{(-)}_{1,\ldots,m+1}=\frac{1}{m+1}\left(1-\sum_{j=2}^{m+1}P_{1,j}\right)P^{(-)}_{2,\ldots,m+1},\quad m=1,\ldots,n-1. \tag{7.1.5}$$

Iterating the above relation yields alternative definition of the projectors

$$P^{(-)}_{1,\ldots,m}=\frac{1}{m!}\sum_{\kappa\in S_m}(-1)^{\text{sign}(\kappa)}P_\kappa,\quad m=2,\ldots,n, \tag{7.1.6}$$

where S_m is the permutation group of m indices, P_κ is a permutation in the group, and sign(κ) is 0 for an even permutation κ and 1 for an odd permutation. We can easily verify that the projection operators satisfy the relations:

$$P^{(-)}_{1,\ldots,m}P^{(-)}_{l,\ldots,m}=P^{(-)}_{l,\ldots,m}P^{(-)}_{1,\ldots,m}=P^{(-)}_{1,\ldots,m},\quad l=1,\ldots,m-1, \tag{7.1.7}$$

$$P_\kappa\, P^{(-)}_{1,\ldots,m}=(-1)^{\text{sign}(\kappa)}\,P^{(-)}_{1,\ldots,m},\quad \forall\kappa\in S_m. \tag{7.1.8}$$

Let us introduce fusion-operators $\{\hat{R}_{\langle 1,\ldots,m\rangle}(-\eta)|m=2,\ldots,n\}$ in terms of the R-matrix via the recursive relation

$$\hat{R}_{\langle 1,\ldots,m+1\rangle}(-\eta)=R_{2,1}(-\eta)R_{3,1}(-2\eta)\ldots R_{m+1,1}(-m\eta)\,\hat{R}_{\langle 2,\ldots,m+1\rangle}(-\eta),$$
$$m=1,\ldots,n-1. \tag{7.1.9}$$

In the present case, these operators are proportional to the projectors (7.1.5), i.e.,

$$\hat{R}_{\langle 1,\dots,m\rangle}(-\eta) = q_m(\eta) P^{(-)}_{1,\dots,m}, \quad q_m(\eta) = (-\eta)^{\frac{m(m-1)}{2}} \prod_{l=1}^{m} l!. \tag{7.1.10}$$

The above relation can be proven by induction as follows. Thanks to (7.1.4), it holds for the case of $m = 2$. Assume that (7.1.10) holds up to m. The recursive relation (7.1.9) implies

$$\begin{aligned}
\hat{R}_{\langle 1,\dots,m+1\rangle}(-\eta) &= q_m(\eta) R_{2,1}(-\eta) R_{3,1}(-2\eta) \dots R_{m+1,1}(-m\eta) P^{(-)}_{2,\dots,m+1} \\
&= (-\eta)^m (m+1)! q_m(\eta) P^{(-)}_{1,\dots,m+1} \\
&= q_{m+1}(\eta) P^{(-)}_{1,\dots,m+1}.
\end{aligned}$$

Hence we have proven the relations (7.1.10).

Now let us turn to fusion of the monodromy matrix. The one-row monodromy matrices are defined as usual

$$T_0(u) = R_{0,N}(u - \theta_N) R_{0,N-1}(u - \theta_{N-1}) \cdots R_{0,1}(u - \theta_1), \tag{7.1.11}$$

$$\hat{T}_0(u) = R_{1,0}(u + \theta_1) \cdots R_{N-1,0}(u + \theta_{N-1}) R_{N,0}(u + \theta_N), \tag{7.1.12}$$

which satisfy the Yang-Baxter relations

$$R_{1,2}(u - v) T_1(u) T_2(v) = T_2(v) T_1(u) R_{1,2}(u - v), \tag{7.1.13}$$

$$R_{1,2}(u - v) \hat{T}_1(u) \hat{T}_2(v) = \hat{T}_2(v) \hat{T}_1(u) R_{1,2}(u - v). \tag{7.1.14}$$

The fundamental Yang-Baxter relation (7.1.13) allows us to derive the following relation by induction

$$\begin{aligned}
&\hat{R}_{\langle 1,\dots,m\rangle}(-\eta)\, T_m(u - (m-1)\eta) \cdots T_1(u) \\
&\quad = T_1(u) \cdots T_m(u - (m-1)\eta)\, \hat{R}_{\langle 1,\dots,m\rangle}(-\eta).
\end{aligned} \tag{7.1.15}$$

From the relation (7.1.10) we have

$$P^{(\;)}_{1,\dots,m}\, T_m(u - (m-1)\eta) \cdots T_1(u) = T_1(u) \cdots T_m(u - (m-1)\eta)\, P^{(\;)}_{1,\dots,m},$$

which by (7.1.7) leads to the following fusion relation

$$\begin{aligned}
&P^{(-)}_{1,\dots,m} T_1(u) \cdots T_m(u - (m-1)\eta)\, P^{(-)}_{1,\dots,m} = T_1(u) \cdots T_m(u - (m-1)\eta)\, P^{(-)}_{1,\dots,m}, \\
&\qquad m = 2, \dots, n.
\end{aligned} \tag{7.1.16}$$

Accordingly, we can introduce the fused monodromy matrices and fused R-matrices as follows:

$$T_{\langle 1,\ldots,m\rangle}(u) = P^{(-)}_{1,\ldots,m} T_1(u)\cdots T_m(u-(m-1)\eta)P^{(-)}_{1,\ldots,m}, \tag{7.1.17}$$

$$R_{\langle 1,\ldots,m\rangle,0}(u) = P^{(-)}_{1,\ldots,m} R_{1,0}(u)\cdots R_{m,0}(u-(m-1)\eta)P^{(-)}_{1,\ldots,m}, \tag{7.1.18}$$

$$R_{0,\langle 1,\ldots,m\rangle}(u) = P^{(-)}_{1,\ldots,m} R_{0,1}(u)\cdots R_{0,m}(u-(m-1)\eta)P^{(-)}_{1,\ldots,m}, \tag{7.1.19}$$

$$m = 2,\ldots,n.$$

The fusion relation (7.1.16) of the projection operator $P^{(-)}_{1,\ldots,m}$ plays an important role in the fusion procedure. It is in fact a generalization of YBE for a multi-particle scattering process in the following sense. The fundamental Yang-Baxter relation (7.1.13) and the fusion properties (7.1.16) imply that

$$R_{\langle\bar{1},\ldots,\bar{m}\rangle,1}(u-v)T_{\langle\bar{1},\ldots,\bar{m}\rangle}(u)T_1(v) = T_1(v)T_{\langle\bar{1},\ldots,\bar{m}\rangle}(u)R_{\langle\bar{1},\ldots,\bar{m}\rangle,1}(u-v). \tag{7.1.20}$$

Let us define further

$$\begin{aligned} R_{\langle 1,\ldots,m\rangle,\langle\bar{1},\ldots,\bar{k}\rangle}(u) &= P^{(-)}_{\bar{1},\ldots,\bar{k}} R_{\langle 1,\ldots,m\rangle,\bar{k}}(u+(k-1)\eta)\cdots R_{\langle 1,\ldots,m\rangle,\bar{1}}(u)P^{(-)}_{\bar{1},\ldots,\bar{k}} \\ &= R_{\langle 1,\ldots,m\rangle,\bar{k}}(u+(k-1)\eta)\cdots R_{\langle 1,\ldots,m\rangle,\bar{1}}(u)P^{(-)}_{\bar{1},\ldots,\bar{k}}, \\ & m,k = 2,\ldots,n. \end{aligned} \tag{7.1.21}$$

With the help of the fused Yang-Baxter relation (7.1.20) and the fusion properties (7.1.16) , we can prove that the following fused Yang-Baxter relation holds:

$$R_{a,b}(u-v)T_a(u)T_b(v) = T_b(v)T_a(u)R_{a,b}(u-v), \tag{7.1.22}$$

where a, b are the indices of the projected subspaces (e.g., $\langle 1,\ldots,m\rangle$). As a direct consequence, the fused R-matrices also satisfy YBE

$$R_{a,b}(u-v)R_{a,c}(u)R_{b,c}(v) = R_{b,c}(v)R_{a,c}(u)R_{a,b}(u-v). \tag{7.1.23}$$

Let us further introduce the fused transfer matrices $t_m(u)$

$$t_m(u) = tr_{1,\ldots,m}\{T_{\langle 1,\ldots,m\rangle}(u)\}, \quad m = 1,\ldots,n, \tag{7.1.24}$$

which include the fundamental transfer matrix $t(u) = tr_0 T_0(u)$ as the first one, i.e., $t(u) = t_1(u)$. The fused YBE (7.1.22) gives rise to

$$[t_m(u), t_k(v)] = 0, \quad m,k = 1,\ldots,n, \tag{7.1.25}$$

indicating that the fused transfer matrices have common eigenstates.

7.1.2 The Quantum Determinant

We note that $t_n(u)$ is in fact the quantum determinant (scalar function) for generic $\{\theta_j\}$, because

$$P^{(-)}_{1,\ldots,n} = |\psi_0\rangle\langle\psi_0|, \tag{7.1.26}$$

and $|\psi_0\rangle$ is the $SU(n)$ singlet state. Let $\mathbf{V}^*$ denote the completely antisymmetric subspace of the tensor space $\mathbf{V}^{\otimes n-1}$ and $\{|i^*\rangle | i = 1, \ldots, n\}$ be its orthonormal basis. The vector $|i^*\rangle$ is given by

$$|i^*\rangle = \frac{1}{\sqrt{(n-1)!}} \sum_{i_1,\ldots,i_{n-1}=1}^{n} \varepsilon_i^{i_1,\ldots,i_{n-1}} |i_1, \ldots, i_{n-1}\rangle, \quad i = 1, \ldots, n, \tag{7.1.27}$$

where $\varepsilon_i^{i_1,\ldots,i_{n-1}}$ is the $(n-1)$th order completely antisymmetric tensor. Thus the $SU(n)$ singlet state $|\psi_0\rangle$ can be expressed as

$$|\psi_0\rangle = \frac{1}{\sqrt{n}} \sum_{i=1}^{n} |i, i^*\rangle. \tag{7.1.28}$$

Taking $m = n$ in (7.1.18), we have

$$\begin{aligned} R_{\langle 1,\ldots,n\rangle,0}(u) &= R_{1,0}(u) \cdots R_{n,0}(u-(n-1)\eta) P^{(-)}_{1,\ldots,n} \\ &= \mathrm{Det}_q\{R(u)\}\, P^{(-)}_{1,\ldots,n}. \end{aligned}$$

Applying the above equation to the singlet state $|\psi_0\rangle$ given by (7.1.28) and comparing the coefficients of the vector components of both sides, we arrive at the equations

$$\sum_{i,\beta=1}^{n} [R(u)]_{i\,\beta}^{k\,\gamma} \left[R^{\mathbf{V}^*,\mathbf{V}}(u-\eta)\right]_{i^*\,\alpha}^{j^*\,\beta} = \mathrm{Det}_q\{R(u)\}\, \delta_j^k \delta_\alpha^\gamma,$$
$$j,\, k,\, \alpha,\, \gamma = 1, \ldots, n, \tag{7.1.29}$$

$$\sum_{i,\beta=1}^{n} \left[R^{\mathbf{V}^*,\mathbf{V}}(u)\right]_{i^*\,\beta}^{j^*\,\gamma} [R(u-(n-1)\eta)]_{i\,\alpha}^{k\,\beta} = \mathrm{Det}_q\{R(u)\}\, \delta_k^j \delta_\alpha^\gamma,$$
$$j,\, k,\, \alpha,\, \gamma = 1, \ldots, n, \tag{7.1.30}$$

where the $(n-1)$th fused R-matrix $R^{\mathbf{V}^*,\mathbf{V}}(u)$ is given by (7.1.18) with $m = n-1$, i.e.,

$$R^{\mathbf{V}^*,\mathbf{V}}_{1,0}(u) = R_{\langle 1,\ldots,n-1\rangle,0}(u).$$

Due to the fact that the vector space $\mathbf{V}$ and its dual space $\mathbf{V}^*$ have the same dimension, (7.1.29) and (7.1.30) can be rewritten in the matrix forms

$$R_{1,2}(u)\left[R_{1,2}^{\mathbf{V}^*,\mathbf{V}}(u-\eta)\right]^{t_1}=\mathrm{Det}_q\{R(u)\}\times \mathrm{id}\otimes\mathrm{id}, \tag{7.1.31}$$

$$\left[R_{1,2}^{\mathbf{V}^*,\mathbf{V}}(u)\right]R_{1,2}^{t_1}(u-(n-1)\eta)=\mathrm{Det}_q\{R(u)\}\times \mathrm{id}\otimes\mathrm{id}. \tag{7.1.32}$$

With the help of (7.1.2) and (7.1.3), we obtain the difference equation for the quantum determinant

$$\frac{\mathrm{Det}_q\{R(u+(n-1)\eta)\}}{\mathrm{Det}_q\{R(u+n\eta)\}}=\frac{\varphi_2(u)}{\varphi_1(u+n\eta)}. \tag{7.1.33}$$

The above difference equation and the asymptotic behavior of $\mathrm{Det}_q\{R(u)\}$ for $u \to \infty$ uniquely determine the quantum determinant, i.e.,

$$\mathrm{Det}_q\{R(u)\}=(u+\eta)\prod_{k=1}^{n-1}(u-k\eta). \tag{7.1.34}$$

Keeping the relation (7.1.16) in mind, finally we have

$$\mathrm{Det}_q\{T(u)\}=\prod_{j=1}^{N}\mathrm{Det}_q\{R(u-\theta_j)\}, \tag{7.1.35}$$

and

$$t_n(u)=\mathrm{Det}_q\{T(u)\}\times\mathrm{id}=\prod_{j=1}^{N}(u-\theta_j+\eta)\prod_{k=1}^{n-1}(u-\theta_j-k\eta)\ \times\mathrm{id}. \tag{7.1.36}$$

7.2 The Periodic $SU(n)$ Spin Chain

7.2.1 Operator Product Identities

Let us consider the product of the one-row monodromy matrices at two special points

$$\begin{aligned}T_{\bar 1}(\theta_j)\,T_{\bar 2}(\theta_j-\eta)=&\,R_{\bar 1,N}(\theta_j-\theta_N)\cdots R_{\bar 1,j}(0)\cdots R_{\bar 1,1}(\theta_j-\theta_1)\\&\times R_{\bar 2,N}(\theta_j-\theta_N-\eta)\cdots R_{\bar 2,j}(-\eta)\cdots R_{\bar 2,1}(\theta_j-\theta_1-\eta)\\=&R_{j,j-1}(\theta_j-\theta_{j-1})\cdots R_{j,1}(\theta_j-\theta_1)R_{\bar 1,N}(\theta_j-\theta_N)\cdots R_{\bar 1,j+1}(\theta_j-\theta_{j+1})\\&\times R_{\bar 2,N}(\theta_j-\theta_N-\eta)\cdots R_{\bar 2,j+1}(\theta_j-\theta_{j+1}-\eta)R_{\bar 2,\bar 1}(-\eta)R_{\bar 1,j}(0)\cdots\end{aligned}$$

$$\begin{aligned}
&\overset{(7.1.16)}{=} R_{j,j-1}(\theta_j-\theta_{j-1})\cdots R_{j,1}(\theta_j-\theta_1)P^{(-)}_{\bar{1},\bar{2}}R_{\bar{1},N}(\theta_j-\theta_N)\cdots\\
&\quad\times R_{\bar{1},j+1}(\theta_j-\theta_{j+1})R_{\bar{2},N}(\theta_j-\theta_N-\eta)\cdots\\
&\quad\times R_{\bar{2},j+1}(\theta_j-\theta_{j+1}-\eta)R_{\bar{2},\bar{1}}(-\eta)R_{\bar{1},j}(0)\cdots\\
&= P^{(-)}_{\bar{1},\bar{2}}T_{\bar{1}}(\theta_j)\,T_{\bar{2}}(\theta_j-\eta).
\end{aligned}$$

Then we have the following useful relations

$$T_1(\theta_j)T_2(\theta_j-\eta) = P^{(-)}_{1,2}\,T_1(\theta_j)T_2(\theta_j-\eta), \quad j=1,\ldots,N, \tag{7.2.1}$$

$$P_{1,2}T_1(\theta_j)T_2(\theta_j-\eta) = -T_1(\theta_j)T_2(\theta_j-\eta), \quad j=1,\ldots,N. \tag{7.2.2}$$

Combining the relations (7.1.18) and (7.2.2), we can show that

$$P_{l,l+1}\,T_1(\theta_j)T_{\langle 2,\ldots,m\rangle}(\theta_j-\eta) = -T_1(\theta_j)T_{\langle 2,\ldots,m\rangle}(\theta_j-\eta), \tag{7.2.3}$$

where $l=1,\ldots,m-1$. The above relations allow us to obtain the following operator identities:

$$\begin{aligned}
T_1(\theta_j)T_{\langle 2,\ldots,m\rangle}(\theta_j-\eta) &= P^{(-)}_{1,\ldots,m}T_1(\theta_j)T_2(\theta_j-\eta)\cdots\\
&\quad\times T_m(\theta_j-(m-1)\eta)\,P^{(-)}_{2,\ldots,m}.
\end{aligned}\tag{7.2.4}$$

Let us evaluate the product of the fundamental transfer matrix and the fused ones at some special points

$$\begin{aligned}
t(\theta_j)t_m(\theta_j-\eta) &= tr_{1,\ldots,m+1}\left\{T_1(\theta_j)T_{\langle 2,\ldots,m+1\rangle}(\theta_j-\eta)\right\}\\
&\overset{(7.2.4)}{=} tr_{1,\ldots,m+1}\left\{P^{(-)}_{1,\ldots,m+1}T_1(\theta_j)T_2(\theta_j-\eta)\cdots T_{m+1}(\theta_j-m\eta)P^{(-)}_{2,\ldots,m+1}\right\}\\
&= tr_{1,\ldots,m+1}\left\{T_1(\theta_j)T_2(\theta_j-\eta)\cdots T_{m+1}(\theta_j-m\eta)P^{(-)}_{1,\ldots,m+1}\right\}\\
&= tr_{1,\ldots,m+1}\left\{T_{\langle 1,\ldots,m+1\rangle}(\theta_j)\right\}.
\end{aligned}\tag{7.2.5}$$

According to the definition (7.1.24), we have the following functional relations among the transfer matrices:

$$t(\theta_j)t_m(\theta_j \quad \eta) = t_{m+1}(\theta_j), \quad m=1,\ldots,n \quad 1, \quad j=1,\ldots,N. \tag{7.2.6}$$

From the initial condition (7.1.1), the fusion condition (7.1.4) and the fundamental properties of the fusion operator $P^{(-)}_{i,j}$ we can easily deduce that

$$t_m(\theta_j+\eta) = t_m(\theta_j+2\eta) = \cdots = t_m(\theta_j+(m-1)\eta) = 0. \tag{7.2.7}$$

For convenience, we introduce the following mutually commutative operators $\{\tau_m(u)\}$ associated with the fused transfer matrices $\{t_m(u)\}$:

$$t_m(u) = \prod_{l=1}^{N}\prod_{k=1}^{m-1}(u - \theta_l - k\eta)\tau_m(u), \tag{7.2.8}$$

$$[\tau_l(u), \tau_m(v)] = 0, \quad l, m = 1, \ldots, n. \tag{7.2.9}$$

Obviously, $\{\tau_m(u)\}$ are degree N polynomials of u. The operator identities (7.2.6) imply the relations

$$\tau(\theta_j)\tau_m(\theta_j - \eta) = \prod_{l=1}^{N}(\theta_j - \theta_l - \eta)\tau_{m+1}(\theta_j), \tag{7.2.10}$$

$$j = 1, \ldots, N, \quad m = 1, \ldots, n-1.$$

Note that $\tau(u) = \tau_1(u)$. From the definition (7.1.24) of the fused transfer matrices, we can deduce the asymptotic behavior

$$\tau_m(u) = \frac{n!}{m!(n-m)!}u^N + \cdots, \quad u \to \infty. \tag{7.2.11}$$

7.2.2 Nested $T - Q$ Relation

Let $\{\Lambda_m(u)\}$ (resp. $\{\bar{\Lambda}_m(u)\}$) be the eigenvalues of $\{t_m(u)\}$ (resp. $\{\tau_m(u)\}$). By applying $t_m(u)$ to a common eigenstate, we obtain the relations:

$$\Lambda(\theta_j)\Lambda_m(\theta_j - \eta) = \Lambda_{m+1}(\theta_j), \tag{7.2.12}$$

$$\Lambda_m(\theta_j + \eta) = \Lambda_m(\theta_j + 2\eta) = \cdots = \Lambda_m(\theta_j + (m-1)\eta) = 0,$$
$$j = 1, \ldots, N, \tag{7.2.13}$$

$$\Lambda_m(u) = \frac{n!}{m!(n-m)!}u^{mN} + \cdots, \quad u \to \infty, \tag{7.2.14}$$

$$\Lambda_n(u) = \prod_{j=1}^{N}(u - \theta_j + \eta)\prod_{k=1}^{n-1}(u - \theta_j - k\eta). \tag{7.2.15}$$

It is remarked that the last Eq. (7.2.15) is a direct consequence of (7.1.36). The definitions (7.2.8) of $\{\tau_m(u)\}$ lead to relations among their eigenvalues and $\{\Lambda_m(u)\}$:

$$\Lambda_m(u) = \prod_{l=1}^{N}\prod_{k=1}^{m-1}(u - \theta_l - k\eta)\,\bar{\Lambda}_m(u), \quad m = 1, \ldots, n. \tag{7.2.16}$$

The properties (7.2.7) (or (7.2.13) for their eigenvalues) imply that each $\bar{\Lambda}_m(u)$ as a function of u is a polynomial of degree N and satisfies the relations (a consequence of (7.2.10))

$$\bar{\Lambda}(\theta_j)\bar{\Lambda}_m(\theta_j-\eta)=\prod_{l=1}^{N}(\theta_j-\theta_l-\eta)\bar{\Lambda}_{m+1}(\theta_j),$$

$$j=1,\ldots,N,\quad m=1,\ldots,n-1, \tag{7.2.17}$$

$$\bar{\Lambda}_m(u)=\frac{n!}{m!(n-m)!}u^N+\cdots,\quad u\to\infty, \tag{7.2.18}$$

$$\bar{\Lambda}_n(u)=\prod_{j=1}^{N}(u-\theta_j+\eta). \tag{7.2.19}$$

To give the nested $T-Q$ relations, we introduce the z-functions

$$z_p^{(l)}(u)=Q_p^{(0)}(u)\frac{Q_p^{(l-1)}(u+\eta)Q_p^{(l)}(u-\eta)}{Q_p^{(l-1)}(u)Q_p^{(l)}(u)},\quad l=1,\ldots,n, \tag{7.2.20}$$

where the Q-functions are given by

$$Q_p^{(0)}(u)=\prod_{j=1}^{N}(u-\theta_j), \tag{7.2.21}$$

$$Q_p^{(r)}(u)=\prod_{l=1}^{L_r}(u-\lambda_l^{(r)}),\quad r=1,\ldots,n-1, \tag{7.2.22}$$

$$Q_p^{(n)}(u)=1, \tag{7.2.23}$$

with $\{L_r|r=1,\ldots,n-1\}$ being the number of the Bethe roots $\{\lambda_l^{(r)}|l=1,\ldots,L_r,$ $r=1,\ldots,n-1\}$. The nested $T-Q$ relations are thus constructed as

$$\Lambda_m(u)=\sum_{1\le i_1<i_2<\cdots<i_m\le n}z_p^{(i_1)}(u)z_p^{(i_2)}(u-\eta)\cdots z_p^{(i_m)}(u-(m-1)\eta). \tag{7.2.24}$$

For example, the eigenvalue $\Lambda(u)$ of the fundamental transfer matrix $t(u)$ is

$$\Lambda(u)=Q_p^{(0)}(u+\eta)\frac{Q_p^{(1)}(u-\eta)}{Q_p^{(1)}(u)}+Q_p^{(0)}(u)\frac{Q_p^{(1)}(u+\eta)Q_p^{(2)}(u-\eta)}{Q_p^{(1)}(u)Q_p^{(2)}(u)}+\cdots$$

$$+Q_p^{(0)}(u)\frac{Q_p^{(n-2)}(u+\eta)Q_p^{(n-1)}(u-\eta)}{Q_p^{(n-2)}(u)Q_p^{(n-1)}(u)}+Q_p^{(0)}(u)\frac{Q_p^{(n-1)}(u+\eta)}{Q_p^{(n-1)}(u)}, \tag{7.2.25}$$

while $\Lambda_n(u)$ is

$$\Lambda_n(u) = Q_p^{(0)}(u+\eta)\prod_{l=1}^{n-1} Q_p^{(0)}(u-l\eta), \tag{7.2.26}$$

which is exactly the quantum determinant given in (7.1.36). It should be emphasized that $\{\Lambda_m(u)\}$ are given in terms of the special form of (7.2.24) so that (7.2.12)–(7.2.14) are satisfied for arbitrary Q_i-function. We should also remark that Eq. (7.2.25) is the same as that derived by the algebraic Bethe Ansatz method in Chap. 2. Note that for the present model we have $n-1$ (determined by the rank of the $SU(n)$ algebra) independent Q-functions. The regularity of the $T-Q$ relation (7.2.25) leads to the associated BAEs

$$\prod_{j=1,\neq l}^{L_r} \frac{\lambda_l^{(r)}-\lambda_j^{(r)}-\eta}{\lambda_l^{(r)}-\lambda_j^{(r)}+\eta} = \prod_{k=1}^{L_{r-1}} \frac{\lambda_l^{(r)}-\lambda_k^{(r-1)}}{\lambda_l^{(r)}-\lambda_k^{(r-1)}+\eta} \prod_{m=1}^{L_{r+1}} \frac{\lambda_l^{(r)}-\lambda_m^{(r+1)}-\eta}{\lambda_l^{(r)}-\lambda_m^{(r+1)}},$$
$$l=1,\ldots,L_r,\quad r=1,\ldots,n-1,\quad L_0=N,\quad L_N=0,\quad \lambda_l^{(0)}=\theta_l. \tag{7.2.27}$$

Moreover, the above BAEs also ensure the regularity of all the eigenvalues $\Lambda_m(u)$ given in (7.2.24). Therefore, the BAEs obtained from all the fused transfer matrices are self-consistent. Putting $\mu_j^{(r)} = \lambda_j^{(r)} - r\eta/2$ and $\theta_j = 0$, the resulting BAEs recover those obtained by other Bethe Ansatz methods [13–16]

$$\prod_{j=1,\neq l}^{L_r} \frac{\mu_l^{(r)}-\mu_j^{(r)}-\eta}{\mu_l^{(r)}-\mu_j^{(r)}+\eta} = \prod_{k=1}^{L_{r-1}} \frac{\mu_l^{(r)}-\mu_k^{(r-1)}-\eta/2}{\mu_l^{(r)}-\mu_k^{(r-1)}+\eta/2} \prod_{m=1}^{L_{r+1}} \frac{\mu_l^{(r)}-\mu_m^{(r+1)}-\eta/2}{\mu_l^{(r)}-\mu_m^{(r+1)}+\eta/2},$$
$$l=1,\ldots,L_r,\quad r=1,\ldots,n-1,\quad L_0=N,\quad L_N=0,\quad \lambda_l^{(0)}=0. \tag{7.2.28}$$

7.3 Fundamental Relations of the Open $SU(n)$ Spin Chain

7.3.1 The Model Hamiltonian

Let us consider a generic $K^-(u)$ satisfying RE [17–20]

$$K^-(u) = \xi + uM, \quad M^2 = 1, \tag{7.3.1}$$

where ξ is a boundary parameter and M is an $n\times n$ constant matrix. Obviously,

$$K^-(0) = \xi, \quad K^-(u) = u\,M + \cdots, \quad u\to\infty. \tag{7.3.2}$$

Note that the eigenvalues of M must be ± 1. Let us denote the number of positive eigenvalues as p and the number of negative eigenvalues as q such that $p+q=n$ and $tr M = p-q$. Accordingly, the generic dual K-matrix $K^+(u)$ reads

$$K^+(u) = \bar{\xi} - \left(u + \frac{n}{2}\eta\right)\bar{M}, \quad \bar{M}^2 = 1, \tag{7.3.3}$$

which satisfies the dual RE

$$\begin{aligned} &R_{1,2}(v-u)K_1^+(u)R_{2,1}(-u-v-n\eta)K_2^+(v) \\ &\quad = K_2^+(v)R_{1,2}(-u-v-n\eta)K_1^+(u)R_{2,1}(v-u), \end{aligned} \tag{7.3.4}$$

and

$$K^+\left(-\frac{n}{2}\eta\right) = \bar{\xi}, \quad K^+(u) = -u\,\bar{M} + \cdots, \quad u \to \infty, \tag{7.3.5}$$

where $\bar{\xi}$ is a boundary parameter and $\bar{M}$ is an $n \times n$ constant matrix with eigenvalues ± 1. Similarly, let us denote $\bar{p}$ as the number of positive eigenvalues and $\bar{q}$ as the number of negative eigenvalues such that $\bar{p}+\bar{q}=n$ and $tr\bar{M} = \bar{p}-\bar{q}$. The Hamiltonian defined in terms of the transfer matrix $t(u) = tr_0\{K_0^+(u)T_0(u)K_0^-(u)\hat{T}_0(u)\}$ thus reads

$$\begin{aligned} H &= \eta \left.\frac{\partial \ln t(u)}{\partial u}\right|_{u=0,\{\theta_j=0\}} \\ &= 2\sum_{j=1}^{N-1} P_{j,j+1} + \eta\frac{tr_0 K_0^{+\prime}(0)}{tr_0 K_0^+(0)} + 2\frac{tr_0[K_0^+(0)P_{0N}]}{tr_0 K_0^+(0)} + \eta\frac{1}{\xi}K_1^{-\prime}(0). \end{aligned} \tag{7.3.6}$$

7.3.2 The Fusion Procedure

To derive the operator product identities of the open boundary system, the fusion technique for K-matrices [21, 22] is needed. Following the standard fusion hierarchy, we introduce the following recursive relations:

$$\begin{aligned} K_{1,\ldots,m}^+(u) &= K_{\langle 2,\ldots,m\rangle}^+(u-\eta)R_{1,m}(-2u-n\eta+(m-1)\eta)\cdots \\ &\quad \times R_{1,2}(-2u-n\eta+\eta)K_1^+(u), \end{aligned} \tag{7.3.7}$$

$$K_{\langle 1,\ldots,m\rangle}^+(u) = P_{1,\ldots,m}^{(-)}K_{1,\ldots,m}^+(u)P_{1,\ldots,m}^{(-)}, \tag{7.3.8}$$

$$\begin{aligned} K_{1,\ldots,m}^-(u) &= K_1^-(u)R_{2,1}(2u-\eta)\cdots R_{m,1}(2u-(m-1)\eta) \\ &\quad \times K_{\langle 2,\ldots,m\rangle}^-(u-\eta), \end{aligned} \tag{7.3.9}$$

$$K_{\langle 1,\ldots,m\rangle}^-(u) = P_{1,\ldots,m}^{(-)}\,K_{1,\ldots,m}^-(u)\,P_{1,\ldots,m}^{(-)}, \tag{7.3.10}$$

$$\mathscr{U}_{1,\ldots,m}(u) = \mathscr{U}_1(u) R_{2,1}(2u-\eta) \cdots R_{m,1}(2u-(m-1)\eta) \\ \times \mathscr{U}_{\langle 2,\ldots,m\rangle}(u-\eta), \tag{7.3.11}$$

$$\mathscr{U}_{\langle 1,\ldots,m\rangle}(u) = P^{(-)}_{1,\ldots,m} \mathscr{U}_{1,\ldots,m}(u) P^{(-)}_{1,\ldots,m}. \tag{7.3.12}$$

The mth fused transfer matrix $t_m(u)$ is thus defined as

$$t_m(u) = tr_{1,\ldots,m}\{K^+_{\langle 1,\ldots,m\rangle}(u)\mathscr{U}_{\langle 1,\ldots,m\rangle}(u)\}, \quad m = 1, \ldots, n, \tag{7.3.13}$$

which includes the fundamental transfer matrix $t(u)$ as the first one, i.e., $t(u) = t_1(u)$.

The YBE and RE indicate the following generalized fusion properties [cf. (7.1.16)]

$$P^{(-)}_{1,2} K^-_1(u) R_{2,1}(2u-\eta) K^-_2(u-\eta) P^{(-)}_{1,2} \\ = K^-_1(u) R_{2,1}(2u-\eta) K^-_2(u-\eta) P^{(-)}_{1,2}, \tag{7.3.14}$$

$$P^{(-)}_{1,2} K^+_2(u-\eta) R_{1,2}(-2u-n\eta+\eta) K^+_1(u) P^{(-)}_{1,2} \\ = K^+_2(u-\eta) R_{1,2}(-2u-n\eta+\eta) K^+_1(u) P^{(-)}_{1,2}, \tag{7.3.15}$$

$$\mathscr{U}_{\langle 1,\ldots,m\rangle}(u) = \mathscr{U}_{1,\ldots,m}(u)\, P^{(-)}_{1,\ldots,m}, \quad m = 1, \ldots, n, \tag{7.3.16}$$

$$K^+_{\langle 1,\ldots,m\rangle}(u) = K^+_{1,\ldots,m}(u)\, P^{(-)}_{1,\ldots,m}, \quad m = 1, \ldots, n, \tag{7.3.17}$$

$$K^-_{\langle 1,\ldots,m\rangle}(u) = K^-_{1,\ldots,m}(u)\, P^{(-)}_{1,\ldots,m}, \quad m = 1, \ldots, n. \tag{7.3.18}$$

The above relations can be proven as follows. The associated RE, with special choice of the spectral parameters, can be written as

$$R_{2,1}(-\eta) K^-_2(u-\eta) R_{1,2}(2u-\eta) K^-_1(u) \\ = K^-_1(u) R_{2,1}(2u-\eta) K^-_2(u-\eta) R_{1,2}(-\eta). \tag{7.3.19}$$

This relation and the fusion property (7.1.4) of the R matrix lead to the relation (7.3.14). With the help of the fusion property (7.1.16), one can check (7.3.18) by induction. Similarly, using the dual RE, one can prove (7.3.15) and its fused version (7.3.17). Combining the fusion properties (7.1.16), (7.3.18) and (7.3.20) (see below), one can also check (7.3.16) by induction.

Similarly as we prove the relation (7.1.16), we can show the following relation holds

$$\hat{T}_{\langle 1,\ldots,m\rangle}(u) = P^{(-)}_{1,\ldots,m}\, \hat{T}_1(u) \cdots \hat{T}_m(u-(m-1)\eta)\, P^{(-)}_{1,\ldots,m} \\ = \hat{T}_1(u) \cdots \hat{T}_m(u-(m-1)\eta)\, P^{(-)}_{1,\ldots,m}. \tag{7.3.20}$$

The above relation and the properties (7.1.16) and (7.3.18) allow us to give the following decomposition of the fused double-row monodromy matrices

$$\mathscr{U}_{\langle 1,\ldots,m\rangle}(u) = T_{\langle 1,\ldots,m\rangle}(u)K^{-}_{\langle 1,\ldots,m\rangle}(u)\hat{T}_{\langle 1,\ldots,m\rangle}(u), \quad m = 1,\ldots,n. \tag{7.3.21}$$

The fused R-matrix (7.1.18) and the fused K-matrix (7.3.18) satisfy the fused RE

$$\begin{aligned}
&R_{\langle 1,\ldots,m\rangle,0}(u-v)K^{-}_{\langle 1,\ldots,m\rangle}(u)R_{0,\langle 1,\ldots,m\rangle}(u+v)K^{-}_{0}(v)\\
&= K^{-}_{0}(v)R_{\langle 1,\ldots,m\rangle,0}(u+v)K^{-}_{\langle 1,\ldots,m\rangle}(u)R_{0,\langle 1,\ldots,m\rangle}(u-v).
\end{aligned}\tag{7.3.22}$$

The above relation can be demonstrated by using YBE and RE repeatedly. To show this clearly, let us consider the $m = 2$ case. From the definitions of the fused R- and K-matrices, we have

$$\begin{aligned}
&R_{\langle 1,2\rangle,0}(u-v)K^{-}_{\langle 1,2\rangle}(u)R_{\langle 1,2\rangle,0}(u+v)K^{-}_{0}(v)\\
&= R_{1,0}(u-v)K^{-}_{1}(u)\left\{R_{2,0}(u-v-\eta)R_{2,1}(2u-\eta)R_{1,0}(u+v)\right\}\\
&\quad \times K^{-}_{2}(u-\eta)R_{2,0}(u+v-\eta)K^{-}_{0}(v)P^{(-)}_{1,2}\\
&= R_{1,0}(u-v)K^{-}_{1}(u)R_{1,0}(u+v)R_{2,1}(2u-\eta)\\
&\quad \times \left\{R_{2,0}(u-v-\eta)K^{-}_{2}(u-\eta)R_{2,0}(u+v-\eta)K^{-}_{0}(v)\right\}P^{(-)}_{1,2}\\
&= K^{-}_{0}(v)R_{1,0}(u+v)K^{-}_{1}(u)\left\{R_{1,0}(u-v)R_{2,1}(2u-\eta)R_{2,0}(u+v-\eta)\right\}\\
&\quad \times K^{-}_{2}(u-\eta)R_{2,0}(u-v-\eta)P^{(-)}_{1,2}\\
&= K^{-}_{0}(v)R_{1,0}(u+v)R_{2,0}(u+v-\eta)K^{-}_{1}(u)R_{2,1}(2u-\eta)\\
&\quad \times K^{-}_{2}(u-\eta)R_{1,0}(u-v)R_{2,0}(u-v-\eta)P^{(-)}_{1,2}\\
&= K^{-}_{0}(v)R_{\langle 1,2\rangle,0}(u+v)K^{-}_{\langle 1,2\rangle}(u)R_{\langle 1,2\rangle,0}(u-v).
\end{aligned}\tag{7.3.23}$$

Similarly, we can demonstrate the relations

$$\begin{aligned}
&R_{a,b}(u-v)K^{-}_{a}(u)R_{b,a}(u+v)K^{-}_{b}(v)\\
&\qquad = K^{-}_{b}(v)R_{a,b}(u+v)K^{-}_{a}(u)R_{b,a}(u-v),
\end{aligned}\tag{7.3.24}$$

$$\begin{aligned}
&R_{a,b}(v-u)K^{+}_{a}(u)R_{b,a}(-u-v-n\eta)K^{+}_{b}(v)\\
&\qquad = K^{+}_{b}(v)R_{a,b}(-u-v-n\eta)K^{+}_{a}(u)R_{b,a}(v-u),
\end{aligned}\tag{7.3.25}$$

and therefore

$$\begin{aligned}
&R_{a,b}(u-v)\mathscr{U}_{a}(u)R_{b,a}(u+v)\mathscr{U}_{b}(v)\\
&\qquad = \mathscr{U}_{b}(v)R_{a,b}(u+v)\mathscr{U}_{a}(u)R_{b,a}(u-v),
\end{aligned}\tag{7.3.26}$$

where $a = \langle 1, \ldots, m \rangle$ and $b = \langle \bar{1}, \ldots, \bar{k} \rangle$. Note that the fused crossing-unitary relation holds:

$$\begin{aligned}
&R^{t_{1,\ldots,m}}_{\langle 1,\ldots,m\rangle,\langle \bar{1},\ldots,\bar{k}\rangle}(u) R^{t_{1,\ldots,m}}_{\langle \bar{1},\ldots,\bar{k}\rangle,\langle 1,\ldots,m\rangle}(-u-n\eta) \\
&= (-1)^{m+\bar{k}} \prod_{l=1}^{m}\prod_{i=1}^{\bar{k}} (u-(l-1)\eta)(u-(l-1-n)\eta)(u-(i-1)\eta) \\
&\quad \times (u-(i-1-n)\eta) P^{(-)}_{1,\ldots,m} P^{(-)}_{\bar{1},\ldots,\bar{k}}. \qquad (7.3.27)
\end{aligned}$$

With the same procedure introduced in Eq. (2.2.11), Eqs. (7.3.24)–(7.3.27) allow us to get the equation

$$[t_m(u),\ t_k(v)] = 0, \quad m,\ k = 1, \ldots, n. \qquad (7.3.28)$$

We remark that $t_n(u)$ is the quantum determinant [23, 24] as in the periodic boundary case

$$t_n(u) = \Delta_q(u) \times \mathrm{id}. \qquad (7.3.29)$$

The coefficient $\Delta_q(u)$ can be calculated as follows. Taking $m = n$ in (7.3.8), (7.3.10) and (7.3.20) yields

$$\begin{aligned}
\hat{T}_{\langle 1,\ldots,n\rangle}(u) &= \mathrm{Det}_q\{\hat{T}(u)\} \times \mathrm{id}, \\
K^{-}_{\langle 1,\ldots,n\rangle}(u) &= \mathrm{Det}_q\{K^{-}(u)\} \times \mathrm{id}, \\
K^{+}_{\langle 1,\ldots,n\rangle}(u) &= \mathrm{Det}_q\{K^{+}(u)\} \times \mathrm{id}.
\end{aligned}$$

Using a similar method to compute $\mathrm{Det}_q\{T(u)\}$, we can easily derive the coefficients $\mathrm{Det}_q\{K^{-}(u)\}$ and $\mathrm{Det}_q\{K^{+}(u)\}$. Therefore

$$\begin{aligned}
\Delta_q(u) &= \mathrm{Det}_q\{T(u)\}\mathrm{Det}_q\{\hat{T}(u)\}\mathrm{Det}_q\{K^{+}(u)\}\mathrm{Det}_q\{K^{-}(u)\} \\
&= \prod_{l=1}^{N}(u-\theta_l+\eta)(u+\theta_l+\eta) \prod_{m=1}^{N}\prod_{k=1}^{n-1}(u-\theta_m-k\eta)(u+\theta_m-k\eta) \\
&\quad \times \prod_{i=1}^{n-1}\prod_{j=1}^{i}(2u-(i+j)\eta)(-2u+(n-2-i-j)\eta)(-1)^{q+\bar{q}} \\
&\quad \times \prod_{k=0}^{\bar{q}-1}\left(-u+\frac{n-2}{2}\eta-\bar{\xi}-k\eta\right) \prod_{k=0}^{\bar{p}-1}\left(-u+\frac{n-2}{2}\eta+\bar{\xi}-k\eta\right) \\
&\quad \times \prod_{k=0}^{q-1}(u-\xi-k\eta) \prod_{k=0}^{p-1}(u+\xi-k\eta). \qquad (7.3.30)
\end{aligned}$$

From YBE and RE, we rewrite the fused K-matrix in another form

$$K^{+}_{\langle 1,\ldots,m\rangle}(u) = P^{(-)}_{1,\ldots,m} K^{+}_{m}(u-(m-1)\eta) R_{m,m-1}(-2u - n\eta + (2m-3)\eta)\cdots \\ \times R_{m,1}(-2u - n\eta + (m-1)\eta) K^{+}_{\langle 1,\ldots,m-1\rangle}(u) P^{(-)}_{1,\ldots,m}. \quad (7.3.31)$$

The initial condition of the R-matrix and the conditions

$$K^{-}(0) = \xi, \quad K^{+}\left(-\frac{n}{2}\eta\right) = \bar{\xi}, \quad (7.3.32)$$

imply that the transfer matrix $t_m(u)$ can be expressed in terms of $\{t_l(u)|l = 1, \ldots, m-1\}$ for the following $2m$ degenerate points of the spectral parameter u:

$$0, \frac{\eta}{2}, \ldots, \frac{m-1}{2}\eta, -\frac{n}{2}\eta + (m-1)\eta, \\ -\frac{n}{2}\eta + (m-1)\eta - \frac{\eta}{2}, \ldots, -\frac{n}{2}\eta + \frac{m-1}{2}\eta, \quad (7.3.33)$$

which provides some necessary conditions to determine the eigenvalue function $\Lambda(u)$.

7.3.3 Operator Product Identities and Functional Relations

Let us first prove some useful relations that are necessary to obtain the operator product identities. Similar to (7.2.1), the following relation holds:

$$\hat{T}_1(-\theta_j)\hat{T}_2(-\theta_j - \eta) = P^{(-)}_{1,2}\, \hat{T}_1(-\theta_j)\hat{T}_2(-\theta_j - \eta), \quad j = 1, \ldots, N. \quad (7.3.34)$$

With the same procedure used in (7.3.23), we can obtain the relations

$$\mathscr{U}_1(\pm\theta_j) R_{2,1}(\pm 2\theta_j - \eta) \mathscr{U}_2(\pm\theta_j - \eta) \\ = P^{(-)}_{1,2} \mathscr{U}_1(\pm\theta_j) R_{2,1}(\pm 2\theta_j - \eta) \mathscr{U}_2(\pm\theta_j - \eta), \quad j = 1, \ldots, N. \quad (7.3.35)$$

Using relations (7.3.16) and (7.3.35), we can derive

$$P_{l,l+1}\, \mathscr{U}_{1,\ldots,m}(\pm\theta_j) = -\mathscr{U}_{1,\ldots,m}(\pm\theta_j), \quad l = 1, \ldots, m-1. \quad (7.3.36)$$

Finally we arrive at the identities

$$\mathscr{U}_{1,\ldots,m}(\pm\theta_j) = P^{(-)}_{1,\ldots,m} \mathscr{U}_{1,\ldots,m}(\pm\theta_j). \quad (7.3.37)$$

Now let us evaluate the operator product

$$
\begin{aligned}
t(\pm\theta_j)t_m(\pm\theta_j-\eta) &= tr_{1,\ldots,m+1}\left\{\mathscr{U}_1^{t_1}(\pm\theta_j)K_1^+(\pm\theta_j)^{t_1}\right.\\
&\quad\left.\times\mathscr{U}_{\langle 2,\ldots,m+1\rangle}(\pm\theta_j-\eta)K^+_{\langle 2,\ldots,m+1\rangle}(\pm\theta_j-\eta)\right\}\\
&=\prod_{k=1}^m\varphi_2^{-1}(\pm 2\theta_j-k\eta)\times tr_{1,\ldots,m+1}\left\{\mathscr{U}_1^{t_1}(\pm\theta_j)K_1^+(\pm\theta_j)^{t_1}\right.\\
&\quad\times R_{1,2}^{t_1}(\mp 2\theta_j+\eta-n\eta)\cdots R_{1,m+1}^{t_1}(\mp 2\theta_j+m\eta-n\eta)R_{1,m+1}^{t_1}(\pm 2\theta_j-m\eta)\\
&\quad\left.\times\cdots\times R_{1,2}^{t_1}(\pm 2\theta_j-\eta)\mathscr{U}_{\langle 2,\ldots,m+1\rangle}(\pm\theta_j-\eta)K^+_{\langle 2,\ldots,m+1\rangle}(\pm\theta_j-\eta)\right\}\\
&=\prod_{k=1}^m\varphi_2^{-1}(\pm 2\theta_j-k\eta)tr_{1,\ldots,m+1}\left\{R_{1,m+1}(\mp 2\theta_j+m\eta-n\eta)\cdots\right.\\
&\quad\left.\times R_{1,2}(\mp 2\theta_j+\eta-n\eta)K_1^+(\pm\theta_j)\mathscr{U}_{1,\ldots,m+1}(\pm\theta_j)K^+_{\langle 2,\ldots,m+1\rangle}(\pm\theta_j-\eta)\right\}\\
&=\prod_{k=1}^m\varphi_2^{-1}(\pm 2\theta_j-k\eta)tr_{1,\ldots,m+1}\left\{K^+_{\langle 2,\ldots,m+1\rangle}(\pm\theta_j-\eta)R_{1,m+1}(\mp 2\theta_j+m\eta-n\eta)\right.\\
&\quad\left.\times\cdots\times R_{1,2}(\mp 2\theta_j+\eta-n\eta)K_1^+(\pm\theta_j)\,\mathscr{U}_{1,\ldots,m+1}(\pm\theta_j)\right\}\\
&=\prod_{k=1}^m\varphi_2^{-1}(\pm 2\theta_j-k\eta)tr_{1,\ldots,m+1}\left\{K^+_{1,\ldots,m+1}(\pm\theta_j)\mathscr{U}_{1,\ldots,m+1}(\pm\theta_j)\right\}\\
&\overset{(7.3.37)}{=}\prod_{k=1}^m\varphi_2^{-1}(\pm 2\theta_j-k\eta)tr_{1,\ldots,m+1}\left\{K^+_{1,\ldots,m+1}(\pm\theta_j)P^{(-)}_{1,\ldots,m+1}\mathscr{U}_{1,\ldots,m+1}(\pm\theta_j)\right\}\\
&=\prod_{k=1}^m\varphi_2^{-1}(\pm 2\theta_j-k\eta)tr_{1,\ldots,m+1}\left\{K^+_{\langle 1,\ldots,m+1\rangle}(\pm\theta_j)\mathscr{U}_{1,\ldots,m+1}(\pm\theta_j)P^{(-)}_{1,\ldots,m+1}\right\}\\
&=\prod_{k=1}^m\varphi_2^{-1}(\pm 2\theta_j-k\eta)tr_{1,\ldots,m+1}\left\{K^+_{\langle 1,\ldots,m+1\rangle}(\pm\theta_j)\mathscr{U}_{\langle 1,\ldots,m+1\rangle}(\pm\theta_j)\right\}.
\end{aligned}
$$

Therefore, the following recursive operator product identities hold:

$$
t(\pm\theta_j)t_m(\pm\theta_j-\eta)=t_{m+1}(\pm\theta_j)\prod_{k=1}^m\varphi_2^{-1}(\pm 2\theta_j-k\eta),\qquad (7.3.38)
$$

$$
j=1,\ldots,N,\quad m=1,\ldots,n-1.
$$

Let $|\Psi\rangle$ be a common eigenstate of $\{t_m(u)\}$ with the eigenvalue $\Lambda_m(u)$, i.e.,

$$
t_m(u)|\Psi\rangle=\Lambda_m(u)|\Psi\rangle,\quad m=1,\ldots,n. \qquad (7.3.39)
$$

We have the following functional relations:

$$\Lambda(\pm\theta_j)\Lambda_m(\pm\theta_j-\eta)=\Lambda_{m+1}(\pm\theta_j)\prod_{k=1}^{m}\varphi_2^{-1}(\pm 2\theta_j-k\eta), \quad (7.3.40)$$
$$j=1,\ldots,N, \quad m=1,\ldots,n-1.$$

Using a similar method as in the periodic case, we rewrite the transfer matrix as

$$t_m(u)=\prod_{i=1}^{m-1}\prod_{j=1}^{i}(2u-i\eta-j\eta)(-2u+(2m-2-n)\eta-i\eta-j\eta)$$
$$\times\prod_{l=1}^{N}\prod_{k=1}^{m-1}(u-\theta_l-k\eta)(u+\theta_l-k\eta)\,\tau_m(u). \quad (7.3.41)$$

Let us introduce the following correspondence:

$$\Lambda_m(u)=\prod_{l=1}^{m-1}\prod_{j=1}^{i}(2u-i\eta-j\eta)(-2u+(2m-2-n)\eta-i\eta-j\eta)$$
$$\times\prod_{l=1}^{N}\prod_{k=1}^{m-1}(u-\theta_l-k\eta)(u+\theta_l-k\eta)\bar{\Lambda}_m(u). \quad (7.3.42)$$

Then we have

$$\bar{\Lambda}(\pm\theta_j)\bar{\Lambda}_m(\pm\theta_j-\eta)=\bar{\Lambda}_{m+1}(\pm\theta_j)\prod_{k=1}^{m}\varphi_2^{-1}(\pm 2\theta_j-k\eta)\varphi_0(\pm\theta_j),$$
$$m=1,\ldots,n-1, \quad j=1,\ldots,N, \quad (7.3.43)$$

where the function $\varphi_0(u)$ is given by

$$\varphi_0(u)=\prod_{l=1}^{N}(u-\theta_l-\eta)(u+\theta_l-\eta)\prod_{k=2}^{m+1}(2u-k\eta)(-2u+k\eta-(n+2)\eta).$$

Obviously, $\bar{\Lambda}_m(u)$ is a degree $2N+2m$ polynomial of u. Note that

$$\bar{\Lambda}_n(u)=(-1)^{q+\bar{q}}\prod_{l=1}^{N}(u-\theta_l+\eta)(u+\theta_l+\eta)\prod_{k=0}^{\bar{q}-1}(-u+\frac{n-2}{2}\eta-\bar{\xi}-k\eta)$$
$$\times\prod_{k=0}^{\bar{p}-1}(u+\frac{n-2}{2}\eta+\xi-k\eta)\prod_{k=0}^{q-1}(u-\xi-k\eta)\prod_{k=0}^{p-1}(u+\xi-k\eta). \quad (7.3.44)$$

7.3.4 Asymptotic Behavior of the Transfer Matrices

We note that the leading order of $\{t_m(u)\}$ is completely determined by the eigenvalues $\{\lambda_l|l=1,\ldots,n\}$ of $\bar{M}M$. With the conditions $M^2=\bar{M}^2=1$ we have the following useful relations for arbitrary integer k:

$$\sum_{l=1}^{n}\lambda_l^k = tr\left\{(\bar{M}M)^k\right\} = tr\left\{(M\bar{M})^k\right\} = tr\left\{(\bar{M}M)^{-k}\right\} = \sum_{l=1}^{n}\lambda_l^{-k}. \quad (7.3.45)$$

In addition,

$$\mathrm{Det}\left(\bar{M}M\right) = \lambda_1\cdots\lambda_n = (-1)^{q+\bar{q}}. \quad (7.3.46)$$

The above equations indicate that the eigenvalues of $M\bar{M}$ must take the form

$$\{\lambda_1,\ldots,\lambda_n\} = \left\{1,\ldots,1,-1,\ldots,-1,e^{-i\vartheta_1},e^{i\vartheta_1},\ldots,e^{-i\vartheta_r},e^{i\vartheta_r}\right\}, \quad (7.3.47)$$

where ϑ_j are some continuous free parameters related to the boundary fields. The maximum number of ϑ_j is $\lfloor\frac{n}{2}\rfloor$. With the relations (7.3.2) and (7.3.5) we can easily deduce the asymptotic behavior

$$\bar{\Lambda}_m(u) = (-1)^m\delta_m\,u^{2N+2m} + \cdots, \quad m=1,\ldots,n, \quad u\to\infty, \quad (7.3.48)$$

with

$$\delta_m = \sum_{1\le i_1<i_2<\cdots<i_m\le n}\lambda_{i_1}\cdots\lambda_{i_m}, \quad m=1,\ldots,n. \quad (7.3.49)$$

As $\bar{\Lambda}_m(u)$ is a degree $2N+2m$ polynomial of u, in addition to the functional relations (7.3.43) and the asymptotic behavior (7.3.48), $\sum_{m=1}^{n-1}2m$ additional conditions at the degenerate points listed in (7.3.33) are also needed to determine $\{\bar{\Lambda}_m(u)|m=1,\ldots,n-1\}$ completely. Note that $\bar{\Lambda}_n(u)$ has already been fixed by (7.3.44). The contributions of the K-matrices to the quantum determinant $\bar{\Lambda}_n(u)$ (7.3.44) can be factorized out by n polynomials of degree 2 $\{K^{(l)}(u)|l=1,\ldots,n\}$ satisfying the relations

$$\prod_{l=1}^{n}K^{(l)}(u-(l-1)\eta) = (-1)^{q+\bar{q}}\prod_{k=0}^{\bar{q}-1}\left(-u+\frac{n-2}{2}\eta-\bar{\xi}-k\eta\right)$$
$$\times\prod_{k=0}^{\bar{p}-1}\left(-u+\frac{n-2}{2}\eta+\bar{\xi}-k\eta\right)\prod_{k=0}^{q-1}(u-\xi-k\eta)\prod_{k=0}^{p-1}(u+\xi-k\eta), \quad (7.3.50)$$

$$K^{(l)}(u)K^{(l)}(-u-l\eta) = K^{(l+1)}(u)K^{(l+1)}(-u-l\eta), \quad (7.3.51)$$

where $l = 1, \ldots, n-1$. The condition (7.3.51) is required to construct self-consistent nested $T-Q$ relations for the eigenvalues $\Lambda_m(u)$ as we shall show in the following sections of this chapter. We remark that although there are several possible choices for $K^{(l)}(u)$ satisfying (7.3.50)–(7.3.51), any choice should be able to give a complete set of solutions of the model, while different choices only give rise to different parameterizations of the eigenvalues as in the $SU(2)$ case.

7.4 Solution of the $SU(3)$ Case

7.4.1 Functional Relations

To show the procedure clearly, let us first consider the $SU(3)$ case. Without losing generality, we take the eigenvalues of $\bar{M}M$ as

$$\{\lambda_1, \lambda_2, \lambda_3\} = \left\{1, e^{-i\vartheta}, e^{i\vartheta}\right\}. \tag{7.4.1}$$

The functional relations (7.3.43) now read

$$\bar{\Lambda}(\pm\theta_j)\bar{\Lambda}_m(\pm\theta_j - \eta) = \bar{\Lambda}_{m+1}(\pm\theta_j)\prod_{k=1}^{m}\varphi_2^{-1}(\pm 2\theta_j - k\eta)\varphi_0(\pm\theta_j), \tag{7.4.2}$$

where $m = 1, 2,\ j = 1, \ldots, N$ and the functions $\varphi_0(u)$ and $\bar{\Lambda}_3(u)$ are given by

$$\varphi_0(u) = \prod_{l=1}^{N}(u-\theta_l-\eta)(u+\theta_l-\eta)\prod_{k=2}^{m+1}(2u-k\eta)(-2u+k\eta-5\eta), \tag{7.4.3}$$

$$\begin{aligned}\bar{\Lambda}_3(u) = &\prod_{l=1}^{N}(u-\theta_l+\eta)(u+\theta_l+\eta)\left(\bar{\xi}+\frac{\eta}{2}-u\right)(\xi+u)\\ &\times\left(\bar{\xi}+\frac{\eta}{2}+u\right)(\xi-u)\left(\bar{\xi}-\frac{\eta}{2}+u\right)(\xi-u+\eta).\end{aligned} \tag{7.4.4}$$

We choose the functions $K^{(l)}(u)$ for the present case as

$$K^{(1)}(u) = \left(\bar{\xi}+\frac{1}{2}\eta-u\right)(\xi+u), \tag{7.4.5}$$

$$K^{(2)}(u) = \left(\bar{\xi}+\frac{3}{2}\eta+u\right)(\xi-u-\eta), \tag{7.4.6}$$

$$K^{(3)}(u) = \left(\bar{\xi}+\frac{3}{2}\eta+u\right)(\xi-u-\eta), \tag{7.4.7}$$

which satisfy the relations

$$K^{(1)}(u)K^{(2)}(u-\eta)K^{(3)}(u-2\eta) = \left(\bar{\xi}+\frac{\eta}{2}-u\right)(\xi+u)$$
$$\times\left(\bar{\xi}+\frac{\eta}{2}+u\right)(\xi-u)\left(\bar{\xi}-\frac{\eta}{2}+u\right)(\xi-u+\eta), \quad (7.4.8)$$
$$K^{(l)}(u)K^{(l)}(-u-l\eta) = K^{(l+1)}(u)K^{(l+1)}(-u-l\eta), \quad l=1,2. \quad (7.4.9)$$

From the definition (7.3.13) of $t_m(u)$ and the asymptotic behavior of $K^{\pm}(u)$, we have

$$\bar{\Lambda}(u)|_{u\to\infty} = -tr(\bar{M}M)u^{2N+2}+\cdots = -\sum_{i=1}^{3}\lambda_i u^{2N+2}+\cdots$$
$$= -(1+2\cos\vartheta)u^{2N+2}+\cdots, \quad (7.4.10)$$
$$\bar{\Lambda}_2(u)|_{u\to\infty} = tr_{1,2}\left\{P^{(-)}_{1,2}(\bar{M}M)_1(\bar{M}M)_2P^{(-)}_{1,2}\right\}u^{2N+4}+\cdots$$
$$= \sum_{1\le i_1<i_2\le 3}\lambda_{i_1}\lambda_{i_2}u^{2N+4}+\cdots = (2\cos\vartheta+1)u^{2N+4}+\cdots. \quad (7.4.11)$$

Moreover, the Eqs. (7.1.1)–(7.1.4), (7.3.2) and (7.3.5) allow us to obtain the relations

$$t(0) = (-1)^N\xi\prod_{l=1}^{N}(\theta_l+\eta)(\theta_l-\eta)tr\{K^+(0)\}\times \mathrm{id}, \quad (7.4.12)$$

$$t\left(-\frac{3}{2}\eta\right) = (-1)^N\bar{\xi}\prod_{l=1}^{N}\left(\theta_l+\frac{3}{2}\eta\right)\left(\theta_l-\frac{3}{2}\eta\right)$$
$$\times tr\left\{K^-\left(-\frac{3}{2}\eta\right)\right\}\times \mathrm{id}, \quad (7.4.13)$$

$$t_2\left(\frac{\eta}{2}\right) = tr_{1,2}\left\{P^{(-)}_{1,2}K_2^+\left(-\frac{\eta}{2}\right)R_{1,2}(-3\eta)K_1^+\left(\frac{\eta}{2}\right)P^{(-)}_{1,2}\right\}\left(\frac{\eta^2}{4}-\xi^2\right)\eta$$
$$\times\prod_{l=1}^{N}\left(\theta_l+\frac{3}{2}\eta\right)\left(\theta_l-\frac{3}{2}\eta\right)\left(\theta_l+\frac{\eta}{2}\right)\left(\theta_l-\frac{\eta}{2}\right)\times \mathrm{id}, \quad (7.4.14)$$

$$t_2(-\eta) = tr_{1,2}\left\{P^{(-)}_{1,2}K_1^-(-\eta)R_{2,1}(-3\eta)K_2^-(-2\eta)P^{(-)}_{1,2}\right\}\left(\frac{\eta^2}{4}-\bar{\xi}^2\right)\eta$$
$$\times\prod_{l=1}^{N}(\theta_l+\eta)(\theta_l-\eta)(\theta_l+2\eta)(\theta_l-2\eta)\times \mathrm{id}, \quad (7.4.15)$$

$$t_2(0) = (-1)^N 2\xi\eta^2 \prod_{l=1}^{N} (\theta_l + \eta)(\theta_l - \eta) tr\{K^+(0)\} t(-\eta), \quad (7.4.16)$$

$$t_2\left(-\frac{\eta}{2}\right) = (-1)^N 2\bar{\xi}\eta^2 \prod_{l=1}^{N} \left(\theta_l + \frac{3}{2}\eta\right)\left(\theta_l - \frac{3}{2}\eta\right) \times tr\left\{K^-\left(-\frac{3}{2}\eta\right)\right\} t\left(-\frac{\eta}{2}\right). \quad (7.4.17)$$

The above six conditions together with the relations (7.4.2) and the asymptotic behavior (7.4.10)–(7.4.11) allow us to determine the eigenvalues $\bar{\Lambda}_m(u)$.

7.4.2 The Nested Inhomogeneous $T - Q$ Relation

For the open chain, we define the corresponding Q-functions as

$$Q^{(0)}(u) = \prod_{j=1}^{N} (u - \theta_j)(u + \theta_j), \quad (7.4.18)$$

$$Q^{(r)}(u) = \prod_{l=1}^{L_r} \left(u - \lambda_l^{(r)}\right)\left(u + \lambda_l^{(r)} + r\eta\right), \quad r = 1, \ldots, n-1, \quad (7.4.19)$$

$$Q^{(n)}(u) = 1, \quad (7.4.20)$$

where $\{L_r | r = 1, \ldots, n-1\}$ are some non-negative integers, and $r = 1, 2$ in the present case. We introduce further three $\tilde{z}(u)$ functions:

$$\tilde{z}_1(u) = z_1(u) + x_1(u), \quad \tilde{z}_2(u) = z_2(u), \quad \tilde{z}_3(u) = z_3(u). \quad (7.4.21)$$

Here $z_m(u)$ is defined as

$$z_m(u) = \frac{u(u + \frac{n}{2}\eta) K^{(m)}(u) Q^{(0)}(u)}{\left(u + \frac{(m-1)}{2}\eta\right)\left(u + \frac{m}{2}\eta\right)} \frac{Q^{(m-1)}(u+\eta) Q^{(m)}(u-\eta)}{Q^{(m-1)}(u) Q^{(m)}(u)}, \quad (7.4.22)$$

with $m = 1, 2, 3$, $n = 3$, and $x_1(u)$ is defined as

$$x_1(u) = u\left(u + \frac{3}{2}\eta\right) Q^{(0)}(u+\eta) Q^{(0)}(u) \frac{F_1(u)}{Q^{(1)}(u)}. \quad (7.4.23)$$

We make the following nested $T-Q$ Ansatz

$$\Lambda(u)=\sum_{i_1=1}^{3}\tilde{z}_{i_1}(u)=\sum_{i_1=1}^{3}z_{i_1}(u)+x_1(u), \tag{7.4.24}$$

$$\Lambda_2(u)=\varphi_2(2u-\eta)\left\{\sum_{1\le i_1<i_2\le 3}\tilde{z}_{i_1}(u)\tilde{z}_{i_2}(u-\eta)-x_1(u)z_2(u-\eta)\right\}, \tag{7.4.25}$$

$$\Lambda_3(u)=\prod_{k=1}^{3}\varphi_2(2u-k\eta)\left\{z_1(u)z_2(u-\eta)z_3(u-2\eta)\right\}. \tag{7.4.26}$$

It should be emphasized that, thanks to $Q^{(0)}(\pm\theta_j)=0$, the above Ansatz (7.4.24)–(7.4.25) makes the functional relations (7.4.2) automatically fulfilled for arbitrary Q-functions and $F_1(u)$. These functions (polynomials) are then fixed by the other conditions required by $\Lambda_i(u)$ as follows. To match the degree of $\Lambda(u)$, $F_1(u)$ is a polynomial of degree $2L_1-2N$. The regularity (7.4.24) at the singular points $\lambda_j^{(1)}$ and $-\lambda_j^{(1)}-\eta$ must give the same equation. This consistency requires

$$F_1(u)=f_1(u)Q^{(2)}(-u-\eta), \tag{7.4.27}$$

with

$$f_1(u)=f_1(-u-\eta). \tag{7.4.28}$$

One can check that the regularity conditions of (7.4.25) indeed coincide with those of $\Lambda(u)$. In order to make (7.4.12)–(7.4.17) satisfied, we need further that all terms with the factor x_1 in those equations must be zero, which allow us to choose the polynomial $f_1(u)$ with a minimal degree as follows

$$f_1(u)=cu\left(u+\frac{1}{2}\eta\right)^2\left(u-\frac{1}{2}\eta\right)\left(u+\frac{3}{2}\eta\right)(u+\eta), \tag{7.4.29}$$

where c is a constant. The non-negative integers L_1 and L_2 satisfy the relation

$$L_1=N+L_2+3, \tag{7.4.30}$$

and

$$c=2(1-\cos\vartheta). \tag{7.4.31}$$

The above relation and (7.3.42) lead to the asymptotic behavior (7.4.10)–(7.4.11) of $\bar{\Lambda}_m(u)$ being automatically satisfied. The regular property (at $u = \lambda_j^{(k)}, -\lambda_j^{(k)} - k\eta$ and $k = 1, 2$) of $\Lambda(u)$ leads to the associated BAEs

$$1 + \frac{\lambda_l^{(1)}}{\lambda_l^{(1)} + \eta} \frac{K^{(2)}(\lambda_l^{(1)})Q^{(0)}(\lambda_l^{(1)})}{K^{(1)}(\lambda_l^{(1)})Q^{(0)}(\lambda_l^{(1)} + \eta)} \frac{Q^{(1)}(\lambda_l^{(1)} + \eta)Q^{(2)}(\lambda_l^{(1)} - \eta)}{Q^{(1)}(\lambda_l^{(1)} - \eta)Q^{(2)}(\lambda_l^{(1)})}$$
$$= -c\,(\lambda_l^{(1)})^2 \left(\lambda_l^{(1)} + \frac{1}{2}\eta\right)^3 (\lambda_l^{(1)} + \eta)(\lambda_l^{(1)} + \frac{3}{2}\eta)(\lambda_l^{(1)} - \frac{1}{2}\eta)$$
$$\times \frac{Q^{(0)}(\lambda_l^{(1)})Q^{(2)}(\lambda_l^{(1)} - \eta)}{K^{(1)}(\lambda_l^{(1)})Q^{(1)}(\lambda_l^{(1)} - \eta)}, \quad l = 1, \ldots, L_1, \tag{7.4.32}$$

$$\frac{\lambda_l^{(2)} + \frac{3}{2}\eta}{\lambda_l^{(2)} + \frac{1}{2}\eta} \frac{K^{(2)}(\lambda_l^{(2)})}{K^{(3)}(\lambda_l^{(2)})} \frac{Q^{(1)}(\lambda_l^{(2)} + \eta)Q^{(2)}(\lambda_l^{(2)} - \eta)}{Q^{(1)}(\lambda_l^{(2)})Q^{(2)}(\lambda_l^{(2)} + \eta)} = -1,$$
$$l = 1, \ldots, L_2. \tag{7.4.33}$$

One may check that the chosen $F_1(u)$ and the BAEs (7.4.32)–(7.4.33) also guarantee the regularity of the Ansatz $\Lambda_2(u)$ given by (7.4.25).

The eigenvalue of the Hamiltonian (7.3.6) for $n = 3$ is given by

$$E = \sum_{l=1}^{L_1} \frac{2\eta^2}{\lambda_l^{(1)}(\lambda_l^{(1)} + \eta)} + 2(N-1) + \eta \frac{\bar{\xi} + \frac{3}{2}\eta - \bar{p}\eta - \xi}{\xi\,(\bar{\xi} + \frac{3}{2}\eta - \bar{p}\eta)} + \frac{2}{3}, \tag{7.4.34}$$

where the parameters $\{\lambda_l^{(1)}\}$ are the roots of the BAEs (7.4.32)–(7.4.33) in the homogeneous limit $\{\theta_j = 0\}$. We note that for $c = 0$, the condition (7.4.30) is no longer necessary.

Numerical results for $N = 2$ shown in Table 7.1 imply that the BAEs indeed give rise to the correct and complete solutions of the model.

7.5 Solution of the $SU(4)$ Case

The $SU(3)$ case is similar to the $SU(2)$ case since there is only one continuous boundary parameter ϑ. The first nontrivial case could be the $SU(4)$ case as it might include two free continuous parameters ϑ_1 and ϑ_2 defined in (7.3.47). Let us consider the case of matrices M and $\bar{M}$ with $p = 2$ and $\bar{p} = 2$. We introduce

$$K^{(1)}(u) = (\xi + u)(\bar{\xi} - u), \tag{7.5.1}$$
$$K^{(2)}(u) = (\xi - u - \eta)(\bar{\xi} + u + \eta), \tag{7.5.2}$$
$$K^{(3)}(u) = (\xi + u + \eta)(\bar{\xi} - u - \eta), \tag{7.5.3}$$
$$K^{(4)}(u) = (\xi - u - 2\eta)(\bar{\xi} + u + 2\eta), \tag{7.5.4}$$

Table 7.1 Solutions of the BAEs (7.4.32)–(7.4.33) for $\eta = 1$, $N = 2$, $\xi = 0.6$, $\bar{\xi} = 1$ and $\vartheta = \pi$

$\lambda_1^{(1)}$	$\lambda_2^{(1)}$	$\lambda_3^{(1)}$	$\lambda_4^{(1)}$	$\lambda_5^{(1)}$			
$-3.24631 - 0.98654i$	$-3.24631 + 0.98654i$	$-3.11950 - 0.00000i$	$-2.48724 - 0.00000i$	$-0.50000 + 0.24330i$			
$-2.95278 - 0.87908i$	$-2.95278 + 0.87908i$	$-2.92063 - 0.00000i$	$-2.51445 + 0.00000i$	$-0.50000 - 0.23261i$			
$-3.17318 - 0.34885i$	$-3.17318 + 0.34885i$	$-3.06506 - 1.38815i$	$-3.06506 + 1.38815i$	$-2.49912 - 0.00000i$			
$-3.17846 - 0.67173i$	$-3.17846 + 0.67173i$	$-2.78686 - 0.00000i$	$-1.19200 + 0.00000i$	$-0.80905 - 0.00000i$			
$-3.32589 - 0.91516i$	$-3.32589 + 0.91516i$	$-3.19981 + 0.00000i$	$-2.49491 + 0.00000i$	$-1.00089 - 0.20677i$			
$-3.39489 - 1.26957i$	$-3.39489 + 1.26957i$	$-3.31805 - 0.29000i$	$-3.31805 + 0.29000i$	$-2.50122 - 0.00000i$			
$-3.00582 - 0.90565i$	$-3.00582 + 0.90565i$	$-2.97538 - 0.00000i$	$-2.50506 + 0.00000i$	$-0.50000 - 0.84595i$			
$-3.15610 - 0.35736i$	$-3.15610 + 0.35736i$	$-3.10262 - 1.42135i$	$-3.10262 + 1.42135i$	$-2.49961 + 0.00000i$			
$-3.03207 - 0.36026i$	$-3.03207 + 0.36026i$	$-2.91597 - 1.54782i$	$-2.91597 + 1.54782i$	$-2.50113 - 0.00000i$			
$\lambda_6^{(1)}$	$\lambda_7^{(1)}$	$\lambda_1^{(2)}$	$\lambda_2^{(2)}$	L_1	L_2	E_n	n
$-0.50000 + 1.03475i$	–	$-1.00000 - 1.08680i$	–	6	1	-3.10819	1
–	–	–	–	5	0	-1.56940	2
$-0.50000 - 0.26257i$	–	$0.08675 + 0.00000i$	–	6	1	-1.27008	3
–	–	–	–	5	0	0.33333	4
$-1.00089 + 0.20677i$	–	$0.40753 - 0.00000i$	–	6	1	1.10819	5
$-1.00759 - 0.33318i$	$-1.00759 + 0.33318i$	$-2.41040 + 0.29871i$	$-2.41040 - 0.29871i$	7	2	1.66667	6
–	–	–	–	5	0	2.90273	7
$-0.50000 - 1.28087i$	–	$-1.00000 - 0.61727i$	–	6	1	3.93675	8
–	–	–	–	5	0	5.00000	9

The symbol n indicates the number of the energy levels, and E_n is the corresponding eigenenergy. The energy E_n calculated from (7.4.34) is the same as that from the exact diagonalization of the Hamiltonian which satisfy the relations (7.3.50)–(7.3.51) with $n = 4$. Suppose that the eigenvalues of $\bar{M}M$ are

$$(\lambda_1, \lambda_2, \lambda_3, \lambda_4) = (e^{i\vartheta_1}, e^{-i\vartheta_1}, e^{i\vartheta_2}, e^{-i\vartheta_2}). \tag{7.5.5}$$

The asymptotic behavior of the eigenvalues of the transfer matrices reads

$$\begin{aligned}
\bar{\Lambda}_1(u)|_{u\to\infty} &= -tr\{\bar{M}M\}u^{2N+2} + \cdots \\
&= -(2\cos\vartheta_1 + 2\cos\vartheta_2)u^{2N+2} + \cdots, \qquad (7.5.6)\\
\bar{\Lambda}_2(u)|_{u\to\infty} &= tr_{1,2}\left\{P^{(-)}_{1,2}(\bar{M}M)_1(\bar{M}M)_2P^{(-)}_{1,2}\right\}u^{2N+4} + \cdots \\
&= \sum_{1\le i_1<i_2\le 4} \lambda_{i_1}\lambda_{i_2}u^{2N+4} + \cdots \\
&= (2 + 4\cos\vartheta_1\cos\vartheta_2)u^{2N+4} + \cdots, \qquad (7.5.7)\\
\bar{\Lambda}_3(u)|_{u\to\infty} &= -tr_{1,2,3}\left\{P^{(-)}_{1,2,3}(\bar{M}M)_1(\bar{M}M)_2(\bar{M}M)_3P^{(-)}_{1,2,3}\right\}u^{2N+6} + \cdots \\
&= -\sum_{1\le i_1<i_2<i_3\le 4} \lambda_{i_1}\lambda_{i_2}\lambda_{i_3}u^{2N+6} + \cdots \\
&= -(2\cos\vartheta_1 + 2\cos\vartheta_2)u^{2N+6} + \cdots. \qquad (7.5.8)
\end{aligned}$$

Moreover, we can derive the relations among the fused transfer matrices:

$$t(0) = (-1)^N\xi\prod_{l=1}^{N}(\theta_l + \eta)(\theta_l - \eta)tr\{K^+(0)\} \times \text{id}, \tag{7.5.9}$$

$$t(-2\eta) = (-1)^N\bar{\xi}\prod_{l=1}^{N}(\theta_l + 2\eta)(\theta_l - 2\eta)tr\{K^-(-2\eta)\} \times \text{id}, \tag{7.5.10}$$

$$t_2(0) = 3(-1)^N\xi\eta^2\prod_{l=1}^{N}(\theta_l + \eta)(\theta_l - \eta)tr\{K^+(0)\}t_1(-\eta), \tag{7.5.11}$$

$$\begin{aligned}
t_2\left(\frac{\eta}{2}\right) &= tr_{1,2}\left\{K^+_{\langle 1,2\rangle}\left(\frac{\eta}{2}\right)\right\}\eta\left(\frac{\eta^2}{4} - \xi^2\right) \\
&\times\prod_{l=1}^{N}\left(\theta_l - \frac{\eta}{2}\right)\left(\theta_l + \frac{\eta}{2}\right)\left(\theta_l - \frac{3}{2}\eta\right)\left(\theta_l + \frac{3}{2}\eta\right) \times \text{id}, \qquad (7.5.12)
\end{aligned}$$

$$\begin{aligned}
t_2\left(-\frac{3}{2}\eta\right) &= \eta\left(\frac{\eta^2}{4} - \bar{\xi}^2\right)\prod_{l=1}^{N}\left(\theta_l - \frac{5}{2}\eta\right)\left(\theta_l + \frac{5}{2}\eta\right) \\
&\times\left(\theta_l - \frac{3}{2}\eta\right)\left(\theta_l + \frac{3}{2}\eta\right)tr_{1,2}\left\{K^-_{\langle 1,2\rangle}\left(-\frac{3}{2}\eta\right)\right\} \times \text{id}, \qquad (7.5.13)
\end{aligned}$$

$$t_2(-\eta) = 3(-1)^N \bar{\xi}\eta^2 \prod_{l=1}^{N}(\theta_l + 2\eta)(\theta_l - 2\eta)tr\{K^-(-2\eta)\}\, t_1(-\eta), \quad (7.5.14)$$

and

$$t_3(0) = 12(-1)^N \xi\eta^4 \prod_{l=1}^{N}(\theta_l + \eta)(\theta_l - \eta)tr\{K^+(0)\}\, t_2(-\eta), \quad (7.5.15)$$

$$t_3(0) = 12(-1)^N \bar{\xi}\eta^4 \prod_{l=1}^{N}(\theta_l + 2\eta)(\theta_l - 2\eta)tr\{K^-(-2\eta)\}\, t_2(0), \quad (7.5.16)$$

$$\begin{aligned} t_3\left(\frac{\eta}{2}\right) &= 12tr_{1,2}\left\{K^+_{\langle 1,2\rangle}\left(\frac{\eta}{2}\right)\right\}\eta^5\left(\frac{\eta^2}{4} - \xi^2\right)t_1\left(-\frac{3}{2}\eta\right) \\ &\quad \times \prod_{l=1}^{N}\left(\theta_l - \frac{\eta}{2}\right)\left(\theta_l + \frac{\eta}{2}\right)\left(\theta_l - \frac{3}{2}\eta\right)\left(\theta_l + \frac{3}{2}\eta\right), \end{aligned} \quad (7.5.17)$$

$$\begin{aligned} t_3\left(-\frac{\eta}{2}\right) &= 12\eta^5\left(\frac{\eta^2}{4} - \bar{\xi}^2\right)tr_{2,3}\left\{K^-_{\langle 2,3\rangle}\left(-\frac{3}{2}\eta\right)\right\}t_1\left(-\frac{\eta}{2}\right) \\ &\quad \times \prod_{l=1}^{N}\left(\theta_l - \frac{5}{2}\eta\right)\left(\theta_l + \frac{5}{2}\eta\right)\left(\theta_l - \frac{3}{2}\eta\right)\left(\theta_l + \frac{3}{2}\eta\right), \end{aligned} \quad (7.5.18)$$

$$\begin{aligned} \frac{\partial}{\partial u}t_3(u)\,|_{u=\eta} &= 4\xi\eta^2(\xi^2 - \eta^2)(-1)^N tr_{1,2,3}\left\{K^+_{\langle 1,2,3\rangle}(\eta)\right\} \\ &\quad \times \prod_{l=1}^{N}\theta_l^2(\theta_l - \eta)(\theta_l + \eta)(\theta_l - 2\eta)(\theta_l + 2\eta) \times \mathrm{id}, \end{aligned} \quad (7.5.19)$$

$$\begin{aligned} \frac{\partial}{\partial u}t_3(u)\,|_{u=-\eta} &= 4\bar{\xi}\eta^2(\eta^2 - \bar{\xi}^2)(-1)^N tr_{1,2,3}\left\{K^-_{\langle 1,2,3\rangle}(-\eta)\right\}\prod_{l=1}^{N}(\theta_l - \eta) \\ &\quad \times (\theta_l + \eta)(\theta_l - 2\eta)(\theta_l + 2\eta)(\theta_l - 3\eta)(\theta_l + 3\eta) \times \mathrm{id}. \end{aligned} \quad (7.5.20)$$

In this case, the $\tilde{z}(u)$ functions read

$$\tilde{z}_{2l-1}(u) = z_{2l-1}(u) + x_{2l-1}(u), \quad (7.5.21)$$

$$\tilde{z}_{2l}(u) = z_{2l}(u), \quad l = 1, 2. \quad (7.5.22)$$

Here the function $z_m(u)$ is defined in (7.4.22) with $m = 1, \ldots, 4$ and $Q^{(4)}(u) \equiv 1$. The functions $x_{2l-1}(u)$ are

$$x_{2l-1}(u) = u(u + 2\eta)Q^{(0)}(u + \eta)Q^{(0)}(u)\frac{F_{2l-1}(u)}{Q^{(2l-1)}(u)}, \quad l = 1, 2. \quad (7.5.23)$$

The rules for constructing the inhomogeneous terms $x_{2l-1}(u)$ are:

- To make (7.5.9)–(7.5.20) fulfilled, all terms in those equations with a factor x_{2l-1} must be zero and therefore are irrelevant to the operator identities (7.3.43), which determines zeros of $f_{2l-1}(u)$ (see below).
- The denominators must be $Q^{(l)}(u)$ or part of them to ensure that no new pole is generated.
- They must preserve the self-consistency and symmetry of the BAEs.

The nested $T-Q$ Ansatz can be constructed as

$$\begin{aligned}
\Lambda(u) &= \sum_{i_1=1}^{4}\tilde{z}_{i_1}(u) = \sum_{i_1=1}^{4} z_{i_1}(u) + x_1(u) + x_3(u),\\
\Lambda_2(u) &= \varphi_2(2u-\eta)\left\{\sum_{1\le i_1<i_2\le 4}\tilde{z}_{i_1}(u)\tilde{z}_{i_2}(u-\eta)\right.\\
&\quad \left. -x_1(u)z_2(u-\eta) - x_3(u)z_4(u-\eta)\right\},\\
\Lambda_3(u) &= \prod_{k=1}^{3}\varphi_2(2u-k\eta)\left\{\sum_{1\le i_1<i_2<i_3\le 4}\tilde{z}_{i_1}(u)\tilde{z}_{i_2}(u-\eta)\tilde{z}_{i_3}(u-2\eta)\right.\\
&\quad -x_1(u)z_2(u-\eta)(\tilde{z}_3(u-2\eta)+\tilde{z}_4(u-2\eta))\\
&\quad \left. -(\tilde{z}_1(u)+\tilde{z}_2(u))x_3(u-\eta)z_4(u-2\eta)\right\}. \qquad (7.5.24)
\end{aligned}$$

With analysis similar to that used in the $SU(3)$ case, we have

$$F_1(u) = f_1(u)Q^{(2)}(-u-\eta), \qquad (7.5.25)$$
$$F_3(u) = f_3(u)Q^{(0)}(-u-2\eta)Q^{(2)}(-u-3\eta), \qquad (7.5.26)$$

with

$$f_1(u) = f_1(-u-\eta) = c_1\prod_{k=1}^{4}\left(u+\frac{k}{2}\eta\right)\left(u+\eta-\frac{k}{2}\eta\right), \qquad (7.5.27)$$

$$f_3(u) = f_3(-u-3\eta) = c_3\prod_{k=1}^{4}\left(u+\eta+\frac{k}{2}\eta\right)\left(u+2\eta-\frac{k}{2}\eta\right). \qquad (7.5.28)$$

The non-negative integers $\{L_1, L_2, L_3\}$ satisfy the relations

$$L_1 = 4 + L_2 + N, \quad L_3 = 4 + 2N + L_2, \qquad (7.5.29)$$

and to ensure automatic satisfaction of the asymptotic behavior (7.5.6)–(7.5.8), the parameters c_1 and c_3 are determined by the eigenvalues (7.5.5) of $\bar{M}M$ through the equations:

$$\begin{cases} -4 + c_1 + c_3 = -2\cos\vartheta_1 - 2\cos\vartheta_2, \\ 4 - 2c_1 - 2c_3 + c_1c_3 = 4\cos\vartheta_1\cos\vartheta_2. \end{cases} \tag{7.5.30}$$

We can easily show that the Ansatz (7.5.24) also fulfills the functional relations (7.3.43). The regular property of $\Lambda(u)$ leads to the associated BAEs

$$1 + \frac{\lambda_l^{(1)}}{\lambda_l^{(1)}+\eta}\frac{K^{(2)}(\lambda_l^{(1)})Q^{(0)}(\lambda_l^{(1)})}{K^{(1)}(\lambda_l^{(1)})Q^{(0)}(\lambda_l^{(1)}+\eta)}\frac{Q^{(1)}(\lambda_l^{(1)}+\eta)Q^{(2)}(\lambda_l^{(1)}-\eta)}{Q^{(1)}(\lambda_l^{(1)}-\eta)Q^{(2)}(\lambda_l^{(1)})}$$
$$+\frac{\lambda_l^{(1)}(\lambda_l^{(1)}+\frac{\eta}{2})Q^{(0)}(\lambda_l^{(1)})}{K^{(1)}(\lambda_l^{(1)})}\frac{F_1(\lambda_l^{(1)})}{Q^{(1)}(\lambda_l^{(1)}-\eta)} = 0, \quad l = 1, \ldots, L_1, \tag{7.5.31}$$

$$\frac{\lambda_l^{(2)}+\frac{\eta}{2}}{\lambda_l^{(2)}+\frac{3}{2}\eta}\frac{K^{(3)}(\lambda_l^{(2)})}{K^{(2)}(\lambda_l^{(2)})}\frac{Q^{(1)}(\lambda_l^{(2)})Q^{(2)}(\lambda_l^{(2)}+\eta)Q^{(3)}(\lambda_l^{(2)}-\eta)}{Q^{(1)}(\lambda_l^{(2)}+\eta)Q^{(2)}(\lambda_l^{(2)}-\eta)Q^{(3)}(\lambda_l^{(2)})}$$
$$= -1, \quad l = 1, \ldots, L_2, \tag{7.5.32}$$

$$1 + \frac{(\lambda_l^{(3)}+\eta)(\lambda_l^{(3)}+\frac{3}{2}\eta)Q^{(0)}(\lambda_l^{(3)}+\eta)}{K^{(3)}(\lambda_l^{(3)})}\frac{Q^{(2)}(\lambda_l^{(3)})F_3(\lambda_l^{(3)})}{Q^{(2)}(\lambda_l^{(3)}+\eta)Q^{(3)}(\lambda_l^{(3)}-\eta)}$$
$$+\frac{\lambda_l^{(3)}+\eta}{\lambda_l^{(3)}+2\eta}\frac{K^{(4)}(\lambda_l^{(3)})}{K^{(3)}(\lambda_l^{(3)})}\frac{Q^{(2)}(\lambda_l^{(3)})Q^{(3)}(\lambda_l^{(3)}+\eta)}{Q^{(2)}(\lambda_l^{(3)}+\eta)Q^{(3)}(\lambda_l^{(3)}-\eta)} = 0,$$
$$l = 1, \ldots, L_3. \tag{7.5.33}$$

It can be shown that the BAEs (7.5.31)–(7.5.33) also guarantee the regularity of $\Lambda_2(u)$ and $\Lambda_3(u)$ given by (7.5.24). Moreover, the Ansatz (7.5.24) indeed satisfies the relations (7.5.9)–(7.5.20).

The eigenvalue of the Hamiltonian for $n = 4$ is given by

$$E = \sum_{l=1}^{L_1}\frac{2\eta^2}{\lambda_l^{(1)}(\lambda_l^{(1)}+\eta)} + 2(N-1) + \eta\frac{[K^{(1)}(u)]'}{K^{(1)}(u)}|_{u\to 0} + \frac{1}{2}, \tag{7.5.34}$$

where the parameters $\{\lambda_l^{(1)}\}$ are the roots of the BAEs (7.5.31)–(7.5.33) in the homogeneous limit $\{\theta_j = 0\}$.

7.6 Solution of the $SU(n)$ Case

By following the same procedure introduced in previous sections, the result for the $SU(n)$-invariant spin chain with general open boundary conditions can be readily derived. The functions $z_m(u)$ are given by Eq. (7.4.22) with $\{K^{(l)}(u)|l = 1, \ldots, n\}$ satisfying (7.3.50)–(7.3.51). In principle, $K^{(l)}(u)$ could be any decomposition of Eq. (7.3.50). For simplicity, we parameterize them, satisfying the relation

$$K^{(l)}(u) = K^{(l+1)}(-u - l\eta), \quad l = 1, \ldots, n-1. \tag{7.6.1}$$

With the same procedure used in the last section, we can obtain the functions $x_m(u)$:

$$\begin{cases} x_{2l-1}(u) = u\left(u + \frac{n}{2}\eta\right) Q^{(0)}(u+\eta)Q^{(0)}(u)\frac{F_{2l-1}(u)}{Q^{(2l-1)}(u)}, \\ x_{2l}(u) = 0, \end{cases} \tag{7.6.2}$$

for $l = 1, 2, \ldots, \lfloor\frac{n}{2}\rfloor$. The functions $\{F_{2l-1}(u)\}$ are thus given by

$$F_1(u) = f_1(u)Q^{(2)}(-u-\eta), \tag{7.6.3}$$

$$F_{2l-1}(u) = f_{2l-1}(u)Q^{(2l-2)}(-u-(2l-1)\eta)$$
$$\times Q^{(2l)}(-u-(2l-1)\eta)Q^{(0)}(-u-2(l-1)\eta), \quad l = 2, \ldots, \lfloor\frac{n}{2}\rfloor, \tag{7.6.4}$$

and

$$f_{2l-1}(u) = c_{2l-1}\prod_{k=1}^{n}\left(u + (l-1+\frac{k}{2})\eta\right)\left(u + (l-\frac{k}{2})\eta\right),$$
$$l = 1, \ldots, \lfloor\frac{n}{2}\rfloor, \tag{7.6.5}$$

which possess the crossing symmetry

$$f_{2l-1}(u) = f_{2l-1}(-u-(2l-1)\eta). \tag{7.6.6}$$

The nested $T-Q$ relation is thus

$$\Lambda(u) = \sum_{i=1}^{n}\tilde{z}_i(u). \tag{7.6.7}$$

All the eigenvalues $\Lambda_m(u)$ can be given by

$$\Lambda_m(u) = \prod_{l=1}^{k}\prod_{k=1}^{m-1}\varphi_2(2u-(k+l-1)\eta){\sum}'_{1\leq i_1<\cdots<i_m\leq n}\big\{\tilde{z}_{i_1}(u)\tilde{z}_{i_2}(u-\eta)\cdots\tilde{z}_{i_m}(u-(m-1)\eta)\big\},\quad m=2,\ldots,n, \tag{7.6.8}$$

where the prime indicates that the terms with factors $x_{2l-1}z_{2l}$ are not included in the summation. The parameters $\{c_{2l-1}\}$ are determined by the asymptotic behavior of the transfer matrices:

$$\bar{\Lambda}_m(u) = (-1)^m \sum_{1\leq i_1<\cdots<i_m\leq n}\lambda_{i_1}\lambda_{i_2}\cdots\lambda_{i_m}u^{2(N+m)}+\cdots. \tag{7.6.9}$$

We remark that the asymptotic behavior of $\Lambda_l(u)$ and $\Lambda_{n-l}(u)$ give the same equation to determine c_{2l-1}. The parameters $\{\lambda_l^{(r)}\}$ satisfy the associated BAEs:

$$\begin{aligned}&\frac{\lambda_j^{(1)}}{\lambda_j^{(1)}+\eta}K^{(2)}(\lambda_j^{(1)})Q^{(0)}(\lambda_j^{(1)})Q^{(1)}(\lambda_j^{(1)}+\eta)\frac{Q^{(2)}(\lambda_j^{(1)}-\eta)}{Q^{(2)}(\lambda_j^{(1)})}\\&\quad+\lambda_j^{(1)}(\lambda_j^{(1)}+\frac{\eta}{2})Q^{(0)}(\lambda_j^{(1)}+\eta)Q^{(0)}(\lambda_j^{(1)})F_1(\lambda_j^{(1)})\\&\quad+K^{(1)}(\lambda_j^{(1)})Q^{(0)}(\lambda_j^{(1)}+\eta)Q^{(1)}(\lambda_j^{(1)}-\eta)=0,\quad j=1,\ldots,L_1,\end{aligned} \tag{7.6.10}$$

$$\begin{aligned}&\frac{2\lambda_k^{(2l)}+(2l+1)\eta}{2\lambda_k^{(2l)}+(2l-1)\eta}\frac{Q^{(2l-1)}(\lambda_k^{(2l)}+\eta)Q^{(2l+1)}(\lambda_k^{(2l)})}{Q^{(2l-1)}(\lambda_k^{(2l)})Q^{(2l+1)}(\lambda_k^{(2l)}-\eta)}\\&\quad=-\frac{K^{(2l+1)}(\lambda_k^{(2l)})}{K^{(2l)}(\lambda_k^{(2l)})}\frac{Q^{(2l)}(\lambda_k^{(2l)}+\eta)}{Q^{(2l)}(\lambda_k^{(2l)}-\eta)},\\&\quad k=1,\ldots,L_{2l},\quad l=1,\ldots,\lfloor\frac{n}{2}\rfloor,\end{aligned} \tag{7.6.11}$$

$$\begin{aligned}&K^{(2s+1)}(\lambda_j^{(2s+1)})Q^{(2s+1)}(\lambda_j^{(2s+1)}-\eta)+\frac{\lambda_j^{(2s+1)}+s\eta}{\lambda_j^{(2s+1)}+(s+1)\eta}K^{(2s+2)}(\lambda_j^{(2s+1)})\\&\times Q^{(2s+1)}(\lambda_j^{(2s+1)}+\eta)\frac{Q^{(2s)}(\lambda_j^{(2s+1)})Q^{(2s+2)}(\lambda_j^{(2s+1)}-\eta)}{Q^{(2s)}(\lambda_j^{(2s+1)}+\eta)Q^{(2s+2)}(\lambda_j^{(2s+1)})}+(\lambda_j^{(2s+1)}+s\eta)\\&\times(\lambda_j^{(2s+1)}+\frac{2s+1}{2}\eta)Q^{(0)}(\lambda_j^{(2s+1)}+\eta)\frac{Q^{(2s)}(\lambda_j^{(2s+1)})F_{2s+1}(\lambda_j^{(2s+1)})}{Q^{(2s)}(\lambda_j^{(2s+1)}+\eta)}=0,\\&j=1,\ldots,L_{2s+1},\quad s=1,\ldots,\lfloor\frac{n}{2}\rfloor-1.\end{aligned} \tag{7.6.12}$$

The eigenvalue of the Hamiltonian (7.3.6) is thus given by

$$E = \sum_{l=1}^{L_1} \frac{2\eta^2}{\lambda_l^{(1)}(\lambda_l^{(1)} + \eta)} + 2(N-1) + \eta \frac{[K^{(1)}(u)]'}{K^{(1)}(u)}|_{u \to 0} + \frac{2}{n}, \qquad (7.6.13)$$

with $\{\lambda_l^{(1)}\}$ being the Bethe roots at $\{\theta_j = 0\}$.

It should be remarked that alternative $T - Q$ relations also exist for the open $SU(n)$ spin chain as in the $SU(2)$ case.

References

1. A. Doikou, Fusion and analytical Bethe Ansatz for the $A_{n-1}^{(1)}$ open spin chain. J. Phys. A **33**, 4755 (2000)
2. A. Doikou, R.I. Nepomechie, Duality and quantum-algebra symmetry of the $A_{N-1}^{(1)}$ open spin chain with diagonal boundary fields. Nucl. Phys. B **530**, 641 (1998)
3. A. Doikou, R.I. Nepomechie, Bulk and boundary S matrices for the $SU(N)$ chain. Nucl. Phys. B **521**, 547 (1998)
4. H.J. de Vega, A. González-Ruiz, Exact solution of the $SU_q(n)$ invariant quantum spin chains. Nucl. Phys. B **417**, 553 (1994)
5. H.J. de Vega, A. Gonzǎlez-Ruiz, Exact Bethe Ansatz solution for A_{n-1} chains with non-$SU_q(n)$ invariant open boundary conditions. Mod. Phys. Lett. **A 09**, 2207 (1994)
6. S. Belliard, N. Crampé, Heisenberg XXX model with general boundaries: eigenvectors from algebraic Bethe Ansatz. SIGMA **9**, 072 (2013)
7. J. Cao, W.-L. Yang, K. Shi, Y. Wang, Nested off-diagonal Bethe Ansatz and exact solutions of the $su(n)$ spin chain with generic integrable boundaries. J. High Energy Phys. **04**, 143 (2014)
8. M. Karowski, On the bound state problem in 1 + 1 dimensional field theories. Nucl. Phys. B **153**, 244 (1979)
9. P.P. Kulish, N.Y. Reshetikhin, E.K. Sklyanin, Yang-Baxter equation and representation theory: I. Lett. Math. Phys. **5**, 393 (1981)
10. P.P. Kulish, E.K. Sklyanin, Quantum spectral transform method: recent developments. Lect. Notes Phys. **151**, 61 (1982)
11. A.N. Kirillov, N.Y. Reshetikhin, Exact solution of the Heisenberg XXZ model of spin s. J. Sov. Math. **35**, 2627 (1986)
12. A.N. Kirillov, N.Y. Reshetikhin, Exact solution of the integrable XXZ Heisenberg model with arbitrary spin. I. The ground state and the excitation spectrum. J. Phys. A **20**, 1565 (1987)
13. B. Sutherland, Model for a multicomponent quantum system. Phys. Rev. B **12**, 3795 (1975)
14. H.J. De Vega, E. Lopes, Exact solution of the Perk-Schultz model. Phys. Rev. Lett. **67**, 489 (1991)
15. E. Lopes, Exact solution of the multi-component generalized six-vertex model. Nucl. Phys. B **370**, 636 (1992)
16. I. Krichever, O. Lipan, P. Wiegmann, A. Zabrodin, Quantum integrable systems and elliptic solutions of classical discrete nonlinear equations. Commun. Math. Phys. **188**, 267 (1997)
17. H. Frahm, N.A. Slavnov, New solutions to the reflection equation and the projecting method. J. Phys. A **32**, 1547 (1999)
18. W. Galleas, M.J. Martins, Solution of the $SU(N)$ vertex model with non-diagonal open boundaries. Phys. Lett. A **335**, 167 (2005)
19. H.J. de Vega, A. González-Ruiz, Boundary K-matrices for the XYZ, XXZ and XXX spin chains. J. Phys. A **27**, 6129 (1994)

20. P.P. Kulish, Yang-Baxter equation and reflection equations in integrable models, arXiv:hep-th/9507070
21. L. Mezincescu, R.I. Nepomechie, Fusion procedure for open chains. J. Phys. A **25**, 2533 (1992)
22. Y.-K. Zhou, Row transfer matrix functional relations for Baxter's eight-vertex and six-vertex models with open boundaries via more general reflection matrices. Nucl. Phys. B **458**, 504 (1996)
23. P.P. Kulish, E.K. Sklyanin, Quantum inverse scattering method and the Heisenberg ferromagnet. Phys. Lett. A **70**, 461 (1979)
24. E.K. Sklyanin, Boundary conditions for integrable quantum systems. J. Phys. A **21**, 2375 (1988)

Chapter 8
The Hierarchical Off-Diagonal Bethe Ansatz

The integrable models confined in higher dimensional quantum space are particularly interesting because of their important applications in quantum field theory. A typical model is the $SU(2)$-invariant spin-s Heisenberg chain which is closely related to the Wess-Zumino-Novikov-Witten (WZNW) models [1–4]. Its anisotropic and graded versions are relevant to the lower dimensional quantum field theories [5, 6] such as the super-symmetric sine-Gordon model [7–9]. In addition, this model has important applications in condensed matter physics such as fractional statistics [10] and the multi-channel Kondo problem [11–13] when it is coupled to an impurity spin. The anisotropic $s = 1$ integrable spin chain was first proposed by Zamolodchikov and Fateev [14] to solve YBE. It was subsequently found that the integrable spin-s chain can be constructed via the fusion techniques [15–19] from the fundamental $s = \frac{1}{2}$ representation of YBE [20, 21]. This method establishes a hierarchical structure of the $SU(2)$-invariant integrable spin chains and allows us to derive the exact spectra of the models in the framework of the algebraic Bethe Ansatz (e.g., see [22–24]). Moreover, the boundary YBE or RE [25, 26] implies that, with generic integrable boundary fields [27, 28], this model is also integrable. Profound results have been obtained for the constraint boundary conditions [29–33] in the past decade. Nevertheless, it was only recently that the model with generic integrable boundary conditions was solved via the ODBA method [34–36].

This chapter is an introduction to the application of ODBA in the high-spin Heisenberg chain with generic integrable boundaries. With the fusion techniques, the R-matrices and K-matrices for the high spin model can be constructed. This allows us to establish a hierarchical structure among the fused transfer matrices. Based on this hierarchy and the operator product identities, the inhomogeneous $T - Q$ relation as well as the corresponding BAEs can be derived in a systematic way. In addition, the exact solution of the alternating-spin chain model, which corresponds to the lattice nonlinear Schrödinger model, will also be introduced.

Y. Wang et al., *Off-Diagonal Bethe Ansatz for Exactly Solvable Models*,
DOI 10.1007/978-3-662-46756-5_8

8.1 The Fusion Procedure

8.1.1 Fusion of the R-Matrices and the K-Matrices

Throughout this chapter, $R_{i,j}^{(l_i,l_j)}(u)$ denotes the spin-(l_i, l_j) R-matrix acting on the tensor space $\mathbf{V}^{l_i} \otimes \mathbf{V}^{l_j}$ and satisfies YBE

$$R_{1,2}^{(l_1,l_2)}(u-v)R_{1,3}^{(l_1,l_3)}(u)R_{2,3}^{(l_2,l_3)}(v) = R_{2,3}^{(l_2,l_3)}(v)R_{1,3}^{(l_1,l_3)}(u)R_{1,2}^{(l_1,l_2)}(u-v). \quad (8.1.1)$$

Explicitly, the fundamental $R_{1,2}^{(\frac{1}{2},s)}(u)$ defined in the tensor product of the two-dimensional auxiliary space and the $(2s+1)$-dimensional quantum space is given by [15–19]

$$R_{1,2}^{(\frac{1}{2},s)}(u) = u + \frac{\eta}{2} + \eta\,\boldsymbol{\sigma}_1 \cdot \mathbf{S}_2, \quad (8.1.2)$$

where $\mathbf{S}$ is the spin-s realization of the $SU(2)$ generators. Namely, all the generators of $SU(2)$ can be realized by $(2s+1) \times (2s+1)$ matrices on $(2s+1)$-dimensional spin-s space as follows

$$S^z|m\rangle = m\,|m\rangle, \quad (8.1.3)$$

$$S^+|m\rangle = \sqrt{(s+1+m)(s-m)}\,|m+1\rangle, \quad (8.1.4)$$

$$S^-|m\rangle = \sqrt{(s+m)(s+1-m)}\,|m-1\rangle, \quad m = s, s-1, \ldots, -s, \quad (8.1.5)$$

where $\{|m\rangle | m = s, s-1, \ldots, -s\}$ is an orthonormal basis of the spin-s space and the Casimir operator $C_2 = (S^z)^2 + \frac{1}{2}(S^-S^+ + S^+S^-)$ acting on the spin-s space is proportional to the identity operator, i.e.,

$$C_2\,|m\rangle = s(s+1)\,|m\rangle, \quad m = s, s-1, \ldots, -s. \quad (8.1.6)$$

For $s = \frac{1}{2}$, the corresponding R-matrix is given by (1.5.2). In particular, with the help of (8.1.6) we can verify that the spin-$(\frac{1}{2}, s)$ R-matrix possesses the unitary relation [cf. (1.5.5)]

$$R_{1,2}^{(\frac{1}{2},s)}(u)R_{2,1}^{(s,\frac{1}{2})}(-u) = -\left[u + \left(\frac{1}{2}+s\right)\eta\right]\left[u - \left(\frac{1}{2}+s\right)\eta\right] \times \text{id}. \quad (8.1.7)$$

We note that the R-matrix (8.1.2) can be obtained by the fusion procedure

$$R_{a,\{1,\ldots,2s\}}^{(\frac{1}{2},s)}(u) = \frac{1}{\prod_{k=1}^{2s-1}(u+(\frac{1}{2}-s+k)\eta)} P_{1,\ldots,2s}^{(+)}$$
$$\times \prod_{k=1}^{2s}\left\{R_{a,k}^{(\frac{1}{2},\frac{1}{2})}(u+(k-\frac{1}{2}-s)\eta)\right\} P_{1,\ldots,2s}^{(+)}, \qquad (8.1.8)$$

with the product being in the order of increasing k from left to right, where $P_{1,\ldots,2s}^{(+)}$ is the symmetric projector given by

$$P_{1,\ldots,2s}^{(+)} = \frac{1}{(2s)!}\prod_{k=1}^{2s}\left(\sum_{l=1}^{k} P_{l,k}\right), \qquad (8.1.9)$$

which possesses the properties

$$P_{1,\ldots,n}^{(+)} = \frac{1}{n}\left(1+\sum_{l=1}^{n-1} P_{l,n}\right) P_{1,\ldots,n-1}^{(+)} = \frac{1}{n} P_{1,\ldots,n-1}^{(+)}\left(1+\sum_{l=1}^{n-1} P_{l,n}\right), \qquad (8.1.10)$$

$$P_{1,\ldots,n}^{(+)} = P_{q_1,\ldots,q_n}^{(+)}, \qquad (8.1.11)$$

$$P_{1,\ldots,n}^{(+)} = P_{1,\ldots,n}^{(+)} P_{q_1,\ldots,q_m}^{(+)} = P_{q_1,\ldots,q_m}^{(+)} P_{1,\ldots,n}^{(+)}, \quad m \le n, \qquad (8.1.12)$$

with $\{q_1,\ldots,q_n\}$ being an arbitrary permutation of $\{1,\ldots,n\}$. Similarly, the spin-(j,s) R-matrix can be constructed from the R-matrix (8.1.2) by the symmetric fusion process

$$R_{\{1,\ldots,2j\},\{1,\ldots,2s\}}^{(j,s)}(u) = P_{1,\ldots,2j}^{(+)} \prod_{k=1}^{2j}\left\{R_{k,\{1,\ldots,2s\}}^{(\frac{1}{2},s)}(u+(k-j-\frac{1}{2})\eta)\right\}$$
$$\times P_{1,\ldots,2j}^{(+)}. \qquad (8.1.13)$$

With the same procedure introduced in Chap. 7, we can easily show that the fused R-matrices (8.1.13) and (8.1.8) satisfy the associated YBE (8.1.1). Particularly, the spin-(s,s) R-matrix can be calculated as [15]

$$R_{1,2}^{(s,s)}(u) = \prod_{j=1}^{2s}(u-j\eta)\sum_{l=0}^{2s}\prod_{k=1}^{l}\frac{u+k\eta}{u-k\eta}\mathbf{P}_{1,2}^{(l)}, \qquad (8.1.14)$$

where $\mathbf{P}_{1,2}^{(l)}$ is a projector defined in the tensor space of two spin-s, which projects the tensor space into the irreducible subspace of spin-l. Explicitly, it reads

$$\mathbf{P}_{1,2}^{(l)} = \prod_{j=0,\neq l}^{2s}\frac{(\mathbf{S}_1+\mathbf{S}_2)^2 - j(j+1)}{l(l+1)-j(j+1)}. \qquad (8.1.15)$$

The R-matrix (8.1.14) possesses the following important properties which are quite useful to derive the closed operator identities:

$$\text{Initial condition}: \; R^{(s,s)}_{1,2}(0) = (2s)!\eta^{2s} P_{1,2}, \tag{8.1.16}$$

$$\text{Antisymmetry}: \; R^{(s,s)}_{1,2}(-\eta) = (-1)^{2s}\eta^{2s}(2s+1)!\, \mathbf{P}^{(0)}_{1,2}. \tag{8.1.17}$$

The projector $\mathbf{P}^{(0)}_{1,2}$ resembles the properties of $P^{(-)}_{1,2}$ for the spin-$\frac{1}{2}$ system and projects the tensor product space of two spin-s into the singlet subspace, i.e.,

$$\mathbf{P}^{(0)} = |\Phi_0\rangle\langle\Phi_0|, \quad |\Phi_0\rangle = \frac{1}{\sqrt{2s+1}}\sum_{l=0}^{2s}(-1)^l|s-l\rangle \otimes |-s+l\rangle. \tag{8.1.18}$$

Moreover, from the explicit expression (8.1.2) of the R-matrix, we can show that [cf.(1.5.9)]

$$R^{(s,\frac{1}{2})}(\pm(\frac{1}{2}+s)\eta) = \pm(2s+1)\eta\, \mathscr{P}^{(s\pm\frac{1}{2})}, \tag{8.1.19}$$

where the projectors $\mathscr{P}^{(s\pm\frac{1}{2})}$ project the tensor product space of the spin-s and the spin-$\frac{1}{2}$ spaces into the subspaces with the spin-$(s\pm\frac{1}{2})$ respectively.

Following the methods developed in [37, 38], the corresponding fused K-matrices can be constructed as

$$\begin{aligned} K^{-(j)}_{\{a\}}(u) &= P^{(+)}_{a_1,\ldots,a_{2j}} \prod_{k=1}^{2j}\left\{\left[\prod_{l=1}^{k-1} R^{(\frac{1}{2},\frac{1}{2})}_{a_l,a_k}(2u+(k+l-2j-1)\eta)\right]\right. \\ &\quad \times \left. K^{-(\frac{1}{2})}_{a_k}(u+(k-j-\frac{1}{2})\eta)\right\} P^{(+)}_{a_1,\ldots,a_{2j}}, \end{aligned} \tag{8.1.20}$$

where $\{a\} \equiv \{a_1,\ldots,a_{2j}\}$. It should be noted that the products in $[\cdots]$ and $\{\cdots\}$ of (8.1.20) are in the order of increasing l and k from left to right. Let us consider the generic non-diagonal spin-$\frac{1}{2}$ K-matrix [27, 28]

$$K^{-(\frac{1}{2})}(u) = \begin{pmatrix} p_- + u & \alpha_- u \\ \alpha_- u & p_- - u \end{pmatrix}, \tag{8.1.21}$$

where p_- and α_- are some boundary parameters. It can be easily demonstrated that $K^{-(j)}_{\{a\}}(u)$ satisfy the RE

$$\begin{aligned} &R^{(j,s)}_{\{a\},\{b\}}(u-v)\, K^{-(j)}_{\{a\}}(u)\, R^{(s,j)}_{\{b\},\{a\}}(u+v)\, K^{-(s)}_{\{b\}}(v) \\ &= K^{-(s)}_{\{b\}}(v)\, R^{(j,s)}_{\{a\},\{b\}}(u+v)\, K^{-(j)}_{\{a\}}(u)\, R^{(s,j)}_{\{b\},\{a\}}(u-v). \end{aligned} \tag{8.1.22}$$

The dual version of the fused K-matrix is given by the duality

$$K^{+(j)}_{\{a\}}(u) = \frac{1}{f^{(j)}(u)} K^{-(j)}_{\{a\}}(-u-\eta)\Big|_{(p_-,\alpha_-)\to(p_+,-\alpha_+)}, \tag{8.1.23}$$

where p_+ and α_+ are two free boundary parameters and the normalization factor $f^{(j)}(u)$ is

$$f^{(j)}(u) = \prod_{l=1}^{2j-1}\prod_{k=1}^{l}[-\varphi(2u+(l+k+1-2j)\eta)], \tag{8.1.24}$$

$$\varphi(u) = (u-\eta)(u+\eta). \tag{8.1.25}$$

The generic fundamental dual K-matrix satisfying the dual RE (5.1.6) is

$$K^{+(\frac{1}{2})}(u) = \begin{pmatrix} p_+ - u - \eta & \alpha_+(u+\eta) \\ \alpha_+(u+\eta) & p_+ + u + \eta \end{pmatrix}. \tag{8.1.26}$$

8.1.2 Fused Transfer Matrices

The fused (or spin-(j,s)) transfer matrix $t^{(j,s)}(u)$ can be constructed by the fused R-matrices and K-matrices as follows [26, 30]

$$t^{(j,s)}(u) = tr_{\{a\}}\left\{K^{+(j)}_{\{a\}}(u)\, T^{(j,s)}_{\{a\}}(u)\, K^{-(j)}_{\{a\}}(u)\, \hat{T}^{(j,s)}_{\{a\}}(u)\right\}, \tag{8.1.27}$$

where $T^{(j,s)}_{\{a\}}(u)$ and $\hat{T}^{(j,s)}_{\{a\}}(u)$ are the fused one-row monodromy matrices defined by

$$T^{(j,s)}_{\{a\}}(u) = R^{(j,s)}_{\{a\},\{b^{[N]}\}}(u-\theta_N)\cdots R^{(j,s)}_{\{a\},\{b^{[1]}\}}(u-\theta_1),$$

$$\hat{T}^{(j,s)}_{\{a\}}(u) = R^{(j,s)}_{\{a\},\{b^{[1]}\}}(u+\theta_1)\cdots R^{(j,s)}_{\{a\},\{b^{[N]}\}}(u+\theta_N). \tag{8.1.28}$$

Here $\{\theta_j|j=1,\ldots,N\}$ are still the inhomogeneous parameters we defined previously. By following the same procedure introduced in Chap. 2, YBE (8.1.1), the RE (8.1.22) and the dual RE lead to

$$\left[t^{(j,s)}(u), t^{(j',s)}(v)\right] = 0. \tag{8.1.29}$$

Therefore, $t^{(j,s)}(u)$ can be treated as the generating functionals of the conserved quantities.

8.2 Operator Identities

Let us fix an $s \in \{\frac{1}{2}, 1, \frac{3}{2}, \ldots\}$ for all the N quantum spaces. Consider the one-row monodromy matrix $T^{(j,s)}_{\{a\}}(u)$ and its corresponding transfer matrix $t^{(j,s)}_p(u)$ given by

$$t^{(j,s)}_p(u) = tr_{\{a\}} T^{(j,s)}_{\{a\}}(u), \quad j = 0, \frac{1}{2}, 1, \ldots, \tag{8.2.1}$$

with the convention $t^{(0,s)}_p(u) = \text{id}$. The fusion procedure (8.1.13) of the R-matrices allows us to express the fused monodromy matrices (8.1.28) in terms of the fundamental $T^{(\frac{1}{2},s)}_a(u)$ as

$$\begin{aligned} T^{(j,s)}_{\{a\}}(u) &= P^{(+)}_{1,\ldots,2j}\, T^{(\frac{1}{2},s)}_1(u-(j-\frac{1}{2})\eta) T^{(\frac{1}{2},s)}_2(u-(j-\frac{1}{2})\eta+\eta) \\ &\quad \times \cdots T^{(\frac{1}{2},s)}_{2j}(u+(j-\frac{1}{2})\eta)\, P^{(+)}_{1,\ldots,2j} \\ &= T^{(\frac{1}{2},s)}_1(u-(j-\frac{1}{2})\eta) T^{(\frac{1}{2},s)}_2(u-(j-\frac{1}{2})\eta+\eta) \\ &\quad \times \cdots T^{(\frac{1}{2},s)}_{2j}(u+(j-\frac{1}{2})\eta)\, P^{(+)}_{1,\ldots,2j}. \end{aligned} \tag{8.2.2}$$

The second equality of the above equation can be proven by the method proving (7.1.16). Similarly, we have

$$\begin{aligned} P^{(-)}_{1,2} T^{(\frac{1}{2},s)}_1(u-\eta) T^{(\frac{1}{2},s)}_2(u) &= P^{(-)}_{1,2} T^{(\frac{1}{2},s)}_1(u-\eta) T^{(\frac{1}{2},s)}_2(u) P^{(-)}_{1,2} \\ &= P^{(-)}_{1,2} T^{(\frac{1}{2},s)}_1(u) T^{(\frac{1}{2},s)}_2(u-\eta) P^{(-)}_{1,2} \\ &= \text{Det}_q\{T^{(\frac{1}{2},s)}(u)\} P^{(-)}_{1,2}, \end{aligned} \tag{8.2.3}$$

where the quantum determinant $\text{Det}_q\{T^{(\frac{1}{2},s)}(u)\}$ is

$$\text{Det}_q\{T^{(\frac{1}{2},s)}(u)\} = \prod_{l=1}^{N} \left(u-\theta_l+\left(\frac{1}{2}+s\right)\eta\right)\left(u-\theta_l-\left(\frac{1}{2}+s\right)\eta\right) \times \text{id}.$$

Let us calculate the product of the fused transfer matrices (8.2.1):

$$\begin{aligned} & t^{(\frac{1}{2},s)}_p(u)\, t^{(j-\frac{1}{2},s)}_p(u-j\eta) \\ &= tr_{1,\ldots,2j}\left\{P^{(+)}_{1,\ldots,2j-1} T^{(\frac{1}{2},s)}_1(u-(2j-1)\eta)\ldots T^{(\frac{1}{2},s)}_{2j}(u) P^{(+)}_{1,\ldots,2j-1}\right\} \\ &= tr_{1,\ldots,2j}\left\{\left[1+(2j-1)P_{2j-1,2j}\right](2j)^{-1} P^{(+)}_{1,\ldots,2j-1} T^{(\frac{1}{2},s)}_1(u-(2j-1)\eta)\right. \end{aligned}$$

$$\times \cdots T_{2j}^{(\frac{1}{2},s)}(u) P^{(+)}_{1,\ldots,2j-1}\Big\} + (2j-1)(2j)^{-1} tr_{1,\ldots,2j}\big\{(1-P_{2j-1,2j})$$

$$\times P^{(+)}_{1,\ldots,2j-1} T_1^{(\frac{1}{2},s)}(u-(2j-1)\eta)\cdots T_{2j}^{(\frac{1}{2},s)}(u) P^{(+)}_{1,\ldots,2j-1}\Big\}$$

$$\overset{(8.2.2)}{=} tr_{1,\ldots,2j}\Big\{\big[1+P_{2j-1,2j}+\cdots+P_{1,2j}\big](2j)^{-1} P^{(+)}_{1,\ldots,2j-1}$$

$$\times T_1^{(\frac{1}{2},s)}(u-(2j-1)\eta)\cdots T_{2j}^{(\frac{1}{2},s)}(u) P^{(+)}_{1,\ldots,2j-1}\Big\} + (2j-1)(2j)^{-1}$$

$$\times tr_{1,\ldots,2j}\Big\{(1-P_{2j-1,2j}) T_1^{(\frac{1}{2},s)}(u-(2j-1)\eta)\cdots T_{2j}^{(\frac{1}{2},s)}(u) P^{(+)}_{1,\ldots,2j-1}\Big\}$$

$$\overset{(8.2.3)}{=} tr_{1,\ldots,2j}\Big\{P^{(+)}_{1,\ldots,2j} T_1^{(\frac{1}{2},s)}(u-(2j-1)\eta)\cdots T_{2j}^{(\frac{1}{2},s)}(u) P^{(+)}_{1,\ldots,2j}\Big\}$$

$$+(2j-1)(2j)^{-1}\mathrm{Det}_q\{T^{(\frac{1}{2},s)}(u)\} tr_{1,\ldots,2j-2}\Big\{T_1^{(\frac{1}{2},s)}(u-(2j-1)\eta)$$

$$\times \cdots T_{2j-2}^{(\frac{1}{2},s)}(u-2\eta) tr_{2j-1,2j}\Big((1-P_{2j-1,2j}) P^{(+)}_{1,\ldots,2j-1}\Big)\Big\}$$

$$= t_p^{(j,s)}(u-(j-\frac{1}{2})\eta) + \mathrm{Det}_q\{T^{(\frac{1}{2},s)}(u)\}\, t_p^{(j-1,s)}(u-(j+\frac{1}{2})\eta). \qquad (8.2.4)$$

To reach the last equality of the above equation, we have used the identity

$$\frac{2j-1}{2j} tr_{2j-1,2j}\Big\{(1-P_{2j-1,2j}) P^{(+)}_{1,\ldots,2j-1}\Big\}$$

$$= tr_{2j-1,2j}\Big\{\frac{(1-P_{2j-1,2j})}{2j}(1+P_{1,2j-1}+\cdots+P_{2j-2,2j-1}) P^{(+)}_{1,\ldots,2j-2}\Big\}$$

$$= P^{(+)}_{1,\ldots,2j-2},$$

and the relation $tr_i\{P_{i,j}\} = \mathrm{id}$.

With a procedure similar to that used in deriving (8.2.4), we may show that the fused double-row transfer matrices $\{t^{(j,s)}(u)\}$ given by (8.1.27) obey the fusion hierarchy relation [29, 30, 37, 38]

$$t^{(\frac{1}{2},s)}(u)\, t^{(j-\frac{1}{2},s)}(u-j\eta) = t^{(j,s)}(u-(j-\frac{1}{2})\eta)$$

$$+\,\delta^{(s)}(u)\, t^{(j-1,s)}(u-(j+\frac{1}{2})\eta), \quad j=\frac{1}{2},1,\frac{3}{2},\ldots, \qquad (8.2.5)$$

with the convention $t^{(0,s)} = \mathrm{id}$ and

$$\begin{aligned}
\delta^{(s)}(u) &= \text{Det}_q\{K^{+(\frac{1}{2})}(u)\}\text{Det}_q\{T^{(\frac{1}{2},s)}(u)\}\text{Det}_q\{K^{-(\frac{1}{2})}(u)\}\text{Det}_q\{\hat{T}^{(\frac{1}{2},s)}(u)\} \\
&= \frac{(2u-2\eta)(2u+2\eta)}{(2u-\eta)(2u+\eta)}((1+\alpha_-^2)u^2-p_-^2)((1+\alpha_+^2)u^2-p_+^2) \\
&\quad \times \prod_{l=1}^{N}(u-\theta_l+(\frac{1}{2}+s)\eta)(u+\theta_l+(\frac{1}{2}+s)\eta) \\
&\quad \times \prod_{l=1}^{N}(u-\theta_l-(\frac{1}{2}+s)\eta)(u+\theta_l-(\frac{1}{2}+s)\eta).
\end{aligned} \tag{8.2.6}$$

From (8.2.5) we can express $t^{(j,s)}(u)$ in terms of $t^{(\frac{1}{2},s)}(u)$ as

$$\begin{aligned}
t^{(j,s)}(u) &= t^{(\frac{1}{2},s)}(u+(j-\frac{1}{2})\eta)\,t^{(\frac{1}{2},s)}(u+(j-\frac{1}{2})\eta-\eta)\cdots t^{(\frac{1}{2},s)}(u-(j-\frac{1}{2})\eta) \\
&-\delta^{(s)}(u+(j-\frac{1}{2})\eta)\,t^{(\frac{1}{2},s)}(u+(j-\frac{1}{2})\eta-2\eta)\cdots t^{(\frac{1}{2},s)}(u-(j-\frac{1}{2})\eta) \\
&-\delta^{(s)}(u+(j-\frac{1}{2})\eta-\eta)\,t^{(\frac{1}{2},s)}(u+(j-\frac{1}{2})\eta)\,t^{(\frac{1}{2},s)}(u+(j-\frac{1}{2})\eta-3\eta) \\
&\times\cdots t^{(\frac{1}{2},s)}(u-(j-\frac{1}{2})\eta) \\
&\vdots \\
&-\delta^{(s)}(u-(j-\frac{1}{2})\eta+\eta)\,t^{(\frac{1}{2},s)}(u+(j-\frac{1}{2})\eta)\cdots t^{(\frac{1}{2},s)}(u-(j-\frac{1}{2})\eta+2\eta) \\
&+\cdots.
\end{aligned} \tag{8.2.7}$$

Here we list the first three:

$$t^{(1,s)}(u) = t^{(\frac{1}{2},s)}(u+\frac{\eta}{2})\,t^{(\frac{1}{2},s)}(u-\frac{\eta}{2}) - \delta^{(s)}(u+\frac{\eta}{2}), \tag{8.2.8}$$

$$\begin{aligned}
t^{(\frac{3}{2},s)}(u) &= t^{(\frac{1}{2},s)}(u+\eta)\,t^{(\frac{1}{2},s)}(u)\,t^{(\frac{1}{2},s)}(u-\eta) - \delta^{(s)}(u+\eta)\,t^{(\frac{1}{2},s)}(u-\eta) \\
&\quad -\delta^{(s)}(u)\,t^{(\frac{1}{2},s)}(u+\eta),
\end{aligned} \tag{8.2.9}$$

$$\begin{aligned}
t^{(2,s)}(u) &= t^{(\frac{1}{2},s)}(u+\frac{3\eta}{2})\,t^{(\frac{1}{2},s)}(u+\frac{\eta}{2})\,t^{(\frac{1}{2},s)}(u-\frac{\eta}{2})\,t^{(\frac{1}{2},s)}(u-\frac{3\eta}{2}) \\
&\quad -\delta^{(s)}(u+\frac{3\eta}{2})\,t^{(\frac{1}{2},s)}(u-\frac{\eta}{2})\,t^{(\frac{1}{2},s)}(u-\frac{3\eta}{2}) \\
&\quad -\delta^{(s)}(u+\frac{\eta}{2})\,t^{(\frac{1}{2},s)}(u+\frac{3\eta}{2})\,t^{(\frac{1}{2},s)}(u-\frac{3\eta}{2}) \\
&\quad -\delta^{(s)}(u-\frac{\eta}{2})\,t^{(\frac{1}{2},s)}(u+\frac{3\eta}{2})\,t^{(\frac{1}{2},s)}(u+\frac{\eta}{2}) \\
&\quad +\delta^{(s)}(u+\frac{3\eta}{2})\,\delta^{(s)}(u-\frac{\eta}{2}).
\end{aligned} \tag{8.2.10}$$

Keeping the generalized initial condition (8.1.16) of $R^{(s,s)}(u)$ and the fusion condition (8.1.19) of $R^{(s,\frac{1}{2})}(u)$ in mind and using the similar procedure to prove the relation (7.2.1) in Sect. 7.2.1, we may show that

$$T_1^{(s,s)}(\theta_j)T_2^{(\frac{1}{2},s)}(\theta_j-(\frac{1}{2}+s)\eta)=\mathscr{P}_{1,2}^{(s-\frac{1}{2})}\,T_1^{(s,s)}(\theta_j)T_2^{(\frac{1}{2},s)}(\theta_j-(\frac{1}{2}+s)\eta),$$

where the projector $\mathscr{P}_{1,2}^{(s-\frac{1}{2})}$ projects the tensor product space of the spin-s and the spin-$\frac{1}{2}$ spaces into the subspace with the spin-$(s-\frac{1}{2})$ (see (8.1.19)). Then with the similar fusion procedure used in Sect. 7.3.3 of Chap. 7, we can derive the identities [34]

$$t^{(s,s)}(\theta_j)\,t^{(\frac{1}{2},s)}(\theta_j-(\frac{1}{2}+s)\eta)=\delta^{(s)}(\theta_j+(\frac{1}{2}-s)\eta)\,t^{(s-\frac{1}{2},s)}(\theta_j+\frac{\eta}{2}),$$
$$j=1,\ldots,N. \tag{8.2.11}$$

Now let us derive some properties of $t^{(\frac{1}{2},s)}(u)$. The R-matrix (8.1.2) and the K-matrices (8.1.21) and (8.1.26) imply that the following relations hold:

$$t^{(\frac{1}{2},s)}(0)=2p_-p_+\prod_{l=1}^{N}(\theta_l+(\frac{1}{2}+s)\eta)(-\theta_l+(\frac{1}{2}+s)\eta)\times\mathrm{id}, \tag{8.2.12}$$

$$t^{(\frac{1}{2},s)}(u)\,|_{u\to\infty}=2(\alpha_-\alpha_+-1)u^{2N+2}\times\mathrm{id}+\cdots, \tag{8.2.13}$$

$$t^{(\frac{1}{2},s)}(-u-\eta)=t^{(\frac{1}{2},s)}(u). \tag{8.2.14}$$

An obvious fact is that $t^{(\frac{1}{2},s)}(u)$, as a function of u, is a polynomial of degree $2N+2$. The fusion hierarchy relation (8.2.5) implies that all the other fused transfer matrices $t^{(j,s)}(u)$ can be expressed in terms of $t^{(\frac{1}{2},s)}(u)$ (see (8.2.7)). Therefore the identities (8.2.11) lead to N constraints on $t^{(\frac{1}{2},s)}(u)$, which together with (8.2.12)–(8.2.14) can completely determine the eigenvalues of $t^{(\frac{1}{2},s)}(u)$ (and as a consequence, also the eigenvalues of all the other transfer matrices $\{t^{(j,s)}(u)\}$).

Let $\Lambda^{(j,s)}(u)$ be the eigenvalue of the fused transfer matrix $t^{(j,s)}(u)$. The fusion hierarchy relation (8.2.5) directly implies

$$\Lambda^{(\frac{1}{2},s)}(u)\,\Lambda^{(j-\frac{1}{2},s)}(u-j\eta)=\Lambda^{(j,s)}(u-(j-\frac{1}{2})\eta)$$
$$+\delta^{(s)}(u)\,\Lambda^{(j-1,s)}(u-(j+\frac{1}{2})\eta),\quad j=\frac{1}{2},1,\frac{3}{2},\ldots, \tag{8.2.15}$$

with $\Lambda^{(0,s)}(u)=1$. The operator identities (8.2.11) also imply that

$$\Lambda^{(s,s)}(\theta_j)\,\Lambda^{(\frac{1}{2},s)}(\theta_j-(\frac{1}{2}+s)\eta)=\delta^{(s)}(\theta_j+(\frac{1}{2}-s)\eta)\,\Lambda^{(s-\frac{1}{2},s)}(\theta_j+\frac{\eta}{2}),$$
$$j=1,\ldots,N. \tag{8.2.16}$$

Equations (8.2.12)–(8.2.14) give rise to the relations

$$\Lambda^{(\frac{1}{2},s)}(0) = 2p_-p_+\prod_{l=1}^{N}(\theta_l + (\frac{1}{2}+s)\eta)(-\theta_l + (\frac{1}{2}+s)\eta), \qquad (8.2.17)$$

$$\Lambda^{(\frac{1}{2},s)}(u)\,|_{u\to\infty} = 2(\alpha_-\alpha_+ - 1)u^{2N+2} + \cdots, \qquad (8.2.18)$$

$$\Lambda^{(\frac{1}{2},s)}(-u-\eta) = \Lambda^{(\frac{1}{2},s)}(u). \qquad (8.2.19)$$

The analyticity of $t^{(\frac{1}{2},s)}(u)$ implies that

$$\Lambda^{(\frac{1}{2},s)}(u), \text{ as a function of } u, \text{ is a polynomial of degree } 2N+2. \qquad (8.2.20)$$

8.3 The Inhomogeneous $T-Q$ Relation

Following the method introduced in previous chapters, we can prove that each eigenvalue $\Lambda^{(\frac{1}{2},s)}(u)$ can be parameterized by the following inhomogeneous $T-Q$ relation:

$$\begin{aligned}\Lambda^{(\frac{1}{2},s)}(u) &= a^{(s)}(u)\frac{Q(u-\eta)}{Q(u)} + d^{(s)}(u)\frac{Q(u+\eta)}{Q(u)}\\ &\quad + c\,u(u+\eta)\frac{F^{(s)}(u)}{Q(u)},\end{aligned} \qquad (8.3.1)$$

where the functions $a^{(s)}(u)$, $d^{(s)}(u)$, $F^{(s)}(u)$ and the constant c are given by

$$\begin{aligned}a^{(s)}(u) &= \frac{2u+2\eta}{2u+\eta}[(1+\alpha_+^2)^{\frac{1}{2}}u + p_+][(1+\alpha_-^2)^{\frac{1}{2}}u + p_-]\\ &\quad\times\prod_{l=1}^{N}(u-\theta_l+(\frac{1}{2}+s)\eta)(u+\theta_l+(\frac{1}{2}+s)\eta),\end{aligned} \qquad (8.3.2)$$

$$d^{(s)}(u) = a^{(s)}(-u-\eta), \qquad (8.3.3)$$

$$F^{(s)}(u) = \prod_{l=1}^{N}\prod_{k=0}^{2s}(u-\theta_l+(\frac{1}{2}-s+k)\eta)(u+\theta_l+(\frac{1}{2}-s+k)\eta), \qquad (8.3.4)$$

$$c = 2[\alpha_-\alpha_+ - 1 - (1+\alpha_-^2)^{\frac{1}{2}}(1+\alpha_+^2)^{\frac{1}{2}}]. \qquad (8.3.5)$$

The Q-function is parameterized by $2sN$ parameters $\{\lambda_j|j=1,\ldots,2sN\}$ as

$$Q(u) = \prod_{j=1}^{2sN}(u-\lambda_j)(u+\lambda_j+\eta) = Q(-u-\eta). \qquad (8.3.6)$$

Proof Let us introduce a function $f(u)$ associated with each solution $\Lambda(u)$ of (8.2.15)–(8.2.20)

$$f(u) = \Lambda^{(\frac{1}{2},s)}(u)Q(u) - a^{(s)}(u)Q(u-\eta) \\ -d^{(s)}(u)Q(u+\eta) - c\,u(u+\eta)F^{(s)}(u). \tag{8.3.7}$$

It follows from its definition that the function $f(u)$, as a function of u, is a polynomial of degree $2(2s+1)N+2$ with the crossing symmetry

$$f(-u-\eta) = f(u). \tag{8.3.8}$$

This property implies that the function can be fixed by its values at $(2s+1)N+2$ different points. It is clear from the relations (8.2.17) and (8.2.18) that

$$f(0) = 0, \quad \text{and} \quad f(u)|_{u\to\infty} = 0 \times u^{2(2s+1)N+2} + \cdots, \tag{8.3.9}$$

which, together with its values at other $(2s+1)N$ independent points, are sufficient to completely determine $f(u)$. Thanks to the fact that $Q(u)$ is also a crossing polynomial (i.e., $Q(-u-\eta) = Q(u)$) of degree $4sN$ with a known coefficient of the term u^{4sN}, for each solution $\Lambda(u)$ of (8.2.15)–(8.2.20) we can always choose the function $Q(u)$ given by (8.3.6) such that the following equations hold:

$$f(\theta_j + (s - \frac{1}{2} - k)\eta) = 0, \quad k = 0, \ldots, 2s, \quad j = 1, \ldots, N. \tag{8.3.10}$$

Then the relations (8.3.8)–(8.3.10) lead to $f(u) = 0$ or that each solution of (8.2.15)–(8.2.20) can be parameterized in terms of the inhomogeneous $T-Q$ relation (8.3.1) with proper choices of the function $Q(u)$. In fact the conditions (8.3.10) are equivalent to the following $(2s+1)N$ linear equations with respect to the values of $Q(u)$ at the $(2s+1)N$ different points $\{\theta_j + (s-\frac{1}{2}-k)\eta | k = 0, \ldots, 2s,\ j = 1, \ldots, N\}$, namely,

$$B^{(j)}\, X^{(j)} = 0, \quad j = 1, \ldots, N, \tag{8.3.11}$$

with each $(2s+1) \times (2s+1)$ matrix $B^{(j)}$ (superscripts of the elements are omitted for simplicity) being given by

$$\begin{pmatrix} \Lambda(\theta_j + (s-\frac{1}{2})\eta) & -a(\theta_j + (s-\frac{1}{2})\eta) & & & \\ -d(\theta_j + (s-\frac{3}{2})\eta) & \Lambda(\theta_j + (s-\frac{3}{2})\eta) & -a(\theta_j + (s-\frac{3}{2})\eta) & & \\ & \ddots & \ddots & \ddots & \\ & & & -d(\theta_j - (s+\frac{1}{2})\eta) & \Lambda(\theta_j - (s+\frac{1}{2})\eta) \end{pmatrix},$$

and the $(2s+1)$-components vector $X^{(j)}$ being given by

$$\begin{pmatrix} Q(\theta_j + (s-\frac{1}{2})\eta) \\ Q(\theta_j + (s-\frac{3}{2})\eta) \\ \vdots \\ Q(\theta_j - (s+\frac{1}{2})\eta) \end{pmatrix}.$$

Direct calculation shows that the identities (8.2.16) give rise to $\det(B^{(j)}) = 0$ which ensures the $(2s+1)N$ linear equations (8.3.11) to have non-zero solutions. In this case, the number of independent linear equations (8.3.11) is reduced to $2sN$ and we can always fix at most the $2sN$ values (up to a scaling factor) of $Q(u)$ at $2sN$ points among $\{\theta_j + (s-\frac{1}{2}-k)\eta | k = 0, 1, \ldots, 2s,\ j = 1, \ldots, N\}$, for an example, $\{\theta_j + (s-\frac{1}{2}-k)\eta | k = 1, \ldots, 2s,\ j = 1, \ldots, N\}$. Therefore, each solution $\Lambda(u)$ of (8.2.15)–(8.2.20) allows us to parameterize it in terms of the inhomogeneous $T-Q$ relation (8.3.1) where $Q(u)$ can be determined either by its values at $2sN$ different points via the Eqs. (8.3.11) or by its roots: $\{\lambda_j | j = 1, \ldots, 2sN\}$ determined by the BAEs (see below). We can check that the $T-Q$ relation (8.3.1) does satisfy the relations (8.2.15)–(8.2.20). □

From (8.3.1) we find that the $T-Q$ relation might have some readily apparent and simple poles at the points

$$\lambda_j,\ -\lambda_j - \eta, \quad j = 1, \ldots, 2sN. \tag{8.3.12}$$

The regularity of the eigenvalue $\Lambda^{(\frac{1}{2},s)}(u)$ requires that its residues at these points must be zero. This requirement leads to the BAEs:

$$\begin{aligned} &a^{(s)}(\lambda_j)Q(\lambda_j - \eta) + d^{(s)}(\lambda_j)Q(\lambda_j + \eta) \\ &\quad + c\,\lambda_j(\lambda_j + \eta)\,F^{(s)}(\lambda_j) = 0, \quad j = 1, \ldots, 2sN. \end{aligned} \tag{8.3.13}$$

Similar to the case of the isotropic spin-$\frac{1}{2}$ chain [39], another simple $T-Q$ relation exists for the spin-s case

$$\begin{aligned} \Lambda^{(\frac{1}{2},s)}(u) = {}& a^{(s)}(u)\frac{Q_1(u-\eta)}{Q_2(u)} + d^{(s)}(u)\frac{Q_2(u+\eta)}{Q_1(u)} \\ & + (-1)^M c\, u(u+\eta)\frac{F^{(s)}(u)}{Q_1(u)\,Q_2(u)}, \end{aligned} \tag{8.3.14}$$

where

$$\begin{aligned} c &= 2(\alpha_-\alpha_+ - 1 - (-1)^M[(1+\alpha_-^2)(1+\alpha_+^2)]^{\frac{1}{2}}), \\ Q_1(u) &= \prod_{j=1}^{M}(u - \mu_j) = Q_2(-u-\eta), \quad M = 2sN. \end{aligned} \tag{8.3.15}$$

The resulting BAEs are

$$d^{(s)}(\mu_j)Q_2(\mu_j)Q_2(\mu_j+\eta)+(-1)^M c\,\mu_j(\mu_j+\eta)\,F^{(s)}(\mu_j)=0,$$
$$j=1,\ldots,M. \tag{8.3.16}$$

Taking the homogeneous limit $\theta_j \to 0$, we can obtain all the eigenvalues $\Lambda^{(j,s)}(u)$ of the fused homogeneous transfer matrices. In particular, the resulting $\Lambda^{(s,s)}(u)$ gives the spectrum of the spin-s XXX Hamiltonian with generic integrable boundary terms.

We note that there are several possibilities for factoring out the functions $a^{(s)}(u)$ and $d^{(s)}(u)$ from the quantum determinant $\Delta^{(s,s)}(u)$. For convenience, let us introduce the notations $\{\varepsilon_i=\pm1|i=1,2,3\}$ and

$$a^{(s)}(u;\varepsilon_1,\varepsilon_2,\varepsilon_3)=\varepsilon_1\frac{2u+2\eta}{2u+\eta}((1+\alpha_+^2)^{\frac{1}{2}}u+\varepsilon_2p_+)((1+\alpha_-^2)^{\frac{1}{2}}u+\varepsilon_3p_-)$$
$$\times\prod_{l=1}^{N}(u-\theta_l+(\frac{1}{2}+s)\eta)(u+\theta_l+(\frac{1}{2}+s)\eta), \tag{8.3.17}$$

$$d^{(s)}(u;\varepsilon_1,\varepsilon_2,\varepsilon_3)=a^{(s)}(-u-\eta;\varepsilon_1,\varepsilon_2,\varepsilon_3),$$
$$c(\varepsilon_1,\varepsilon_2,\varepsilon_3)=2(\alpha_-\alpha_+-1-\varepsilon_1[(1+\alpha_-^2)(1+\alpha_+^2)]^{\frac{1}{2}}). \tag{8.3.18}$$

Similar to the work in [40], the three discrete variables $\{\varepsilon_i\}$ must satisfy the relation

$$\varepsilon_1\varepsilon_2\varepsilon_3=1, \tag{8.3.19}$$

as required by the initial condition (8.2.17). An arbitrary choice of the discrete parameters $\{\varepsilon_i\}$ satisfying the constraint (8.3.19) should give the complete set of the spectrum of the transfer matrix by replacing $a^{(s)}(u)$, $d^{(s)}(u)$ and c in the $T-Q$ relation (8.3.1) with $a^{(s)}(u;\varepsilon_1,\varepsilon_2,\varepsilon_3)$, $d^{(s)}(u;\varepsilon_1,\varepsilon_2,\varepsilon_3)$ and $c(\varepsilon_1,\varepsilon_2,\varepsilon_3)$, respectively. It is particularly important that for $\alpha_-=-\alpha_+$, $K^{\pm}(u)$ can be diagonalized simultaneously and the algebraic Bethe Ansatz method is applicable [33]. In this case, if we choose $\varepsilon_1=-1$, then $\varepsilon_2\varepsilon_3=-1$, the parameter $c(\varepsilon_1,\varepsilon_2,\varepsilon_3)=0$ and the corresponding $T-Q$ relation (8.3.1) is naturally reduced to a conventional one [33]

$$\Lambda^{(\frac{1}{2},s)}(u)=a^{(s)}(u)\frac{\bar{Q}(u-\eta)}{\bar{Q}(u)}+d^{(s)}(u)\frac{\bar{Q}(u+\eta)}{\bar{Q}(u)}, \tag{8.3.20}$$

with

$$\bar{Q}(u)=\prod_{j=1}^{\bar{M}}(u-\lambda_j)(u+\lambda_j+\eta)=\bar{Q}(-u-\eta),\quad \bar{M}=0,1,\ldots. \tag{8.3.21}$$

The corresponding BAEs thus read

$$\frac{a^{(s)}(\lambda_j)}{d^{(s)}(\lambda_j)} = -\frac{\bar{Q}(\lambda_j + \eta)}{\bar{Q}(\lambda_j - \eta)}. \tag{8.3.22}$$

We remark that for other choices of $\{\varepsilon_i\}$, even for $\alpha_- = -\alpha_+$, $c(\varepsilon_1, \varepsilon_2, \varepsilon_3) \neq 0$ and the $T - Q$ relation cannot take the conventional form (8.3.20).

8.4 Completeness of the Solutions

Let us illustrate the completeness of the Bethe Ansatz solutions introduced in the previous section with the $s = 1$ case as an example. In terms of the basis $\{|l\rangle | l = 1, 0, -1\}$ given by

$$\begin{aligned}
|1\rangle &= |\tfrac{1}{2}\rangle \otimes |\tfrac{1}{2}\rangle, \\
|0\rangle &= \frac{1}{\sqrt{2}}\left(|\tfrac{1}{2}\rangle \otimes |-\tfrac{1}{2}\rangle + |-\tfrac{1}{2}\rangle \otimes |\tfrac{1}{2}\rangle\right), \\
|-1\rangle &= |-\tfrac{1}{2}\rangle \otimes |-\tfrac{1}{2}\rangle,
\end{aligned}$$

the explicit form of the fused R-matrix $R^{(1,1)}(u)$ defined in (8.1.13) reads

$$R^{(1,1)}(u) = \left(\begin{array}{ccc|ccc|ccc}
h_3(u) & & & & & & & & \\
& h_2(u) & & h_5(u) & & & & & \\
& & h_4(u) & & h_7(u) & & h_6(u) & & \\
\hline
& h_5(u) & & h_2(u) & & & & & \\
& & h_7(u) & & h_1(u) & & h_7(u) & & \\
& & & & & h_2(u) & & h_5(u) & \\
\hline
& & h_6(u) & & h_7(u) & & h_4(u) & & \\
& & & & & h_5(u) & & h_2(u) & \\
& & & & & & & & h_3(u)
\end{array}\right), \tag{8.4.1}$$

with the non-vanishing entries

$$\begin{aligned}
&h_1(u) = u(u+\eta) + 2\eta^2, \quad h_2(u) = u(u+\eta), \quad h_3(u) = (u+\eta)(u+2\eta), \\
&h_4(u) = u(u-\eta), \quad h_5(u) = 2\eta(u+\eta), \quad h_6(u) = 2\eta^2, \quad h_7(u) = 2u\eta.
\end{aligned} \tag{8.4.2}$$

In addition, the fused K-matrix defined by (8.1.20), in terms of the basis $\{|l\rangle | l = 1, 0, -1\}$, is given by

$$K^{-(1)}(u) = (2u+\eta)\begin{pmatrix} x_1(u) & x_4(u) & x_6(u) \\ x_4(u) & x_2(u) & x_5(u) \\ x_6(u) & x_5(u) & x_3(u) \end{pmatrix}, \tag{8.4.3}$$

where the matrix elements are

$$\begin{aligned}
x_1(u) &= (p_- + u + \frac{\eta}{2})(p_- + u - \frac{\eta}{2}) + \frac{\alpha_-^2}{2}\,\eta\,(u - \frac{\eta}{2}), \\
x_2(u) &= (p_- + u - \frac{\eta}{2})(p_- - u + \frac{\eta}{2}) + \alpha_-^2\,(u + \frac{\eta}{2})(u - \frac{\eta}{2}), \\
x_3(u) &= (p_- - u - \frac{\eta}{2})(p_- - u + \frac{\eta}{2}) + \frac{\alpha_-^2}{2}\,\eta\,(u - \frac{\eta}{2}), \\
x_4(u) &= \sqrt{2}\,\alpha_-\,u\,(p_- + u - \frac{\eta}{2}), \\
x_5(u) &= \sqrt{2}\,\alpha_-\,u\,(p_- - u + \frac{\eta}{2}), \\
x_6(u) &= \alpha^2\,u\,(u - \frac{\eta}{2}),
\end{aligned} \tag{8.4.4}$$

while $K^{+(1)}(u)$ can be given through the duality (8.1.23).

The corresponding spin-1 Hamiltonian in terms of the transfer matrix $t^{(1,1)}(u)$ is thus given by

$$\begin{aligned}
H &= \frac{\partial}{\partial u}\left\{\ln[u(u+\eta)\,t^{(1,1)}(u)]\right\}\Big|_{u=0,\{\theta_j=0\}} \\
&= \frac{1}{\eta^2}\sum_{j=1}^{N-1}\left[\mathbf{S}_j\cdot\mathbf{S}_{j+1} - (\mathbf{S}_j\cdot\mathbf{S}_{j+1})^2\right] \\
&\quad + \frac{1}{p_-^2 - \frac{1}{4}(1+\alpha_-^2)\eta^2}\Big[2p_-\alpha_- S_1^x + 2p_- S_1^z + \frac{1}{2}(\alpha_-^2\eta - 2\eta)(S_1^z)^2 \\
&\qquad - \frac{1}{2}\alpha_-^2\eta((S_1^x)^2 - (S_1^y)^2) - \alpha_-\eta(S_1^z S_1^x + S_1^x S_1^z)\Big] \\
&\quad + \frac{1}{(3p_+^2 - \frac{3}{4}(1+\alpha_+^2)\eta^2)\eta^2}\Big[6p_+\alpha_+\eta S_N^x - 6p_+\eta S_N^z \\
&\qquad + 3\alpha_+\eta^2(S_N^x S_N^z + S_N^z S_N^x) - (2p_+^2 - \frac{3}{2}(1-\alpha_+^2)\eta^2)(S_N^x)^2 \\
&\qquad - (2p_+^2 - \frac{3}{2}(1+\alpha_+^2)\eta^2)(S_N^y)^2 - (2p_+^2 + \frac{3}{2}(1-\alpha_+^2)\eta^2)(S_N^z)^2\Big] \\
&\quad + \frac{\eta(1+\alpha_+^2)}{3p_+^2 - \frac{3}{4}(1+\alpha_+^2)\eta^2} + \frac{\eta}{p_-^2 - \frac{1}{4}(1+\alpha_-^2)\eta^2} + 3N\frac{1}{\eta^2} + \frac{4}{\eta},
\end{aligned} \tag{8.4.5}$$

and the associated eigenvalue of the Hamiltonian reads

$$E = \sum_{j=1}^{2N} \frac{4\eta}{(\lambda_j + \frac{3\eta}{2})(\lambda_j - \frac{\eta}{2})} + E_0, \tag{8.4.6}$$

$$E_0 = -\frac{1}{\eta}\left\{3N + \frac{8}{3} + \frac{2(1+\alpha_+^2)^{\frac{1}{2}} p_+ \eta}{p_+^2 - \frac{\eta^2}{4}(1+\alpha_+^2)} + \frac{2(1+\alpha_-^2)^{\frac{1}{2}} p_- \eta}{p_-^2 - \frac{\eta^2}{4}(1+\alpha_-^2)}\right\}, \tag{8.4.7}$$

with the Bethe roots $\{\lambda_j\}$ determined by (8.3.13) in the homogeneous limit $\{\theta_j = 0\}$ or

$$E = -\sum_{k=1}^{2N} \frac{4(\mu_k + \eta)}{(\mu_k + \frac{\eta}{2})(\mu_k + \frac{3\eta}{2})} + E_0, \tag{8.4.8}$$

with the Bethe roots $\{\mu_k\}$ determined by (8.3.16) in the homogeneous limit $\{\theta_j = 0\}$.

Numerical solutions of the BAEs (8.3.13) and (8.3.16) in the homogeneous limit $\{\theta_j = 0\}$ as well as the exact diagonalization of the Hamiltonian (8.4.5) are performed in [34] for $N = 2$ and randomly chosen boundary parameters. We list those results in Tables 8.1 and 8.2, respectively. The eigenvalues $\Lambda^{(\frac{1}{2},1)}(u)$ of the transfer matrix $t^{(\frac{1}{2},1)}(u)$ are shown in Fig. 8.1. Again, the curves of $\Lambda^{(\frac{1}{2},1)}(u)$ calculated

Table 8.1 Solutions of the BAEs (8.3.13) for $N = 2, s = 1, \eta = 1, \{\theta_j = 0\}, p_+ = 0.1, p_- = 0.2, \alpha_+ = 0.3$ and $\alpha_- = 0.4$

λ_1	λ_2	λ_3	λ_4	E_n	n
0.02022	0.15565 − 0.56301i	0.15565 + 0.56301i	1.28344	−2.82985	1
0.01436 − 0.14539i	0.01436 + 0.14539i	1.02580 − 0.23475i	1.02580 + 0.23475i	0.74454	2
0.00579 − 0.12153i	0.00579 + 0.12153i	0.95719 − 0.17885i	0.95719 + 0.17885i	1.84509	3
−0.50000 + 0.46805i	−0.09323	0.93690	1.18821	3.99277	4
0.06934 − 0.91728i	0.06934 + 0.91728i	1.06778 − 0.60960i	1.06778 + 0.60960i	4.36850	5
−0.50000 + 0.16632i	−0.18832	0.82026	1.19558	5.34163	6
−0.09561	0.89614	1.31281 − 0.54820i	1.31281 + 0.54820i	7.59257	7
−0.25439	0.03756	0.73530 − 0.09425i	0.73530 + 0.09425i	9.12855	8
−0.18554	0.81124	1.29199 − 0.51363i	1.29199 + 0.51363i	9.43905	9

The symbol n indicates the number of the energy levels, and E_n is the corresponding eigenenergy. The energy E_n calculated from (8.4.6) is the same as that from the exact diagonalization of the Hamiltonian (8.4.5) (reproduced from [34])

Table 8.2 Solutions of the BAEs (8.3.16) for $N = 2, s = 1, \eta = 1, \{\theta_j = 0\}, p_+ = 0.1, p_- = 0.2, \alpha_+ = 0.3$ and $\alpha_- = 0.4$

μ_1	μ_2	μ_3	μ_4	E_n	n
−2.01449	−1.03956	−0.20728 − 0.16066*i*	−0.20728 + 0.16066*i*	−2.82985	1
−1.02843	−0.99277	0.08448	9.58424	0.74454	2
−1.24529	−0.75827	0.29385	6.34119	1.84509	3
−0.90627	−0.50220 − 0.45722*i*	−0.50220 + 0.45722*i*	8.15074	3.99277	4
−6.63560	−1.41141	−0.58859	0.74744	4.36850	5
−0.80847	−0.50016 − 0.16539*i*	−0.50016 + 0.16539*i*	6.62545	5.34163	6
−4.94473	−0.90422	2.81138	35.70597	7.59257	7
−4.56687	−0.88866	−0.82335	0.40258	9.12855	8
−4.49000	−0.81430	2.38972	32.53964	9.43905	9

The symbol n indicates the number of the energy levels, and E_n is the corresponding eigenenergy. The energy E_n calculated from (8.4.8) is the same as that from the exact diagonalization of the Hamiltonian (8.4.5) (reproduced from [34])

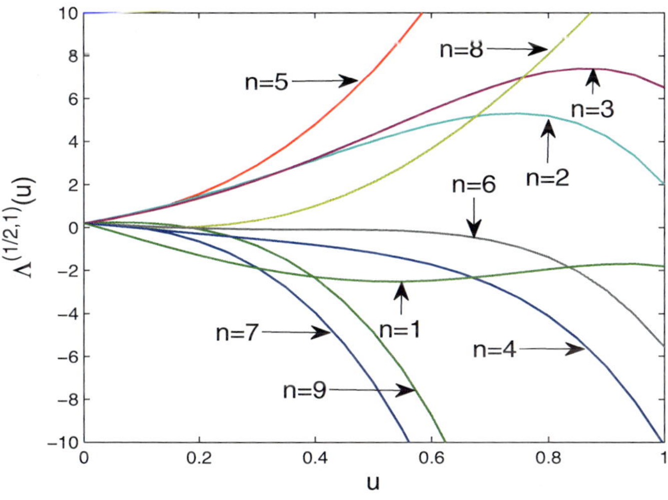

Fig. 8.1 $\Lambda^{(\frac{1}{2},1)}(u)$ versus u calculated from both the $T - Q$ relations and exact diagonalization of $t^{(\frac{1}{2},1)}(u)$ for $N = 2$, $\eta = 1$, $\{\theta_j = 0\}$, $p_+ = 0.1$, $p_- = 0.2$, $\alpha_+ = 0.3$ and $\alpha_- = 0.4$. The correspondence is indicated by the number $n = 1$ to 9 (reproduced from [34])

from the BAEs and the $T - Q$ relations coincide exactly with those from the exact diagonalization of the transfer matrix $t^{(\frac{1}{2},1)}(u)$. The numerical results strongly suggest that any choice of the $T - Q$ relations satisfying the closed functional relations (8.2.15)–(8.2.20) is sufficient to give the complete spectrum of the transfer matrix.

8.5 The Nonlinear Schrödinger Model

An interesting feature of the high-spin Heisenberg chain is its duality to some infinite dimensional models. It has been noted that the XXX spin chain corresponds to the lattice version of the nonlinear Schrödinger model [41] or the Lieb-Liniger model [42, 43], while the XXZ spin chain corresponds to the lattice version of the sine-Gordon model [44, 45]. This correspondence allows us to derive the exact solutions of the associated continuum models by taking a proper limit $s \to \infty$. In this section, by employing the non-linear Schrödinger model as an example, we show that the ODBA method can also be applied to the infinite dimensional models. The method can be naturally generalized to the integrable models associated with cyclic representations such as the lattice sine-Gordon model [44, 45], the τ_2 model [46], the relativistic quantum Toda chain [47] and the chiral Potts model [48–51] with generic integrable boundaries.

8.5.1 The Model Correspondence

The Hamiltonian of the quantum lattice nonlinear Schrödinger model is [41]

$$\begin{aligned} H = &-\frac{4}{3g\Delta^3}\sum_{n=1}^{N}\left(t_n + t_n^\dagger + \frac{8-g\Delta}{8-2g\Delta}\right) \\ &+ \frac{4}{3\Delta^2\left(1-\frac{\Delta^2 g^2}{16}\right)}\sum_{n=1}^{N}\Delta\psi_n^\dagger\psi_n, \end{aligned} \tag{8.5.1}$$

where

$$\begin{aligned} t_n = &\left(\alpha^\dagger(n+2)\alpha(n+1)\right)^{-1}\left\{\left(\alpha^\dagger(n)\alpha(n-1)\right)^{-1}\left(\alpha^\dagger(n+1)\alpha(n)\right)^{-1}\right. \\ &\times\left.\left(\alpha^\dagger(n+1)\sigma^z\alpha(n-1)\right)\right\}\left(\alpha^\dagger(n+2)\alpha(n+1)\right), \end{aligned} \tag{8.5.2}$$

for odd n and

$$\begin{aligned} t_n = &\left(\alpha^\dagger(n-1)\alpha(n-2)\right)^{-1}\left\{\left(\alpha^\dagger(n+1)\alpha(n)\right)^{-1}\left(\alpha^\dagger(n)\alpha(n-1)\right)^{-1}\right. \\ &\times\left.\left(\alpha^\dagger(n+1)\sigma^z\alpha(n-1)\right)\right\}\left(\alpha^\dagger(n-1)\alpha(n-2)\right), \end{aligned} \tag{8.5.3}$$

for even n with α being a two-component vector defined by the elements

$$\alpha_1(n) = -i\left(\frac{g}{2}\right)^{\frac{1}{2}} \Delta\psi_n^\dagger, \tag{8.5.4}$$

$$\alpha_2(n) = \left[2 + ((-1)^n - 1)\frac{g\Delta}{4} + \frac{g\Delta^2}{2}\psi_n^\dagger\psi_n\right]^{\frac{1}{2}}, \tag{8.5.5}$$

where Δ and g are the lattice constant and the coupling constant, respectively, and $\psi_n^\dagger$ (ψ_n) is the creation (annihilation) operator of bosons. The continuum model can be obtained by taking the limit of $\Delta \to 0$, $N \to \infty$ and the length of the system $L = N\Delta$ finite with the correspondence

$$\psi_n = \frac{1}{\Delta}\int_{(n-1)\Delta}^{n\Delta} \psi(x)dx, \quad [\psi(x), \psi^\dagger(y)] = \delta(x-y). \tag{8.5.6}$$

The resulting continuum Hamiltonian thus reads

$$H = \int_0^L \left\{\frac{\partial\psi^\dagger(x)}{\partial x}\frac{\partial\psi(x)}{\partial x} + g\psi^\dagger(x)\psi^\dagger(x)\psi(x)\psi(x)\right\} dx. \tag{8.5.7}$$

To solve the above Hamiltonian, let us first consider its lattice version (8.5.1). The corresponding L-operator is [41]

$$L_n(u) = \begin{pmatrix} 1 + (-1)^n\frac{g\Delta}{4} - \frac{iu\Delta}{2} + \frac{g\Delta^2}{2}\psi_n^\dagger\psi_n & -i\left(\frac{g}{2}\right)^{\frac{1}{2}}\Delta\psi_n^\dagger\rho_n \\ i\left(\frac{g}{2}\right)^{\frac{1}{2}}\Delta\rho_n\psi_n & 1 + (-1)^n\frac{g\Delta}{4} + \frac{iu\Delta}{2} + \frac{g\Delta^2}{2}\psi_n^\dagger\psi_n \end{pmatrix},$$

$$\rho_n = \left[2 + (-1)^n\frac{g\Delta}{2} + \frac{g\Delta^2}{2}\psi_n^\dagger\psi_n\right]^{\frac{1}{2}}, \tag{8.5.8}$$

and the transfer matrix is defined as

$$t(u) = tr\{L_N(u)L_{N-1}(u)\cdots L_1(u)\}. \tag{8.5.9}$$

The Hamiltonian (8.5.1) can be obtained from the transfer matrix as

$$H = -\frac{8i}{3g\Delta^4}\frac{\partial}{\partial u}\ln[t(u)t_0^{-1}(u)]|_{u=u_0} + \frac{8i}{3g\Delta^4}\frac{\partial}{\partial u}\ln[t(u)t_0^{-1}(u)]|_{u=-u_0}$$
$$+\sum_{n=1}^{N}\frac{4\psi_n^\dagger\psi_n}{3\Delta(1-\frac{\Delta^2 g^2}{16})}, \tag{8.5.10}$$

with

$$u_0 = -\frac{2i}{\Delta} + \frac{ig}{2}, \tag{8.5.11}$$

and

$$t_0(u) = (1 + \frac{g\Delta}{4} - \frac{iu\Delta}{2})^{N/2}(1 - \frac{g\Delta}{4} - \frac{iu\Delta}{2})^{N/2}$$
$$+(1 + \frac{g\Delta}{4} + \frac{iu\Delta}{2})^{N/2}(1 - \frac{g\Delta}{4} + \frac{iu\Delta}{2})^{N/2}. \quad (8.5.12)$$

Let us introduce the Holstein-Primakoff transformations

$$S_n^+ = i\left(\frac{2}{g}\right)^{\frac{1}{2}} \rho_n \psi_n, \qquad S_n^- = i\left(\frac{2}{g}\right)^{\frac{1}{2}} \psi_n^+ \rho_n,$$
$$S_n^z = s_n - \Delta\psi_n^\dagger \psi_n, \quad (8.5.13)$$

where the spin s_n takes the following alternating values for even sites and odd sites, respectively

$$s_n = -\frac{2}{g\Delta} - \frac{1}{2}(-1)^n. \quad (8.5.14)$$

The L-operator (8.5.8) can be written as

$$L_n(u) = -\frac{i\Delta}{2}\sigma_0^z R_{0,n}^{(\frac{1}{2},s_n)}(u - \frac{\eta}{2}), \quad (8.5.15)$$

where $R_{0,n}^{(\frac{1}{2},s_n)}(u)$ is given by (8.1.2) and $\eta = -ig$. For convenience, in the following text let us set

$$s_n = \begin{cases} \bar{s} = -\frac{2}{g\Delta} + \frac{1}{2}, \text{ for odd } n, \\ s = -\frac{2}{g\Delta} - \frac{1}{2}, \text{ for even } n. \end{cases} \quad (8.5.16)$$

The corresponding spin chain is then called the alternating-spin chain.

8.5.2 Operator Identities

Let us fix an even number N and introduce the fundamental monodromy matrix for an alternating-spin chain as

$$T_0^{(\frac{1}{2},(\bar{s},s))}(u) = \sigma_0^z R_{0,N}^{(\frac{1}{2},s_N)}(u - \theta_N) \cdots \sigma_0^z R_{0,1}^{(\frac{1}{2},s_1)}(u - \theta_1). \quad (8.5.17)$$

The transfer matrix $t(u)$ of the lattice nonlinear Schrödinger model thus reads

$$t(u) = (-\frac{i\Delta}{2})^N tr_0 T_0^{(\frac{1}{2},(\bar{s},s))}(u - \frac{\eta}{2})|_{\{\theta_j=0\}}. \tag{8.5.18}$$

Obviously, the following YBE holds

$$\begin{aligned} &R_{1,2}^{(\frac{1}{2},\frac{1}{2})}(u-v)T_1^{(\frac{1}{2},(\bar{s},s))}(u)T_2^{(\frac{1}{2},(\bar{s},s))}(v) \\ &= T_2^{(\frac{1}{2},(\bar{s},s))}(v)T_1^{(\frac{1}{2},(\bar{s},s))}(u)R_{1,2}^{(\frac{1}{2},\frac{1}{2})}(u-v), \end{aligned} \tag{8.5.19}$$

which leads to $[t(u), t(v)] = 0$. Similarly, we can introduce the fused monodromy matrices (with $\{a\} = 1, \ldots, 2j$)

$$\begin{aligned} T_{\{a\}}^{(j,(\bar{s},s))}(u) = P_{1,\ldots,2j}^{(+)} &T_1^{(\frac{1}{2},(\bar{s},s))}(u - (j - \frac{1}{2})\eta) \\ &\times \cdots T_{2j}^{(\frac{1}{2},(\bar{s},s))}(u + (j - \frac{1}{2})\eta)P_{1,\ldots,2j}^{(+)}, \end{aligned} \tag{8.5.20}$$

and the mutually commutative fused transfer matrices

$$t^{(j,(\bar{s},s))}(u) = tr_{\{a\}}T_{\{a\}}^{(j,(\bar{s},s))}(u), \tag{8.5.21}$$

$$[t^{(k,(\bar{s},s))}(u)\,,\, t^{(l,(\bar{s},s))}(v)] = 0. \tag{8.5.22}$$

With a procedure similar to that used to prove the relation (8.2.4), we obtain the operator recursion equations

$$\begin{aligned} t^{(\frac{1}{2},(\bar{s},s))}(u)\,t^{(j-\frac{1}{2},(\bar{s},s))}(u - j\eta) &= t^{(j,(\bar{s},s))}(u - (j - \frac{1}{2})\eta) \\ &+\Delta^{(\bar{s},s)}(u)t^{(j-1,(\bar{s},s))}(u - (j + \frac{1}{2})\eta)\,, \end{aligned} \tag{8.5.23}$$

with $t^{(0,(\bar{s},s))} = \text{id}$ and

$$\Delta^{(\bar{s},s)}(u) = a^{(\bar{s},s)}(u)d^{(\bar{s},s)}(u - \eta), \tag{8.5.24}$$

$$a^{(\bar{s},s)}(u) = \prod_{l=1}^{N/2}(u - \theta_{2l-1} + (\bar{s} + \frac{1}{2})\eta)(u - \theta_{2l} + (s + \frac{1}{2})\eta), \tag{8.5.25}$$

$$d^{(\bar{s},s)}(u) = \prod_{l=1}^{N/2}(u - \theta_{2l-1} + (\frac{1}{2} - \bar{s})\eta)(u - \theta_{2l} + (\frac{1}{2} - s)\eta). \tag{8.5.26}$$

With the similar fusion procedure used in Sect. 8.2, we have the operator identities

$$t^{(\bar{s},(\bar{s},s))}(\theta_{2l-1})\,t^{(\bar{s},(\bar{s},s))}(\theta_{2l-1}-\eta) = \Delta_1^{(\bar{s},s)}(\theta_{2l-1}) \times \mathrm{id}, \tag{8.5.27}$$

$$t^{(s,(\bar{s},s))}(\theta_{2l})\,t^{(s,(\bar{s},s))}(\theta_{2l}-\eta) = \Delta_2^{(\bar{s},s)}(\theta_{2l}) \times \mathrm{id}, \quad l = 1, \ldots, \frac{N}{2}. \tag{8.5.28}$$

The functions $\Delta_i^{(\bar{s},s)}(u)$ are given by

$$\Delta_1^{(\bar{s},s)}(u) = \prod_{k=0}^{2\bar{s}-1} \Delta^{(\bar{s},s)}(u-(\bar{s}-\frac{1}{2})\eta+k\eta), \tag{8.5.29}$$

$$\Delta_2^{(\bar{s},s)}(u) = \prod_{k=0}^{2s-1} \Delta^{(\bar{s},s)}(u-(s-\frac{1}{2})\eta+k\eta). \tag{8.5.30}$$

The expression (8.5.18) implies the asymptotic behavior

$$t^{(\frac{1}{2},(\bar{s},s))}(u) = 2u^N + \cdots, \quad u \to \infty. \tag{8.5.31}$$

8.5.3 $T - Q$ Relation

Following the method in Sect. 8.3, we obtain the $T - Q$ relation

$$\Lambda^{(\frac{1}{2},(\bar{s},s))}(u) = a^{(\bar{s},s)}(u)\frac{Q(u-\eta)}{Q(u)} + d^{(\bar{s},s)}(u)\frac{Q(u+\eta)}{Q(u)}, \tag{8.5.32}$$

where the Q-function is parameterized by M Bethe roots $\{\lambda_j\}$ as

$$Q(u) = \prod_{j=1}^{M}(u-\lambda_j), \quad M = 0, 1, \ldots.$$

From (8.5.18) we conclude that the eigenvalue of the transfer matrix of the lattice nonlinear Schrödinger model, denoted by $\Lambda_{NS}(u)$, is given by

$$\begin{aligned}\Lambda_{NS}(u+\frac{\eta}{2}) = (-\frac{\Delta^2}{4})^{\frac{N}{2}} \Big\{ &(u+(\bar{s}+\frac{1}{2})\eta)^{\frac{N}{2}}(u+(s+\frac{1}{2})\eta)^{\frac{N}{2}}\frac{Q(u-\eta)}{Q(u)} \\ &+ (u+(\frac{1}{2}-\bar{s})\eta)^{\frac{N}{2}}(u+(\frac{1}{2}-s)\eta)^{\frac{N}{2}}\frac{Q(u+\eta)}{Q(u)}\Big\},\end{aligned} \tag{8.5.33}$$

with the associated BAEs

$$\frac{(\lambda_j + (\bar{s} + \frac{1}{2})\eta)^{\frac{N}{2}}(\lambda_j + (s + \frac{1}{2})\eta)^{\frac{N}{2}}}{(\lambda_j + (\frac{1}{2} - \bar{s})\eta)^{\frac{N}{2}}(\lambda_j + (\frac{1}{2} - s)\eta)^{\frac{N}{2}}} = -\frac{Q(\lambda_j + \eta)}{Q(\lambda_j - \eta)}. \tag{8.5.34}$$

After a redefinition of the parameters: $\lambda_j \to \lambda_j - \frac{\eta}{2}$, the above BAEs become

$$\frac{(\lambda_j + \bar{s}\eta)^{\frac{N}{2}}(\lambda_j + s\eta)^{\frac{N}{2}}}{(\lambda_j - \bar{s}\eta)^{\frac{N}{2}}(\lambda_j - s\eta)^{\frac{N}{2}}} = -\frac{Q(\lambda_j + \eta)}{Q(\lambda_j - \eta)}. \tag{8.5.35}$$

Then taking the limit $\Delta \to 0$, $N\Delta \to L$, the above BAEs are reduced to

$$e^{i\lambda_j L} = -\prod_{l=1}^{M} \frac{\lambda_j - \lambda_l + ig}{\lambda_j - \lambda_l - ig}, \quad j = 1, \ldots, M, \tag{8.5.36}$$

which are the same BAEs obtained via the coordinate Bethe Ansatz [42, 43].

References

1. J. Wess, B. Zumino, Consequences of anomalous Ward identities. Phys. Lett. B **37**, 95 (1971)
2. S.P. Novikov, The hamiltonian formalism and a many-valued analogue of the Morse theory. Usp. Math. Nauk. **37**, 3 (1982)
3. E. Witten, Non-abelian bosonization in two dimensions. Commun. Math. Phys. **92**, 455 (1984)
4. R. Thomale, S. Rachel, P. Schmitteckert, M. Greiter, Family of spin-S chain representations of $SU(2)_k$ Wess-Zumino-Witten models. Phys. Rev. B **85**, 195149 (2012)
5. R. Shankar, E. Witten, The S-matrix of the supersymmetric nonlinear sigma model. Phys. Rev. D **17**, 2134 (1978)
6. C. Ahn, D. Bernard, A. Leclair, Fractional supersymmetries in peturded coset CFTs and integrable soliton theory. Nucl. Phys. B **346**, 409 (1990)
7. T. Inami, S. Odake, Y.-Z. Zhang, Spersymmetric extension of the sine-Gordon theory with integrable boundary interations. Phys. Lett. B **359**, 118 (1995)
8. R.I. Nepomechie, The boundary supersymmetric sine-Gordon model revisited. Phys. Lett. B **509**, 183 (2001)
9. Z. Bajnok, L. Palla, G. Takacs, Spectrum of boundary states in $N = 1$ SUSY sine-Gordon theory. Nucl. Phys. B **644**, 509 (2002)
10. H. Frahm, M. Stahlsmeier, Spinon statistics in integrable spin-S Heisenberg chains. Phys. Lett. A **250**, 293 (1998)
11. N. Andrei, C. Destri, Solution of the multichannel Kondo problem. Phys. Rev. Lett. **52**, 364 (1984)
12. A.M. Tsvelick, P.B. Wiegmann, Solution of two-channel Kondo problem (scaling and integrability). Z. Physik B **54**, 201 (1984)
13. J. Dai, Y. Wang, U. Eckern, Ghost spins and quantum critical behavior in a spin chain with local bond deformation. Phys. Rev. B **60**, 6594 (1999)
14. A.B. Zamolodchikov, V.A. Fateev, Model factorized S matrix and an integrable Heisenberg chain with spin 1. Sov. J. Nucl. Phys. **32**, 298 (1980)

15. P.P. Kulish, E.K. Sklyanin, Quantum spectral transform method recent developments. Lect. Notes Phys. **151**, 61 (1982)
16. P.P. Kulish, N.Y. Reshetikhin, E.K. Sklyanin, Yang-Baxter equation and representation theory. I. Lett. Math. Phys. **5**, 393 (1981)
17. P.P. Kulish, N.Y. Reshetikhin, Quantum linear problem for the sine-Gordon equation and higher representation. J. Sov. Math. **23**, 2435 (1983)
18. A.N. Kirillov, N.Y. Reshetikhin, Exact solution of the Heisenberg XXZ model of spin s. J. Sov. Math. **35**, 2627 (1986)
19. A.N. Kirillov, NYu. Reshetikhin, Exact solution of the integrable XXZ Heisenberg model with arbitrary spin I. The ground state and the excitation spectrum. J. Phys. A **20**, 1565 (1987)
20. C.N. Yang, Some exact results for the many-body problem in one dimension with repulsive delta-function interaction. Phys. Rev. Lett. **19**, 1312 (1967)
21. R.J. Baxter, *Exactly Solved Models in Statistical Mechanics* (Academic Press, London, 1982)
22. H.M. Babujian, Exact solution of the isotropic Heisenberg chain with arbitrary spins: thermodynamics of the model. Nucl. Phys. B **215**, 317 (1983)
23. L.A. Takhtajan, The picture of low-lying excitations in the isotropic Heisenberg chain of arbitrary spins. Phys. Lett. A **87**, 479 (1982)
24. H.M. Babujian, Exact solution of the one dimensional isotropic Heisenberg chain with arbitrary spins s. Phys. Lett. A **90**, 479 (1982)
25. I.V. Cherednik, Factorizing particles on a half line and root sytsems. Theor. Math. Phys. **61**, 977 (1984)
26. E.K. Sklyanin, Boundary conditions for integrable quantum systems. J. Phys. A **21**, 2375 (1988)
27. H.J. de Vega, A. González Ruiz, Boundary K-matrices for the XYZ, XXZ and XXX spin chains. J. Phys. A **27**, 6129 (1994)
28. S. Ghoshal, A.B. Zamolodchikov, Boundary S-matrix and boundary state in two-dimensional integrable quantum field theory. Int. J. Mod. Phys. A **9**, 3841 (1994)
29. L. Mezincescu, R.I. Nepomechie, V. Rittenberg, Bethe Ansatz solution of the Fateev-Zamolodchikov quantum spin chain with boundary terms. Phys. Lett. A **147**, 70 (1990)
30. L. Frappat, R.I. Nepomechie, E. Ragoucy, Complete Bethe Ansatz solution of the open spin-s XXZ chain with general integrable boundary terms, J. Stat. Mech. **P09008** (2007)
31. R. Murgan, Bethe Ansatz of the open spin-s XXZ chain with nondiagonal boundary terms. J. High Energy Phys. **04**, 076 (2009)
32. R. Baiyasi, R. Murgan, Generalized T-Q relations and the open spin-s XXZ chain with nondiagonal boundary terms, J. Stat. Mech. **P10003** (2012)
33. C.S. Melo, G.A.P. Ribeiro, M.J. Martins, Bethe Ansatz for the XXX-S chain with non-diagonal open boundaries. Nucl. Phys. B **711**, 565 (2005)
34. J. Cao, S. Cui, W.L. Yang, K. Shi, Y. Wang, Exact spectrum of the spin-s Heisenberg chain with generic non-diagonal boundaries. J. High Energy Phys. **02**, 036 (2015)
35. J. Cao, W.-L. Yang, K. Shi, Y. Wang, Exact solution of the XXZ alternating spin chain with generic non-diagonal boundaries. arXiv:1409.3646
36. J. Cao, W.-L. Yang, K. Shi, Y. Wang, On the complete-spectrum characterization of quantum integrable spin chains via the inhomogeneous T-Q relation. arXiv:1409.5303
37. L. Mezincescu, R.I. Nepomechie, Fusion procedure for open chains. J. Phys. A **25**, 2533 (1992)
38. Y.-K. Zhou, Row transfer matrix functional relations for Baxter's eight-vertex and six-vertex models with open boundaries via more general reflection matrices. Nucl. Phys. B **458**, 504 (1996)
39. J. Cao, W.-L. Yang, K. Shi, Y. Wang, Off-diagonal Bethe Ansatz solution of the XXX spin chain with arbitrary boundary conditions. Nucl. Phys. B **875**, 152 (2013)
40. W.-L. Yang, R.I. Nepomechie, Y.-Z. Zhang, Q-operator and T-Q relation from the fusion hierarchy. Phys. Lett. B **633**, 664 (2006)
41. V.E. Korepin, N.M. Bogoliubov, A.G. Izergin, *Quantum Inverse Scattering Method and Correlation Function* (Cambridge University Press, Cambridge, 1993)
42. E.H. Lieb, W. Liniger, Exact analysis of an interacting Bose gas I. The general solution and the ground state. Phys. Rev. **130**, 1605 (1963)

43. E.H. Lieb, Exact analysis of an interacting Bose gas II. The excitation spectrum. Phys. Rev. **130**, 1616 (1963)
44. A.G. Izergin, V.E. Korepin, The lattice quantum sine-Gordon model. Lett. Math. Phys. **5**, 199 (1981)
45. N.M. Bogoliubov, A.G. Izergin, Lattice sine-Gordon model with local hamiltonian. Theor. Math. Phys. **61**, 1195 (1985)
46. V.V. Bazhanov, Yu G. Stroganov, Chiral Potts models as a descendant of the six-vertex models. J. Stat. Phys. **59**, 799 (1990)
47. S.N. Ruijsenaars, Relativistic Toda systems. Commun. Math. Phys. **133**, 217 (1990)
48. S. Howes, L.P. Kadanoff, M. den Nijs, Quantum model for commensurate—incommensurate transitions. Nucl. Phys. B **215**, 169 (1983)
49. G. von Gehlen, V. Rittenberg, $Z(n)$-symmetric quantum chains with an infinite set of conserved charges and $Z(n)$ zero modes. Nucl. Phys. B **257**, 351 (1985)
50. H. Au-Yang, B.M. McCoy, J.H.H. Perk, S. Tang, M.-L. Yan, Commuting transfer matrices in the chiral Potts models: solutions of star-triangle equations with genus > 1. Phys. Lett. A **123**, 219 (1987)
51. R.J. Baxter, J.H.H. Perk, H. Au-Yang, New solutions of the star-triangle relations for the chiral Potts model. Phys. Lett. A **128**, 138 (1988)

Chapter 9
The Izergin-Korepin Model

The integrable models can be classified into several series such as A_n-, B_n-, C_n- and D_n-types [1–3], associated with different Lie algebras [4]. Most of the models we studied in the previous chapters belong to the $A^{(1)}_{n-1}$-type. In fact, the ODBA method might be applied to other types of the quantum integrable models, among which the Izergin-Korepin (IK) model [5] associated with the twisted affine algebra $A^{(2)}_2$ is the simplest one. It has been realized that this model is closely related to a low-dimensional relativistic quantum field model, i.e., the Dodd-Bullough-Mikhailov or Zhiber-Mikhailov-Shabat model [6, 7]. On the other hand, with an open boundary, this model corresponds to the loop model [8] in statistical physics and describes the self-avoiding walks at a boundary [9]. The periodic IK model and the open IK chain with diagonal boundaries have been extensively studied and fruitful results have been obtained in the past several decades [10–22]. However, the exact solution of this model with generic boundaries was derived only recently via the ODBA method [23], even though the most general integrable boundary conditions (corresponding to the non-diagonal reflection matrices) were known long before [13, 19, 21].

In this chapter, by employing the IK model with generic integrable boundaries as an example, we show how the ODBA method can also be applied to the quantum integrable models beyond the A-type. This example, together with the nested ODBA introduced in Chap. 7 and the hierarchical ODBA introduced in Chap. 8, might form a systematic method to approach the eigenvalue problem of general quantum integrable models.

9.1 The Model with Generic Open Boundaries

Throughout this chapter, $\mathbf{V}$ denotes a three-dimensional linear space with the basis $\{|i\rangle | i = 1, 2, 3\}$. The R-matrix of the IK model [5] in terms of this basis is expressed as

Y. Wang et al., *Off-Diagonal Bethe Ansatz for Exactly Solvable Models*,
DOI 10.1007/978-3-662-46756-5_9

$$R(u)=\left(\begin{array}{ccc|ccc|ccc}
h_3(u) & & & & & & & & \\
 & h_2(u) & & e(u) & & & & & \\
 & & h_4(u) & & g(u) & & f(u) & & \\
\hline
 & \bar{e}(u) & & h_2(u) & & & & & \\
 & & \bar{g}(u) & & h_1(u) & & g(u) & & \\
 & & & & & h_2(u) & & e(u) & \\
\hline
 & & \bar{f}(u) & & \bar{g}(u) & & h_4(u) & & \\
 & & & & & \bar{e}(u) & & h_2(u) & \\
 & & & & & & & & h_3(u)
\end{array}\right), \tag{9.1.1}$$

with the non-zero entries

$$\begin{aligned}
h_1(u) &= \sinh(u-3\eta)-\sinh 5\eta+\sinh 3\eta+\sinh\eta,\\
h_2(u) &= \sinh(u-3\eta)+\sinh 3\eta,\\
h_3(u) &= \sinh(u-5\eta)+\sinh\eta, \quad h_4(u)=\sinh(u-\eta)+\sinh\eta,\\
e(u) &= -2e^{-\frac{u}{2}}\sinh 2\eta\cosh\left(\frac{u}{2}-3\eta\right), \quad \bar{e}(u)=-2e^{\frac{u}{2}}\sinh 2\eta\cosh\left(\frac{u}{2}-3\eta\right),\\
f(u) &= -2e^{-u+2\eta}\sinh\eta\sinh 2\eta-e^{-\eta}\sinh 4\eta,\\
\bar{f}(u) &= 2e^{u-2\eta}\sinh\eta\sinh 2\eta-e^{\eta}\sinh 4\eta,\\
g(u) &= 2e^{-\frac{u}{2}+2\eta}\sinh\frac{u}{2}\sinh 2\eta, \quad \bar{g}(u)=-2e^{\frac{u}{2}-2\eta}\sinh\frac{u}{2}\sinh 2\eta.
\end{aligned} \tag{9.1.2}$$

The R-matrix satisfies YBE and possesses the following properties:

$$\text{Initial condition: } R_{1,2}(0)=(\sinh\eta-\sinh 5\eta)P_{1,2}, \tag{9.1.3}$$

$$\text{Unitary relation: } R_{1,2}(u)R_{2,1}(-u)=\varphi_1(u)\times \mathrm{id}, \tag{9.1.4}$$

$$\text{Crossing relation: } R_{1,2}(u)=V_1R_{1,2}^{t_2}(-u+6\eta+i\pi)V_1^{-1}, \tag{9.1.5}$$

$$\text{PT-symmetry: } R_{2,1}(u)=R_{1,2}^{t_1t_2}(u), \tag{9.1.6}$$

$$\text{Periodicity: } R_{1,2}(u+2i\pi)=R_{1,2}(u), \tag{9.1.7}$$

where $R_{2,1}(u)=P_{1,2}R_{1,2}(u)P_{1,2}$, and the function $\varphi_1(u)$ and the crossing matrix V are given by

$$\begin{aligned}
\varphi_1(u) = &-4\sinh\left(\frac{u}{2}-2\eta\right)\sinh\left(\frac{u}{2}+2\eta\right)\\
&\times\cosh\left(\frac{u}{2}-3\eta\right)\cosh\left(\frac{u}{2}+3\eta\right),
\end{aligned} \tag{9.1.8}$$

$$V=\begin{pmatrix} & & -e^{-\eta}\\ & 1 & \\ -e^{\eta} & & \end{pmatrix}, \quad V^2=1. \tag{9.1.9}$$

From the relations (9.1.4), (9.1.5) and (9.1.8), (9.1.9) we can deduce that the R-matrix possesses the crossing-unitary property

$$R_{1,2}^{t_1}(u)\,\mathscr{M}_1\,R_{2,1}^{t_1}(-u+12\eta)\,\mathscr{M}_1^{-1}=\varphi_2(u)\times\mathrm{id},\tag{9.1.10}$$

$$\mathscr{M}=V^t\,V=\begin{pmatrix}e^{2\eta}&&\\&1&\\&&e^{-2\eta}\end{pmatrix},\tag{9.1.11}$$

$$\varphi_2(u)=-4\cosh\left(\frac{u}{2}-5\eta\right)\cosh\left(\frac{u}{2}-\eta\right)\sinh\frac{u}{2}\sinh\left(\frac{u}{2}-6\eta\right),\tag{9.1.12}$$

and the following useful relations

$$\begin{aligned}&\mathscr{M}_1\,\mathscr{M}_2\,R_{1,2}(u)\,\mathscr{M}_1^{-1}\,\mathscr{M}_2^{-1}=R_{1,2}(u),\\&R_{1,2}(u)\,R_{2,1}(v)=R_{2,1}(v)\,R_{1,2}(u).\end{aligned}\tag{9.1.13}$$

The one-row monodromy matrices $T(u)$ and $\hat{T}(u)$ are still defined as usual, as 3×3 matrices with operator-valued elements acting on $\mathbf{V}^{\otimes N}$:

$$T_0(u)=R_{0,N}(u-\theta_N)R_{0,N-1}(u-\theta_{N-1})\cdots R_{0,1}(u-\theta_1),\tag{9.1.14}$$

$$\hat{T}_0(u)=R_{1,0}(u+\theta_1)R_{2,0}(u+\theta_2)\cdots R_{N,0}(u+\theta_N).\tag{9.1.15}$$

We consider the type II (classified in [19, 21]) generic non-diagonal K-matrices $K^-(u)$ and $K^+(u)$ found in [13]

$$\begin{aligned}K^-(u)&=\begin{pmatrix}1+2e^{-u-\varepsilon}\sinh\eta&0&2e^{-\varepsilon+\varsigma}\sinh u\\0&1-2e^{-\varepsilon}\sinh(u-\eta)&0\\2e^{-\varepsilon-\varsigma}\sinh u&0&1+2e^{u-\varepsilon}\sinh\eta\end{pmatrix},\\K^+(u)&=\mathscr{M}K^-(-u+6\eta+i\pi)\big|_{(\varepsilon,\varsigma)\to(\varepsilon',\varsigma')}.\end{aligned}\tag{9.1.16}$$

The former satisfies RE

$$\begin{aligned}&R_{1,2}(u-v)K_1^-(u)R_{2,1}(u+v)K_2^-(v)\\&\quad=K_2^-(v)R_{1,2}(u+v)K_1^-(u)R_{2,1}(u-v),\end{aligned}\tag{9.1.17}$$

and the latter satisfies the dual RE

$$\begin{aligned}&R_{1,2}(v-u)K_1^+(u)\mathscr{M}_1^{-1}R_{2,1}(-u-v+12\eta)\mathscr{M}_1K_2^+(v)\\&\quad=K_2^+(v)\mathscr{M}_2^{-1}R_{1,2}(-u-v+12\eta)\mathscr{M}_2K_1^+(u)R_{2,1}(v-u).\end{aligned}\tag{9.1.18}$$

Apart from the crossing parameter η, the corresponding transfer matrix $t(u)$ given by $t(u)=tr_0\{K_0^+(u)T_0(u)K_0^-(u)\hat{T}_0(u)\}$ has four other free parameters $\{\varepsilon,\ \varsigma,\ \varepsilon',\ \varsigma'\}$ describing the boundary fields.

The IK model Hamiltonian with generic integrable boundaries is thus defined in terms of the transfer matrix as

$$
\begin{aligned}
H &= \left.\frac{\partial \ln t(u)}{\partial u}\right|_{u=0,\{\theta_j=0\}} \\
&= \sum_{j=1}^{N-1} \frac{2P_{j,j+1}R'_{j,j+1}(0)}{\sinh\eta - \sinh 5\eta} + \frac{tr K^{+\prime}(0)}{tr K^{+}(0)} + \frac{2tr_0[K_0^{+}(0)P_{N,0}R'_{N,0}(0)]}{(\sinh\eta - \sinh 5\eta)tr K^{+}(0)} \\
&\quad + \frac{K_1^{-\prime}(0)}{1+2e^{-\varepsilon}\sinh\eta} \\
&= \frac{2}{\sinh\eta - \sinh 5\eta}\sum_{j=1}^{N-1}\Big\{\cosh 5\eta(E_j^{11}E_{j+1}^{11} + E_j^{33}E_{j+1}^{33}) + \sinh 2\eta \\
&\quad \times(\sinh 3\eta - \cosh 3\eta)(E_j^{11}E_{j+1}^{22} + E_j^{22}E_{j+1}^{33}) + (\sinh 3\eta + \cosh 3\eta) \\
&\quad \times \sinh 2\eta(E_j^{22}E_{j+1}^{11} + E_j^{33}E_{j+1}^{22}) + \cosh\eta(E_j^{13}E_{j+1}^{31} + E_j^{31}E_{j+1}^{13}) \\
&\quad +2\sinh\eta\sinh 2\eta(e^{-2\eta}E_j^{11}E_{j+1}^{33} + e^{2\eta}E_j^{33}E_{j+1}^{11}) - e^{-2\eta}\sinh 2\eta \\
&\quad \times(E_j^{12}E_{j+1}^{32} + E_j^{21}E_{j+1}^{23}) + e^{2\eta}\sinh 2\eta(E_j^{23}E_{j+1}^{21} + E_j^{32}E_{j+1}^{12}) \\
&\quad +\cosh 3\eta(E_j^{12}E_{j+1}^{21} + E_j^{21}E_{j+1}^{12} + E_j^{22}E_{j+1}^{22} + E_j^{23}E_{j+1}^{32} + E_j^{32}E_{j+1}^{23})\Big\} \\
&\quad -2e^{-\varepsilon}[1+2e^{-\varepsilon}\sinh\eta]^{-1}\Big[\sinh\eta(E_1^{11} - E_1^{33}) + \cosh\eta E_1^{22} - e^{\varsigma}E_1^{13} \\
&\quad -e^{-\varsigma}E_1^{31}\Big] + 2\Big[(\sinh\eta - \sinh 5\eta)(2e^{-\varepsilon'}\sinh 5\eta - 4e^{-\varepsilon'}\sinh\eta \\
&\quad \times\cosh 4\eta + 1 + 2\cosh 2\eta)\Big]^{-1}\Big\{\Big[(e^{2\eta} - 2e^{-4\eta-\varepsilon'}\sinh\eta)\cosh 5\eta \\
&\quad +(1+2e^{-\varepsilon'}\sinh 5\eta)\sinh 2\eta(\sinh 3\eta - \cosh 3\eta) + 2(e^{-4\eta} - 2e^{2\eta-\varepsilon'} \\
&\quad \times\sinh\eta)\sinh\eta\sinh 2\eta\Big]E_N^{11} + \Big[(e^{2\eta} - 2e^{-4\eta-\varepsilon'}\sinh\eta)\sinh 2\eta \\
&\quad \times(\sinh 3\eta + \cosh 3\eta) + (1+2e^{-\varepsilon'}\sinh 5\eta)\cosh 3\eta \\
&\quad +(e^{-2\eta} - 2e^{4\eta-\varepsilon'}\sinh\eta)\sinh 2\eta(\sinh 3\eta - \cosh 3\eta)\Big]E_N^{22} \\
&\quad +\Big[2(e^{4\eta} - 2e^{-2\eta-\varepsilon'}\sinh\eta)\sinh\eta\sinh 2\eta + (1+2e^{-\varepsilon'}\sinh 5\eta) \\
&\quad \times\sinh 2\eta(\sinh 3\eta + \cosh 3\eta) + (e^{-2\eta} - 2e^{4\eta-\varepsilon'}\sinh\eta)\cosh 5\eta\Big] \\
&\quad \times E_N^{33} - 2e^{-\varepsilon'}\sinh 6\eta\cosh\eta[e^{2\eta+\varsigma'}E_N^{13} + e^{-2\eta-\varsigma'}E_N^{31}]\Big\} \\
&\quad +\frac{2e^{-\varepsilon'}(2\sinh\eta\sinh 4\eta - \cosh 5\eta)}{2\cosh 2\eta - 4e^{-\varepsilon'}\sinh\eta\cosh 4\eta + 1 + 2e^{-\varepsilon'}\sinh 5\eta},
\end{aligned}
\tag{9.1.19}
$$

where $E_j^{\mu\nu}$ ($\mu, \nu = 1, 2, 3$) is the Weyl matrix or the Hubbard operator in Dirac notations

$$E^{\mu\nu} = |\mu\rangle\langle\nu|,$$

and

$$R'_{i,j}(u) = \frac{\partial}{\partial u} R_{i,j}(u), \quad K_j^{\pm\prime}(u) = \frac{\partial}{\partial u} K_j^{\pm}(u). \tag{9.1.20}$$

The Hamiltonian (9.1.19) is hermitian for real η, ε, ε' and imaginary ς and $\varsigma' + 2\eta$.

9.2 Operator Product Identities

To solve the eigenvalue problem of the transfer matrix, we adopt the fusion techniques [24–29] for both R-matrices [30] and K-matrices [16, 31] to derive the desired operator identities determining the spectrum of the transfer matrix $t(u)$. Let us introduce the following vectors in the tensor space $\mathbf{V} \otimes \mathbf{V}$:

$$|\chi_0\rangle = \frac{1}{\sqrt{2\cosh 2\eta + 1}} \left(e^{-\eta}|1\rangle \otimes |3\rangle - |2\rangle \otimes |2\rangle + e^{\eta}|3\rangle \otimes |1\rangle\right), \tag{9.2.1}$$

$$|\chi_1\rangle = \frac{1}{\sqrt{2\cosh 2\eta}} \left(e^{-\eta}|1\rangle \otimes |2\rangle - e^{\eta}|2\rangle \otimes |1\rangle\right), \tag{9.2.2}$$

$$|\chi_2\rangle = \frac{1}{\sqrt{2\cosh 2\eta}} \left(|1\rangle \otimes |3\rangle - 2\sinh\eta|2\rangle \otimes |2\rangle - |3\rangle \otimes |1\rangle\right), \tag{9.2.3}$$

$$|\chi_3\rangle = \frac{1}{\sqrt{2\cosh 2\eta}} \left(e^{-\eta}|2\rangle \otimes |3\rangle - e^{\eta}|3\rangle \otimes |2\rangle\right), \tag{9.2.4}$$

and the associated projectors

$$P_{1,2}^{(1)} = |\chi_0\rangle\langle\chi_0|, \quad P_{1,2}^{(3)} = \sum_{i=1}^{3} |\chi_i\rangle\langle\chi_i|. \tag{9.2.5}$$

The projectors satisfy the properties

$$\{P_{1,2}^{(i)}\}^2 = P_{1,2}^{(i)}, \quad i = 1, 3. \tag{9.2.6}$$

Direct calculation shows that the R-matrix at some degenerate points is proportional to the projectors,

$$R_{1,2}(6\eta + i\pi) = -2\cosh 3\eta \sinh 2\eta (2\cosh 2\eta + 1)\, P_{1,2}^{(1)} \times \gamma_{1,2}^{(1)}, \tag{9.2.7}$$

$$R_{1,2}(4\eta) = 2\cosh\eta \sinh 4\eta\, P_{1,2}^{(3)} \times \gamma_{1,2}^{(3)}, \tag{9.2.8}$$

where $\gamma_{1,2}^{(i)}$ are some non-degenerate diagonal matrices $\in \mathrm{End}(\mathbf{V}\otimes\mathbf{V})$

$$\gamma_{1,2}^{(1)} = \mathrm{Diag}\{1, 1, e^{2\eta}, 1, 1, 1, e^{-2\eta}, 1, 1\},$$
$$\gamma_{1,2}^{(3)} = \mathrm{Diag}\{1, e^{2\eta}, 1, e^{-2\eta}, -1, e^{2\eta}, 1, e^{-2\eta}, 1\}.$$

Using the same procedure from (2.4.19) to (2.4.21) in Chap. 2, we can show that

$$P_{1,2}^{(1)} R_{2,3}(u)\, R_{1,3}(u+6\eta+i\pi)\, P_{1,2}^{(1)} = \varphi_1(u) P_{1,2}^{(1)} \otimes \mathrm{id}, \tag{9.2.9}$$
$$P_{1,2}^{(1)} R_{3,1}(u)\, R_{3,2}(u+6\eta+i\pi)\, P_{1,2}^{(1)} = \varphi_1(u) P_{1,2}^{(1)} \otimes \mathrm{id}, \tag{9.2.10}$$
$$P_{1,2}^{(3)} R_{2,3}(u)\, R_{1,3}(u+4\eta)\, P_{1,2}^{(3)} = \varphi_3(u)\, R_{[1,2],3}(u+2\eta+i\pi), \tag{9.2.11}$$
$$P_{1,2}^{(3)} R_{3,1}(u)\, R_{3,2}(u+4\eta)\, P_{1,2}^{(3)} = \varphi_3(u)\, R_{3,[1,2]}(u+2\eta+i\pi), \tag{9.2.12}$$

where the function $\varphi_3(u)$ is

$$\varphi_3(u) = -2\sinh\left(\frac{u}{2}+2\eta\right)\cosh\left(\frac{u}{2}-3\eta\right). \tag{9.2.13}$$

Noting that the fused space $\mathbf{V}_{[1,2]}$ is defined by

$$\mathbf{V}_{[1,2]} = P_{1,2}^{(3)}\mathbf{V}\otimes\mathbf{V}, \tag{9.2.14}$$

and its dimension is three, after the following isomorphism: $C : \mathbf{V} \to \mathbf{V}_{[1,2]}$

$$C|i\rangle = |\chi_i\rangle, \quad i = 1, 2, 3, \tag{9.2.15}$$

we obtain that the fused R-matrix $R_{[1,2],3}(u)$ is isomorphic to $R_{1,3}(u)$ by direct calculation, i.e.,

$$C^{-1}\, R_{[1,2],3}(u)\, C \overset{\mathbf{V}_{[1,2]}\to\mathbf{V}_1}{=} R_{1,3}(u). \tag{9.2.16}$$

Then the fusion relations (9.2.9)–(9.2.12) allow us to derive the following results:

$$P_{1,2}^{(1)} T_2(u)\, T_1(u+6\eta+i\pi) P_{1,2}^{(1)} = \prod_{l=1}^{N} \varphi_1(u-\theta_l)\, P_{1,2}^{(1)} \otimes \mathrm{id}, \tag{9.2.17}$$

$$P_{1,2}^{(1)} \hat{T}_1(u)\, \hat{T}_2(u+6\eta+i\pi) P_{1,2}^{(1)} = \prod_{l=1}^{N} \varphi_1(u+\theta_l)\, P_{1,2}^{(1)} \otimes \mathrm{id}, \tag{9.2.18}$$

$$P_{1,2}^{(3)} T_2(u)\, T_1(u+4\eta) P_{1,2}^{(3)} = \prod_{l=1}^{N} \varphi_3(u-\theta_l)\, T_{[1,2]}(u+2\eta+i\pi), \tag{9.2.19}$$

$$P_{1,2}^{(3)} \hat{T}_1(u)\, \hat{T}_2(u+4\eta) P_{1,2}^{(3)} = \prod_{l=1}^{N} \varphi_3(u+\theta_l) \hat{T}_{[1,2]}(u+2\eta+i\pi). \tag{9.2.20}$$

We note that in contrast to most of the rational integrable models, here $P^{(i)}_{2,1} = P_{1,2}P^{(i)}_{1,2}P_{1,2} \neq P^{(i)}_{1,2}$.

Similarly, after some tedious calculation we can obtain that

$$\begin{aligned}
&P^{(1)}_{2,1}K^-_1(u)R_{2,1}(2u+6\eta+i\pi)K^-_2(u+6\eta+i\pi)P^{(1)}_{1,2}\\
&\qquad = \mathrm{Det}_q\{K^-(u)\}P^{(1)}_{2,1}\,P_{1,2}, && (9.2.21)\\
&P^{(1)}_{1,2}K^+_2(u+6\eta+i\pi)\mathcal{M}_1R_{1,2}(-2u+6\eta+i\pi)\mathcal{M}_1^{-1}K^+_1(u)P^{(1)}_{2,1}\\
&\qquad = \mathrm{Det}_q\{K^+(u)\}P^{(1)}_{1,2}\,P_{1,2}, && (9.2.22)\\
&P_{1,2}\,P^{(3)}_{2,1}K^-_1(u)R_{2,1}(2u+4\eta)K^-_2(u+4\eta)P^{(3)}_{1,2}\\
&\qquad = f_-(u)\,K^-_{[1,2]}(u+2\eta+i\pi), && (9.2.23)\\
&P^{(3)}_{1,2}K^+_2(u+4\eta)\mathcal{M}_1R_{1,2}(-2u+8\eta)\mathcal{M}_1^{-1}K^+_1(u)P^{(3)}_{2,1}\,P_{1,2}\\
&\qquad = f_+(u)\,K^+_{[1,2]}(u+2\eta+i\pi), && (9.2.24)
\end{aligned}$$

where the functions $f_\pm(u)$ and $\mathrm{Det}_q\{K^\pm(u)\}$ are

$$\begin{aligned}
&f_-(u) = -2(1-2e^{-\varepsilon}\sinh(u-\eta))\cosh(u-\eta)\sinh(u+4\eta), && (9.2.25)\\
&f_+(u) = 2(1-2e^{-\varepsilon'}\sinh(u-\eta))\cosh(u-\eta)\sinh(u-6\eta), && (9.2.26)\\
&\mathrm{Det}_q\{K^-(u)\} = tr_{1,2}\left\{P^{(1)}_{2,1}K^-_1(u)R_{2,1}(2u+6\eta+i\pi)\right.\\
&\qquad \left.\times K^-_2(u+6\eta+i\pi)P^{(1)}_{1,2}P_{1,2}\right\} = -2(1-2e^{-\varepsilon}\sinh(u-\eta))\\
&\qquad \times(1+2e^{-\varepsilon}\sinh(u+\eta))\sinh(u+6\eta)\cosh(u+\eta), && (9.2.27)\\
&\mathrm{Det}_q\{K^+(u)\} = tr_{1,2}\left\{P^{(1)}_{1,2}K^+_2(u+6\eta+i\pi)\mathcal{M}_1R_{1,2}(-2u+6\eta+i\pi)\right.\\
&\qquad \left.\times\mathcal{M}_1^{-1}K^+_1(u)P^{(1)}_{2,1}P_{1,2}\right\} = 2(1-2e^{-\varepsilon'}\sinh(u-\eta))\\
&\qquad \times(1+2e^{-\varepsilon'}\sinh(u+\eta))\sinh(u-6\eta)\cosh(u-\eta). && (9.2.28)
\end{aligned}$$

By using the relations (9.2.17)–(9.2.24) and the fusion procedure introduced in Chap. 7, it is found that the transfer matrix satisfies the operator identities

$$t(\pm\theta_j)t(\pm\theta_j+6\eta+i\pi) = \left.\frac{\delta_1(u)\times \mathrm{id}}{\varphi_1(2u)}\right|_{u=\pm\theta_j}, \qquad (9.2.29)$$

$$t(\pm\theta_j)t(\pm\theta_j+4\eta) = \left.\frac{\delta_2(u)\times t(u+2\eta+i\pi)}{\varphi_2(-2u+8\eta)}\right|_{u=\pm\theta_j}, \qquad (9.2.30)$$

where $j = 1, \ldots, N$ and the functions $\delta_1(u)$ and $\delta_2(u)$ are

$$\delta_1(u) = \mathrm{Det}_q\{K^-(u)\}\mathrm{Det}_q\{K^+(u)\}\prod_{l=1}^{N}\varphi_1(u-\theta_l)\varphi_1(u+\theta_l), \quad (9.2.31)$$

$$\delta_2(u) = f_-(u)f_+(u)\prod_{l=1}^{N}\varphi_3(u-\theta_l)\varphi_3(u+\theta_l). \quad (9.2.32)$$

9.3 Crossing Symmetry and Asymptotic Behavior

As is the case for most of the integrable open boundary models, the transfer matrix of the present model possesses the crossing symmetry

$$t(u) = t(-u+6\eta+i\pi). \quad (9.3.1)$$

It is easy to show that the following relations hold:

$$K^-(u)K^-(-u) = \Delta_-(u) \times \mathrm{id}, \quad (9.3.2)$$
$$V^t\, K^+(-u+6\eta+i\pi)\, V V^t\, K^+(u+6\eta+i\pi)V = \Delta_+(u) \times \mathrm{id}, \quad (9.3.3)$$

where $\Delta_\pm(u)$ are

$$\Delta_-(u) = (1-2e^{-\varepsilon}\sinh(u-\eta))(1+2e^{-\varepsilon}\sinh(u+\eta)), \quad (9.3.4)$$
$$\Delta_+(u) = (1-2e^{-\varepsilon'}\sinh(u-\eta))(1+2e^{-\varepsilon'}\sinh(u+\eta)). \quad (9.3.5)$$

For convenience, we introduce the matrices

$$\bar{K}_1^-(u) = tr_2\left\{P_{1,2}\, R_{2,1}(-2u)\, V_2\, K_2^{-t_2}(u+6\eta+i\pi)\, V_2^{t_2}\right\}, \quad (9.3.6)$$
$$\bar{K}_1^+(u+6\eta+i\pi) = tr_2\left\{P_{1,2}\, R_{1,2}(2u)\, K_2^{+t_2}(u)\right\}. \quad (9.3.7)$$

Substituting the expressions (9.1.16) into the above equations, we find that the K-matrices satisfy the crossing relations

$$\bar{K}^-(u) = \frac{\mathrm{Det}_q\{K^-(u)\}}{\Delta_-(u)}K^-(-u), \quad (9.3.8)$$

$$\bar{K}^+(u+6\eta+i\pi) = \frac{\mathrm{Det}_q\{K^+(u)\}}{\Delta_+(u)}V^t\, K^+(-u+6\eta+i\pi)\, V. \quad (9.3.9)$$

Note that the crossing relation (9.1.5) and PT-symmetry (9.1.6) allow us to derive

$$R_{2,1}(u) = V_1^{t_1}\, R_{1,2}^{t_1}(-u+6\eta+i\pi)\, V_1^{t_1}, \tag{9.3.10}$$

$$R_{2,1}(u) = V_2\, R_{1,2}^{t_2}(-u+6\eta+i\pi)\, V_2. \tag{9.3.11}$$

These relations together with (9.1.5) and (9.1.9) give rise to the following dualities between $T(u)$ and $\hat{T}(u)$:

$$T_0^{t_0}(-u+6\eta+i\pi) = V_0^{t_0}\, \hat{T}_0(u)\, V_0^{t_0}, \tag{9.3.12}$$

$$\hat{T}_0^{t_0}(-u+6\eta+i\pi) = V_0\, T_0(u)\, V_0. \tag{9.3.13}$$

With a procedure similar to that introduced in Chap. 2, we conclude that the crossing symmetry (9.3.1) holds.

In addition, since

$$K^{\pm}(u+2i\pi) - K^{\pm}(u),$$

the following periodicity of the transfer matrix $t(u)$ holds:

$$t(u+2i\pi) = t(u). \tag{9.3.14}$$

Moreover, from the definition of the transfer matrix we can easily deduce the asymptotic behavior

$$\lim_{u\to\pm\infty} t(u) = \left(\frac{1}{2}\right)^{2N} e^{\pm 2(N+1)(u-3\eta)-\varepsilon-\varepsilon'} \times[1+2\cosh(\varsigma'-\varsigma+2\eta)]\times \mathrm{id}+\cdots. \tag{9.3.15}$$

Note that

$$K^-(0) = (1+2e^{-\varepsilon}\sinh\eta)\times \mathrm{id},$$

$$K^+(6\eta+i\pi) = (1+2e^{-\varepsilon'}\sinh\eta)\mathscr{M}, \tag{9.3.16}$$

$$K^-(i\pi) = (1-2e^{-\varepsilon}\sinh\eta)\times \mathrm{id},$$

$$K^+(6\eta) = (1-2e^{-\varepsilon'}\sinh\eta)\mathscr{M}, \tag{9.3.17}$$

From the unitary condition (9.1.4) and the crossing unitary condition (9.1.10) we deduce the relations

$$T_0(u)\hat{T}_0(-u) = \prod_{l-1}^{N}\varphi_1(u-\theta_l)\times \mathrm{id}, \tag{9.3.18}$$

$$T_0^{t_0}(u)\mathscr{M}_0\hat{T}_0^{t_0}(-u+12\eta)\mathscr{M}_0^{-1} = \prod_{l=1}^{N}\varphi_2(u-\theta_l)\times \mathrm{id}, \tag{9.3.19}$$

which imply the identities

$$t(0) = t(6\eta + i\pi) = (1 + 2e^{-\varepsilon} \sinh \eta) tr\{K^+(0)\} \prod_{l=1}^{N} \varphi_1(-\theta_l) \times \text{id}, \quad (9.3.20)$$

$$t(i\pi) = t(6\eta) = (1 - 2e^{-\varepsilon} \sinh \eta) tr\{K^+(i\pi)\} \prod_{l=1}^{N} \varphi_1(i\pi - \theta_l) \times \text{id}. \quad (9.3.21)$$

9.4 The Inhomogeneous $T - Q$ Relation

As we introduced in previous chapters, the eigenvalue $\Lambda(u)$ of the transfer matrix should satisfy the same relations held by $t(u)$. The operator identities (9.2.29) and (9.2.30) indicate that

$$\Lambda(\theta_j)\Lambda(\theta_j + 6\eta + i\pi) = \left.\frac{\delta_1(u)}{\varphi_1(2u)}\right|_{u=\theta_j}, \quad (9.4.1)$$

$$\Lambda(\theta_j)\Lambda(\theta_j + 4\eta) = \left.\frac{\delta_2(u) \times \Lambda(u + 2\eta + i\pi)}{\varphi_2(-2u + 8\eta)}\right|_{u=\theta_j}, \quad (9.4.2)$$

where $j = 1, \ldots, N$ and the functions $\delta_i(u)$ are given by (9.2.31) and (9.2.32). Equation (9.3.1) ensures the crossing symmetry

$$\Lambda(u) = \Lambda(-u + 6\eta + i\pi), \quad (9.4.3)$$

while (9.3.14), (9.3.15) and (9.3.20), (9.3.21) give rise to

$$\Lambda(u + 2i\pi) = \Lambda(u), \quad (9.4.4)$$

$$\Lambda(0) = (1 + 2e^{-\varepsilon} \sinh \eta) tr\{K^+(0)\} \prod_{l=1}^{N} \varphi_1(-\theta_l), \quad (9.4.5)$$

$$\Lambda(i\pi) = (1 - 2e^{-\varepsilon} \sinh \eta) tr\{K^+(i\pi)\} \prod_{l=1}^{N} \varphi_1(i\pi - \theta_l), \quad (9.4.6)$$

$$\lim_{u \to \pm\infty} \Lambda(u) = \left(\frac{1}{2}\right)^{2N} e^{\pm 2(N+1)(u-3\eta) - \varepsilon - \varepsilon'} \times [1 + 2\cosh(\varsigma' - \varsigma + 2\eta)] + \cdots. \quad (9.4.7)$$

By a simple analysis we conclude that

$$\Lambda(u), \text{ as an entire function of } \tfrac{u}{2}, \text{ is a trigonometric polynomial of degree } 4N + 4. \quad (9.4.8)$$

The relations (9.4.1)–(9.4.8) provide $4N + 5$ conditions to determine the $4N + 5$ coefficients of the $\Lambda(u)$ polynomial.

Let us introduce the $T - Q$ relation

$$\Lambda(u) = a(u)\frac{Q_1(u+4\eta)}{Q_2(u)} + d(u)\frac{Q_2(u-6\eta+i\pi)}{Q_1(u-2\eta+i\pi)}$$
$$+b(u)\frac{Q_1(u+2\eta+i\pi)Q_2(u-4\eta)}{Q_2(u-2\eta+i\pi)Q_1(u)} + \frac{1}{\cosh(u-3\eta)}\left[\frac{c(u)Q_1(u+2\eta+i\pi)}{Q_1(u)Q_2(u)}\right.$$
$$\left.-\frac{c(-u+6\eta+i\pi)Q_2(u-4\eta)}{Q_1(u-2\eta+i\pi)Q_2(u-2\eta+i\pi)}\right], \tag{9.4.9}$$

where the functions $a(u)$, $b(u)$, $c(u)$, $d(u)$ and $Q_i(u)$ are given by

$$a(u) = \prod_{l=1}^{N} h_3(u-\theta_l)h_3(u+\theta_l)(1-2e^{-\varepsilon}\sinh(u-\eta))(1-2e^{-\varepsilon'}\sinh(u-\eta))$$
$$\times\frac{\sinh(u-6\eta)\cosh(u-\eta)}{\sinh(u-2\eta)\cosh(u-3\eta)}, \tag{9.4.10}$$

$$d(u) = \prod_{l=1}^{N} h_4(u-\theta_l)h_4(u+\theta_l)(1-2e^{-\varepsilon}\sinh(u-5\eta))(1-2e^{-\varepsilon'}\sinh(u-5\eta))$$
$$\times\frac{\sinh u\cosh(u-5\eta)}{\sinh(u-4\eta)\cosh(u-3\eta)}, \tag{9.4.11}$$

$$b(u) = \prod_{l=1}^{N} h_2(u-\theta_l)h_2(u+\theta_l)(1+2e^{-\varepsilon}\sinh(u-3\eta))(1+2e^{-\varepsilon'}\sinh(u-3\eta))$$
$$\times\frac{\sinh u\sinh(u-6\eta)}{\sinh(u-2\eta)\sinh(u-4\eta)}, \tag{9.4.12}$$

$$c(u) = c_0(u)4^{1-N}\sinh u\sinh(u-6\eta)$$
$$\times\prod_{l=1}^{N} h_3(u-\theta_l)h_3(u+\theta_l)h_4(u-\theta_l)h_4(u+\theta_l), \tag{9.4.13}$$

$$Q_1(u) = \prod_{k=1}^{\bar{N}}\sinh\left[\frac{u-\lambda_k}{2}-\eta\right], \quad \bar{N} = 4N+2(n-1), \tag{9.4.14}$$

$$Q_2(u) = \prod_{k=1}^{N}\sinh\left[\frac{u+\lambda_k}{2}-\eta\right]. \tag{9.4.15}$$

Here $c_0(u)$ is a trigonometric polynomial with a degree n obeying the relation $c_0(u) = c_0(-u+4\eta)$ to match the asymptotic behavior (9.4.7). We choose the minimal $T - Q$ relation with $n = 0$ and

$$c_0(u) = c_0 = -2\frac{\cosh(\varsigma' - \varsigma + 2\eta) - \cosh\left(\bar{N}\eta - \sum_{j=1}^{\bar{N}}\lambda_j\right)}{e^{\varepsilon+\varepsilon'}\cosh\left(\frac{\bar{N}\eta}{2} - \frac{1}{2}\sum_{j=1}^{\bar{N}}\lambda_j\right)}, \quad (9.4.16)$$

$$\bar{N} = 4N - 2. \quad (9.4.17)$$

Note that the functions $Q_i(u)$ possess the periodic and crossing properties

$$Q_i(u + 2i\pi) = Q_i(u), \quad i = 1, 2, \qquad Q_2(u) = Q_1(-u + 4\eta). \quad (9.4.18)$$

Indeed, the $T - Q$ Ansatz (9.4.9) satisfies the relations (9.4.3)–(9.4.6). In addition, since

$$\begin{aligned} h_2(0) = h_4(0) &= h_3(6\eta + i\pi) = h_2(6\eta + i\pi) \\ &= h_3(4\eta) = h_4(2\eta + i\pi) = 0, \end{aligned} \quad (9.4.19)$$

the $T - Q$ Ansatz (9.4.9) also satisfies the functional relations (9.4.1) and (9.4.2).

In the $T - Q$ Ansatz (9.4.9), there exist some readily apparent and simple poles:

$$2\eta, \ 4\eta, \ 3\eta + i\frac{\pi}{2}, \quad \mathrm{mod}(i\pi), \quad (9.4.20)$$

and

$$\lambda_j + 2\eta, \ -\lambda_j + 2\eta, \ \lambda_j + 4\eta + i\pi, \quad -\lambda_j + 4\eta + i\pi, \ \mathrm{mod}(2i\pi), \quad (9.4.21)$$

where $j = 1, \ldots, \bar{N}$. The regularity of the $T - Q$ Ansatz at these poles leads to the corresponding BAEs of the present model

$$\begin{aligned} &\frac{(1 + 2e^{-\varepsilon}\sinh(\lambda_j - \eta))(1 + 2e^{-\varepsilon'}\sinh(\lambda_j - \eta))\cosh(\lambda_j - \eta)}{4\sinh\lambda_j\sinh(\lambda_j - 2\eta)} \\ &= -\prod_{l=1}^{N}\sinh\left[\frac{\lambda_j - \theta_l}{2} - \eta\right]\sinh\left[\frac{\lambda_j + \theta_l}{2} - \eta\right]\cosh\left[\frac{\lambda_j - \theta_l}{2}\right]\cosh\left[\frac{\lambda_j + \theta_l}{2}\right] \\ &\times\frac{c_0 Q_2(\lambda_j + i\pi)}{Q_2(\lambda_j - 2\eta)Q_2(\lambda_j + 2\eta)}, \quad j = 1, \ldots, \bar{N}. \end{aligned} \quad (9.4.22)$$

The above BAEs ensure that the $\Lambda(u)$ given by (9.4.9) is an eigenvalue of the transfer matrix.

Apart from the above solutions, there is another special solution $\bar{N} = 0$ and

$$\Lambda(u) = a(u) + b(u) + d(u). \quad (9.4.23)$$

This solution corresponds to an isolated vacuum state which is independent of the boundary fields and does not contain any Bethe root.

The eigenvalue of the Hamiltonian (9.1.19) reads

$$
\begin{aligned}
E &= \frac{\partial}{\partial u} \ln \Lambda(u)|_{u=0} \\
&= \sum_{j=1}^{\bar{N}} \coth\left(\eta - \frac{\lambda_j}{2}\right) + 2N \frac{\cosh(5\eta)}{\sinh\eta - \sinh(5\eta)} \\
&\quad -2\frac{e^{-\varepsilon'}\cosh\eta}{1+2e^{-\varepsilon'}\sinh\eta} - 2\frac{e^{-\varepsilon}\cosh\eta}{1+2e^{-\varepsilon}\sinh\eta} \\
&\quad +\frac{\sinh(6\eta)\cosh\eta\cosh(5\eta) - \sinh(2\eta)\cosh(3\eta)\cosh(7\eta)}{\sinh(6\eta)\cosh\eta\sinh(2\eta)\cosh(3\eta)},
\end{aligned}
\tag{9.4.24}
$$

with the Bethe roots $\{\lambda_j | j = 1, \ldots, \bar{N}\}$ determined by the BAEs (9.4.22) for $\{\theta_j = 0\}$.

Numerical simulations are performed for $N = 2$ and randomly chosen boundary parameters by keeping a hermitian Hamiltonian. The numerical results shown in Table 9.1 and Fig. 9.1 imply that the $T-Q$ relation (9.4.9) and the BAEs (9.4.22) indeed give the complete solutions of the eigenvalues of the transfer matrix. How-

Table 9.1 One set solutions of the BAEs (9.4.22) for $N = 2$, $\eta = 0.2$, $\{\theta_j = 0\}$, $\varepsilon = 2$, $\varepsilon' = 1$, $\zeta = 0.7$ and $\zeta' = 0.6$

λ_1	λ_2	λ_3
$-3.86730 - 0.76144i\pi$	$-3.86730 + 0.76144i\pi$	$0.40014 - 0.55129i\pi$
$-4.31619 + 0.00000i$	$-3.11935 - 1.00000i\pi$	$0.39986 - 0.55166i\pi$
$-1.19960 + 0.00000i$	$-0.35935 - 0.51198i\pi$	$-0.35935 + 0.51198i\pi$
$-5.21818 - 1.00000i\pi$	$-2.92697 + 0.00000i$	$0.00014 - 0.44834i\pi$
$-5.58665 - 0.00000i$	$-3.11378 + 0.00000i$	$-0.00014 - 0.44871i\pi$
$-0.47543 - 0.24271i\pi$	$-0.47543 + 0.24271i\pi$	$0.39999 - 0.89627i\pi$
$-0.78903 + 1.00000i\pi$	$-0.00004 - 0.13296i\pi$	$-0.00004 + 0.13296i\pi$
$-6.34007 - 1.00000i\pi$	$0.00047 - 0.10337i\pi$	$0.00047 + 0.10337i\pi$
$-0.75861 - 0.44544i\pi$	$-0.75861 + 0.44544i\pi$	$0.00045 - 0.10340i\pi$

λ_4	λ_5	λ_6	E_n	n
$0.40014 + 0.55129i\pi$	$3.51378 + 1.00000i\pi$	$5.98665 - 1.00000i\pi$	-5.43010	1
$0.39986 + 0.55166i\pi$	$3.32697 + 1.00000i\pi$	$5.61818 - 0.00000i$	-5.42987	2
$1.36235 + 1.00000i\pi$	$1.72287 - 0.50381i\pi$	$1.72287 + 0.50381i\pi$	-4.90174	3
$0.00014 + 0.44834i\pi$	$3.51935 - 0.00000i$	$4.71619 + 1.00000i\pi$	-4.58712	4
$-0.00014 + 0.44871i\pi$	$4.26730 - 0.23856i\pi$	$4.26730 + 0.23856i\pi$	-4.58375	5
$0.39999 + 0.89627i\pi$	$0.48065 - 0.24340i\pi$	$0.48065 + 0.24340i\pi$	-3.16027	6
$0.39510 - 0.55436i\pi$	$0.39510 + 0.55436i\pi$	$0.78826 \quad 1.00000i\pi$	0.21408	7
$0.40537 - 0.55059i\pi$	$0.40537 + 0.55059i\pi$	$6.39766 - 0.00000i$	0.66418	8
$0.00045 + 0.10340i\pi$	$1.56521 - 0.44347i\pi$	$1.56521 + 0.44347i\pi$	0.66882	9

The symbol n indicates the number of the energy levels, and E_n is the corresponding eigenenergy. The energy E_n calculated from (9.4.24) is the same as that from the exact diagonalization of the Hamiltonian

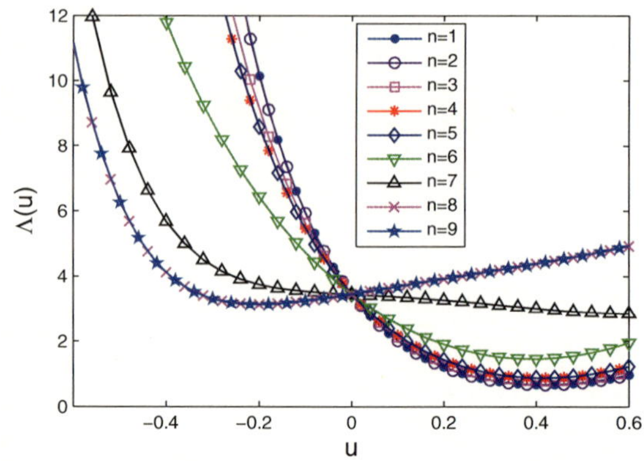

Fig. 9.1 $\Lambda(u)$ versus u for $N = 2$, $\eta = 0.2$, $\{\theta_j = 0\}$, $\varepsilon = 2$, $\varepsilon' = 1$, $\zeta = 0.7$ and $\zeta' = 0.6$. The *curves* calculated from exact diagonalization of the transfer matrix coincide exactly with those from the $T - Q$ relation (9.4.9) and the BAEs (9.4.22)

ever, there could be several sets of Bethe roots corresponding to one $\Lambda(u)$ curve. Such behavior also indicates the multiple parameterizations of the eigenvalues. For a given $\Lambda(u)$, the operator identities give rise to $2N$ independent equations for the coefficients of the $Q(u)$ functions, among which N of them are linear and N of them are quadratic. According to Bézout's theorem [32], the maximum number of possible $Q(u)$ functions that parameterize the given $\Lambda(u)$ correctly, is 2^N. Therefore, a given $\Lambda(u)$ may not correspond to a unique pair of $Q_{1,2}(u)$.

We note that the generalization of the above approaches to the type I non-diagonal K-matrices [19, 21] is straightforward. It should be remarked that there is almost no arbitrariness in the choice of the $T - Q$ relation, as long as $c_0(u)$ is fixed, for the present model. For example, a $T - Q$ relation with one unique Q-function like (8.3.1) or with three Q-functions like (3.2.61), allowed for other rank-1 models, does not exist for the IK model. Such a phenomenon could be a common feature of the integrable models beyond the $A^{(1)}_{n-1}$-type.

9.5 Reduced $T - Q$ Relation for Constrained Boundaries

As in the case of the anisotropic spin-$\frac{1}{2}$ chains introduced in Chap. 5, here $c_0 = 0$ leads to the constraints among the boundary parameters and η. In such cases we might find a proper reference state, enabling application of the conventional Bethe Ansatz methods [33–35].

The BAEs (9.4.22) require that the Bethe roots $\{\lambda_j\}$ form two types of pairs:

$$(\lambda_j, -\lambda_j), \quad (\lambda_j, -\lambda_j + 4\eta), \tag{9.5.1}$$

which imply the constraint condition for a generic η

$$\varsigma' - \varsigma = -4k\eta \mod (2i\pi), \qquad k \in Z. \tag{9.5.2}$$

The resulting $T - Q$ relation (9.4.9) is thus reduced to a conventional one

$$\begin{aligned}\Lambda(u) &= a(u)\frac{Q(u+4\eta)}{Q(u)} + d(u)\frac{Q(u-6\eta-i\pi)}{Q(u-2\eta+i\pi)} \\ &+b(u)\frac{Q(u+2\eta+i\pi)Q(u-4\eta)}{Q(u-2\eta+i\pi)Q(u)},\end{aligned} \tag{9.5.3}$$

where

$$Q(u) = \prod_{j=1}^{M}\sinh[\frac{u-\lambda_j}{2} - \eta]\sinh[\frac{u+\lambda_j}{2} - \eta], \tag{9.5.4}$$

with $M = N-k$ for $k \leq -N$ $M = N+k+1$ for $k \geq N+1$ and $M = N-k,\ N+k-1$ for $1 - N \leq k \leq N$, respectively.

The reduced $T - Q$ relation is also valid if η takes the following discrete values

$$\eta = \frac{\varsigma - \varsigma'}{4N - 4M} + \frac{2i\pi m}{4N - 4M}, \quad m, M \in Z, \text{ and } 0 \leq M. \tag{9.5.5}$$

The resulting BAEs are

$$\begin{aligned}&\prod_{l=1}^{N}\frac{\sinh\left[\frac{\lambda_j-\theta_l}{2} - \eta\right]\sinh\left[\frac{\lambda_j+\theta_l}{2} - \eta\right]}{\sinh\left[\frac{\lambda_j-\theta_l}{2} + \eta\right]\sinh\left[\frac{\lambda_j+\theta_l}{2} + \eta\right]}\frac{(1-2e^{-\varepsilon}\sinh(\lambda_j+\eta))}{(1+2e^{-\varepsilon}\sinh(\lambda_j-\eta))} \\ &= -\frac{(1+2e^{-\varepsilon'}\sinh(\lambda_j-\eta))}{(1-2e^{-\varepsilon'}\sinh(\lambda_j+\eta))}\frac{\sinh(\lambda_j+2\eta)\cosh(\lambda_j-\eta)}{\sinh(\lambda_j-2\eta)\cosh(\lambda_j+\eta)} \\ &\times\frac{Q(\lambda_j-2\eta)Q(\lambda_j+4\eta+i\pi)}{Q(\lambda_j+6\eta)Q(\lambda_j+i\pi)}, \quad j = 1, \ldots, M.\end{aligned} \tag{9.5.6}$$

We remark that the Hamiltonian (9.1.19) is normally not hermitian under the above constraint conditions.

For the diagonal K-matrices

$$K^-(u) = \text{id}, \quad K^+(u) = \mathscr{M}, \tag{9.5.7}$$

the $T - Q$ Ansatz (9.5.3) is reduced to the one [16] obtained by the analytic Bethe Ansatz method

$$\Lambda(u) = \bar{a}(u)\frac{Q(u+4\eta)}{Q(u)} + \bar{d}(u)\frac{Q(u-6\eta+i\pi)}{Q(u-2\eta+i\pi)} + \bar{b}(u)\frac{Q(u-4\eta)Q(u+2\eta+i\pi)}{Q(u-2\eta+i\pi)Q(u)}, \tag{9.5.8}$$

where $\bar{a}(u) = a(u)|_{\varepsilon,\varepsilon'\to+\infty}$, $\bar{d}(u) = d(u)|_{\varepsilon,\varepsilon'\to+\infty}$, $\bar{b}(u) = b(u)|_{\varepsilon,\varepsilon'\to+\infty}$ and the function $Q(u)$ is still given by (9.5.4) but with $M = 0, 1, \ldots, 2N$, because in this case the $U(1)$-symmetry is recovered [16].

9.6 Periodic Boundary Case

For the periodic IK model, the ODBA process is natural and simple. In this case, the transfer matrix reads

$$t(u) = tr_0 T_0(u). \tag{9.6.1}$$

Similarly, using the method in Sect. 9.2, and with the help of the relations (9.2.17) and (9.2.19), we may obtain that the following operator identities hold:

$$t(\theta_j)t(\theta_j+6\eta+i\pi) = \tilde{a}(\theta_j)\tilde{d}(\theta_j+6\eta+i\pi) \times \mathrm{id}, \tag{9.6.2}$$

$$t(\theta_j)t(\theta_j+4\eta) = \tilde{\delta}(\theta_j)t(\theta_j+2\eta+i\pi), \quad j = 1, \ldots, N, \tag{9.6.3}$$

where

$$\tilde{a}(u) = \prod_{l=1}^{N} h_3(u-\theta_l), \quad \tilde{d}(u) = \prod_{l=1}^{N} h_4(u-\theta_l), \tag{9.6.4}$$

$$\tilde{\delta}(\theta_j) = (-2)^N \prod_{l=1}^{N} \cosh\left(\frac{\theta_j-\theta_l}{2} - 3\eta\right) \sinh\left(\frac{\theta_j-\theta_l}{2} + 2\eta\right). \tag{9.6.5}$$

In addition, the asymptotic behavior of $t(u)$ for $u \to \infty$ reads

$$\lim_{u\to\pm\infty} t(u) = 2^{-N} e^{\pm N(u-3\eta)} \left[1 + 2\cosh(2\hat{M}\eta)\right] \times \mathrm{id} + \cdots, \tag{9.6.6}$$

where $\hat{M} = \sum_{j=1}^{N}(E_j^{11} - E_j^{33})$ is related to $U(1)$ charge operator.

The Eqs. (9.6.2)–(9.6.6) completely determine the eigenvalue $\Lambda(u)$ of $t(u)$, which can be given by the $T-Q$ relation

$$\Lambda(u) = \tilde{a}(u)\frac{Q(u+4\eta)}{Q(u)} + \tilde{d}(u)\frac{Q(u-6\eta+i\pi)}{Q(u-2\eta+i\pi)} + \tilde{b}(u)\frac{Q(u-4\eta)Q(u+2\eta+i\pi)}{Q(u-2\eta+i\pi)Q(u)}, \tag{9.6.7}$$

with

$$\tilde{b}(u) = \prod_{l=1}^{N} h_2(u - \theta_l), \tag{9.6.8}$$

$$Q(u) = \prod_{k=1}^{M} \sinh\left[\frac{u - \lambda_k}{2} - \eta\right], \quad M = 0, \ldots, 2N. \tag{9.6.9}$$

The BAEs in this case read

$$\prod_{l=1}^{N} \frac{\sinh\left[\frac{\lambda_j - \theta_l}{2} - \eta\right]}{\sinh\left[\frac{\lambda_j - \theta_l}{2} + \eta\right]} = -\frac{Q(\lambda_j - 2\eta)Q(\lambda_j + 4\eta + i\pi)}{Q(\lambda_j + 6\eta)Q(\lambda_j + i\pi)},$$
$$j = 1, \ldots, M. \tag{9.6.10}$$

References

1. V.V. Bazhanov, Trigonometric solutions of the triangle equations and classical Lie algebras. Phys. Lett. B **159**, 321 (1985)
2. M. Jimbo, Quantum R matrix for the generalized Toda system. Commun. Math. Phys. **102**, 537 (1986)
3. V.V. Bazhanov, Integrable quantum systems and classical Lie algebras. Commun. Math. Phys. **113**, 471 (1987)
4. V.G. Kac, *Infinite Dimensional Lie Algebras* (Birkhauser, Basel, 1983)
5. A.G. Izergin, V.E. Korepin, The inverse scattering method approach to the quantum Shabat-Mikhailov model. Commun. Math. Phys. **79**, 303 (1981)
6. R.K. Dodd, R.K. Bullough, Polynomial conserved densities for the sine-Gordon equations. Proc. R. Soc. Lond. A **352**, 481 (1977)
7. A.V. Zhiber, A.B. Shabat, Klein-Gordon equations with a nontrivial group. Dokl. Akad. Nauk. SSSR **247**, 1103 (1979)
8. C.M. Yung, M.T. Batchelor, Integrable vertex and loop models on the square lattice with open boundaries via reflection matrices. Nucl. Phys. B **435**, 430 (1995)
9. M.T. Batchelor, C.M. Yung, Exact results for the adsorption of a flexible self-avoiding polymer chain in two dimensions. Phys. Rev. Lett. **74**, 2026 (1995)
10. N.Yu. Reshetikhin, The functional equation method in the theory of exactly soluble quantum systems. Sov. Phys. JETP 57, 691 (1983)
11. V.A. Tarasov, Algebraic Bethe Ansatz for the Izergin-Korepin matrix. Theor. Math. Phys. **56**, 793 (1988)
12. S.O. Warnaar, M.T. Batchelor, B. Nienhuis, Critical properties of the Izergin-Korepin and solvable O(n) models and their related quantum spin chains. J. Phys. A **25**, 3077 (1992)
13. J.D. Kim, Boundary K-matrix for the quantum Mikhailov-Shabat model. arXiv:hep-th/9412192
14. M.J. Martins, The exact solution and the finite-size behaviour of the $Osp(1|2)$-invariant spin chain. Nucl. Phys. B **450**, 768 (1995)
15. C.M. Yung, M.T. Batchelor, O(n) model on the honeycomb lattice via reflection matrices: surface critical behaviour. Nucl. Phys. B **453**, 552 (1995)
16. L. Mezincescu, R.I. Nepomechie, Analytical Bethe Ansatz for quantum algebra invariant spin chains. Nucl. Phys. B **372**, 597 (1992)

17. H. Fan, Bethe Ansatz for the Izergin-Korepin model. Nucl. Phys. B **488**, 409 (1997)
18. B.-Y. Hou, W.-L. Yang, Y.-Z. Zhang, The twisted quantum affine algebra $U_q(A_2^{(2)})$ and correlation functions of the Izergin-Korepin model. Nucl. Phys. B **556**, 485 (1999)
19. A. Lima-Santos, Reflection K-matrices for 19-vertex models. Nucl. Phys. B **558**, 637 (1999)
20. W.-L. Yang, Y.-Z. Zhang, Izergin-Korepin model with a boundary. Nucl. Phys. B **596**, 495 (2001)
21. R.I. Nepomechie, Boundary quantum group generators of type A. Lett. Math. Phys. **62**, 83 (2002)
22. G.-L. Li, K.-J. Shi, R.-H. Yue, The algebraic Bethe Ansatz for the Izergin-Korepin model with open boundary conditions. Nucl. Phys. B **670**, 401 (2003)
23. K. Hao, J. Cao, G.-L. Li, W.-L. Yang, K. Shi, Y. Wang, Exact solution of the Izergin-Korepin model with general non-diagonal boundary terms. J. High Energy Phys. **06**, 128 (2014)
24. P.P. Kulish, N. Yu, Reshetikhin, E.K. Sklyanin, Yang-Baxter equation and representation theory: I. Lett. Math. Phys. **5**, 393 (1981)
25. P.P. Kulish, E.K. Sklyanin, Quantum spectral transform method: recent developments. Lect. Notes Phys. **151**, 61 (1982)
26. A.N. Kirillov, N.Yu. Reshetikhin, Exact solution of the Heisenberg XXZ model of spin s. J. Sov. Math. **35**, 2627 (1986)
27. A.N. Kirillov, N.Yu. Reshetikhin, Exact solution of the integrable XXZ Heisenberg model with arbitrary spin. I. The ground state and the excitation spectrum. J. Phys. A **20**, 1565 (1987)
28. P.P. Kulish, N.Yu. Reshetikhin, Quantum linear problem for the sine-Gordon equation and higher representation. J. Sov. Math. **23**, 2435 (1983)
29. L. Mezincescu, R.I. Nepomechie, Fusion procedure for open chains. J. Phys. A **25**, 2533 (1992)
30. M. Karowski, On the bound state problem in 1+1 dimensional field theories. Nucl. Phys. B **153**, 244 (1979)
31. Y.-K. Zhou, Row transfer matrix functional relations for Baxter's eight-vertex and six-vertex models with open boundaries via more general reflection matrices. Nucl. Phys. B **458**, 504 (1996)
32. P. Griffiths, J. Harris, *Principles of Algebraic Geometry* (Wiley Classics Library, New York, 1994)
33. J. Cao, H.-Q. Lin, K.-J. Shi, Y. Wang, Exact solution of XXZ spin chain with unparallel boundary fields. Nucl. Phys. B **663**, 487 (2003)
34. W.-L. Yang, Y.-Z. Zhang, T-Q relation and exact solution for the XYZ chain with general nondiagonal boundary terms. Nucl. Phys. B **744**, 312 (2006)
35. W.-L. Yang, Y.-Z. Zhang, On the second reference state and complete eigenstates of the open XXZ chain. J. High Energy Phys. **04**, 044 (2007)

Index

Y. Wang et al., *Off-Diagonal Bethe Ansatz for Exactly Solvable Models*,
DOI 10.1007/978-3-662-46756-5

Printed by Printforce, the Netherlands